ENVIRONMENTAL POLLUTION IMPACT ON PLANTS

Survival Strategies under Challenging Conditions

ENVIRONMENTAL POLLUTION IMPACT ON PLANTS

Survival Strategies under Challenging Conditions

Edited by

Tariq Aftab, PhD

Khalid Rehman Hakeem, PhD

AAP | APPLE ACADEMIC PRESS

First edition published 2023

Apple Academic Press Inc.
1265 Goldenrod Circle, NE,
Palm Bay, FL 32905 USA

760 Laurentian Drive, Unit 19,
Burlington, ON L7N 0A4, CANADA

CRC Press
6000 Broken Sound Parkway NW,
Suite 300, Boca Raton, FL 33487-2742 USA

4 Park Square, Milton Park,
Abingdon, Oxon, OX14 4RN UK

Library and Archives Canada Cataloguing in Publication

Title: Environmental pollution impact on plants : survival strategies under challenging conditions / edited by Tariq Aftab, PhD, Khalid Rehman Hakeem, PhD.
Names: Aftab, Tariq, editor. | Hakeem, Khalid Rehman, editor.
Description: First edition. | Includes bibliographical references and index.
Identifiers: Canadiana (print) 20220449066 | Canadiana (ebook) 20220449074 | ISBN 9781774911242 (hardcover) | ISBN 9781774911259 (softcover) | ISBN 9781003304210 (ebook)
Subjects: LCSH: Plant ecophysiology. | LCSH: Plants—Effect of pesticides on. | LCSH: Plants—Effect of pollution on.
Classification: LCC QK717 .E58 2023 | DDC 581.7—dc23

Library of Congress Cataloging-in-Publication Data

CIP data on file with US Library of Congress

ISBN: 978-1-77491-124-2 (hbk)
ISBN: 978-1-77491-125-9 (pbk)
ISBN: 978-1-00330-421-0 (ebk)

About the Editors

Tariq Aftab, PhD

Tariq Aftab, PhD, is currently Assistant Professor at Department of Botany from the Aligarh Muslim University, India, where he earned his PhD. He is the recipient of a prestigious Leibniz-DAAD fellowship from Germany, Raman Fellowship from the Government of India, and Young Scientist Awards from the State Government of Uttar Pradesh (India) and Government of India. After completing his doctorate, he worked as a Research Fellow at the National Bureau of Plant Genetic Resources, New Delhi, and as Postdoctoral Fellow at Jamia Hamdard, New Delhi, India. Dr. Aftab also worked as a visiting scientist at the Leibniz Institute of Plant Genetics and Crop Plant Research (IPK), Gatersleben, Germany, and in the Department of Plant Biology, Michigan State University, USA. He is a member of various scientific associations in India and abroad.

He has edited 14 books with international publishers, including Elsevier, Springer Nature, and CRC Press (Taylor & Francis Group), coauthored several book chapters, and published over 65 research papers in peer-reviewed international journals. His research interests include physiological, proteomic, and molecular studies on medicinal and crop plants.

Khalid Rehman Hakeem, PhD

Khalid Rehman Hakeem, PhD is a Professor at the King Abdulaziz University, Jeddah, Saudi Arabia. After completing his doctorate (Botany; specialization in Plant Ecophysiology and Molecular Biology) from Jamia Hamdard, New Delhi, India, in 2011, he worked as a lecturer at the University of Kashmir, Srinagar, India, for a short period. Later, he joined the Universiti Putra Malaysia, Selangor, Malaysia, and worked there as Postdoctoral Fellow in 2012 and Fellow Researcher (Associate Prof.) from 2013 to 2016. Dr. Hakeem has more than 12 years of teaching and research experience in plant ecophysiology, biotechnology and molecular biology, medicinal plant research, plant-microbe-soil interactions as well as in environmental studies. He is the recipient of several fellowships at both national and international levels; also, he has served as a visiting scientist at Jinan University, Guangzhou, China. Currently, he is involved with a number of international research projects with different government organizations.

To date, Dr. Hakeem has authored and edited more than 70 books with international publishers, including Springer Nature, Academic Press (Elsevier), and CRC Press. He also has to his credit more than 180 research publications in peer-reviewed international journals and 65 book chapters in edited volumes with international publishers.

At present, Dr. Hakeem serves as an editorial board member and reviewer of several high-impact international scientific journals from Elsevier, Springer Nature, Taylor and Francis, Cambridge, and John Wiley. He is included in the advisory board of Cambridge Scholars Publishing, UK. He is also a fellow of the Royal Society of Biology, London, fellow of the Plantae Group of the American Society of Plant Biologists, member of the World Academy of Sciences, member of the International Society for Development and Sustainability, Japan, and member of Asian Federation of Biotechnology, Korea. Dr. Hakeem has been listed in Marquis Who's Who in the World, 2014–2019. Currently, Dr. Hakeem is engaged in studying the plant processes at ecophysiological as well as molecular levels.

Contents

Contributors

Muhammad Aasim
Department of Plant Protection, Faculty of Agricultural Sciences and Technologies, Sivas University of Science and Technology, Sivas, Turkey; E-mail: mshazim@gmail.com

Ismail Abiola Adebayo
Microbiology and Immunology Department, Faculty of Biomedical Sciences, Kampala International University, Western Campus, P. O. Box 71, Ishaka-Bushenyi, Uganda; E-mail: ibnahmad507@gmail.com

Muhammad Adnan
College of Agriculture, University of Sargodha, Sargodha 40100, Pakistan

Hamaad Raza Ahmad
Institute of Soil and Environmental Sciences, University of Agriculture Faisalabad, Pakistan

Rana Imtiaz Ahmed
Regional Agricultural Research Institute, Department of Agriculture, Government of Punjab, Bahawalpur 63100, Punjab, Pakistan

Shazia Qurat Ul Aien
Department of Zoology, University of Kashmir Srinagar 190006, Jammu and Kashmir, India

Ozlem Akgur
Department of Plant Protection, Faculty of Agricultural Sciences and Technologies, Sivas University of Science and Technology, Sivas, Turkey

Lal Hussain Akhter
Regional Agricultural Research Institute, Department of Agriculture, Government of Punjab, Bahawalpur 63100, Punjab, Pakistan

Tanveer Alam
Natural and Medical Sciences Research Center, University of Nizwa, Birkat Al Mawz, Sultanate of Oman

Abid Ali
Regional Agricultural Research Institute, Department of Agriculture, Government of Punjab, Bahawalpur 63100, Punjab, Pakistan

Ahmad Ali
Department of Life Sciences, University of Mumbai, Vidyanagari, India

Iftikhar Ali
Department of Soil and Environmental Sciences, The University of Agriculture Peshawar, KPK, Pakistan

Liaquat Ali
Regional Agricultural Research Institute, Department of Agriculture, Government of Punjab, Bahawalpur 63100, Punjab, Pakistan

Wajiha Anum
Regional Agricultural Research Institute, Department of Agriculture, Government of Punjab, Bahawalpur 63100, Punjab, Pakistan

Hasni Arsad
Advanced Medical and Dental Institute, Universiti Sains Malaysia, Bertam, 13200 Kepala Batas, Penang, Malaysia

Muhammad Zohaib Aslam
Institute of Soil and Environmental Sciences, University of Agriculture Faisalabad, Pakistan

Muhammad Ashar Ayub
Institute of Soil and Environmental Sciences, University of Agriculture Faisalabad, Pakistan
Plant Root Biology Lab, Indian River Research Centre, Institute of Food and Agriculture Sciences, University of Florida, Fort Pierce, FL, USA; E-mail: muhammadasharayub@gmail.com

Uzma Azeem
Government College Malerkotla, Malerkotla-148023, Punjab, India;
E-mail: uzmaazeem2@gmail.com

Gurpaul Singh Dhingra
Department of Botany, Punjabi University, Patiala 147002, Punjab, India

Neal Samuel Eash
Biosystems Engineering & Soil Science, The University of Tennessee, Tennessee, USA

Ruchira Ghosh
Centre for Sustainable Technologies, Belfast School of Architecture and the Built Environment, Ulster University, Northern Ireland, United Kingdom

Tajamul Islam
Department of Botany, University of Kashmir Srinagar 190006, Jammu and Kashmir, India

Mohd Nazri Ismail
Analytical Biochemistry Research Centre, Universiti Sains Malaysia (USM), 11800 Pulau Pinang, Malaysia

Noor Zafirah Ismail
Advanced Medical and Dental Institute, Universiti Sains Malaysia, Bertam, 13200 Kepala Batas, Penang, Malaysia

Hanuman Singh Jatav
Sri Karan Narendra Agriculture University Jobner-Jaipur, Rajasthan, India

Parneet Kaur
School of Biotechnology, Faculty of Applied Sciences and Biotechnology, Shoolini University, Solan, Himachal Pradesh, India

Nawab Khan
Climate, Energy and Water Research Institute (CEWRI), National Agricultural Research Centre, Islamabad, Pakistan

Saurabh Kulshreshtha
School of Biotechnology, Faculty of Applied Sciences and Biotechnology, Shoolini University, Solan, Himachal Pradesh, India

Zemran Mustafa
Department of Plant Production and Technology, Faculty of Agricultural Sciences and Technologies, Sivas University of Science and Technology, Sivas, Turkey

Lubna Najam
DAV (PG) College Muzaffarnagar, CCS University, Meerut, UP, India; E-mail: najam.lubna@gmail.com

Subzar Ahmad Nanda
Department of Botany, University of Kashmir Srinagar 190006, Jammu and Kashmir, India

Yusuf Opeyemi Oyebamiji
Department of Biological Sciences and Biotechnology, Faculty of Science and Technology, Universiti Kebangsaan Malaysia, 43600 Bangi, Selangor, Malaysia

Jessica Pandohee
Centre for Crop and Disease Management, School of Molecular and Life Sciences, Curtin University, Bentley, WA 6102, Australia; E-mail: jessica.pandohee@gmail.com

Muhammad Farhan Qadir
Institute of Soil and Environmental Sciences, University of Agriculture, Faisalabad, Punjab, Pakistan

Asad-ur-Rahman
Regional Agricultural Research Institute, Department of Agriculture, Government of Punjab, Bahawalpur 63100, Punjab, Pakistan

Khalid Raza
Department of Computer Science, Jamia Millia Islamia, New Delhi, India

Taqi Raza
Land Resources Research Institute, National Agricultural Research Centre, Islamabad, Pakistan; E-mail: taqiraza85@gmail.com

Ishfaq Ul Rehman
Department of Botany, University of Kashmir Srinagar 190006, Jammu and Kashmir, India

Muhammad Zia ur Rehman
Institute of Soil and Environmental Sciences, University of Agriculture Faisalabad, Pakistan

Summia Rehman
Department of Botany, University of Kashmir Srinagar 190006, Jammu and Kashmir, India; E-mail: sumaiyarehman348@gmail.com

Umair Riaz
Soil and Water Testing Laboratory for Research, Department of Agriculture, Government of Punjab, Bahawalpur 63100, Punjab, Pakistan; E-mail: umairbwp3@gmail.com

Syed Saleemullah
Department of Bioscience, Jamia Millia Islamia, New Delhi, India

Zoya Shaikh
Department of Life Sciences, University of Mumbai, Vidyanagari, India

Noraziyah Abd Aziz Shamsuddin
Department of Biological Sciences and Biotechnology, Faculty of Science and Technology, Universiti Kebangsaan Malaysia, 43600 Bangi, Selangor, Malaysia

Kayenat Sheikh
Department of Computer Science, Jamia Millia Islamia, New Delhi, India; E-mail: kayenat996@gmail.com

Richa Shri
Department of Phramaceutical Sciences and Drug Research, Punjabi University, Patiala 147002, Punjab, India

Priti Sohal
School of Biotechnology, Faculty of Applied Sciences and Biotechnology, Shoolini University, Solan, Himachal Pradesh, India

Wajid Umar
Doctoral School of Environmental Sciences, Szent Istvan University, Godollo, Hungary

Abbreviations

AMF	arbuscular mycorrhizal fungi
APP	acute pesticide poisoning
APX	ascorbate peroxidase
As	arsenic
BC	British Columbia
BHA	butylated hydroxy anisole
BHT	butylated hydroxytoluene
BOD	biological oxygen demand
BTEX	benzene, toluene, ethyl benzene, xylene
CAT	catalase
CMC	critical micelle concentration
COD	chemical oxygen demand
DDT	dichloro-diphenyl-trichloroethane
D	dust
DHAR	deydroascorbate reductase
EC	emulsifiable concentrate
E	emulsion
ENMs	engineered nanomaterials
EPA	Environmental Protection Agency
EPS	extracellular polymeric substances
EU	Europe
FAO	Food and Agriculture Organization
FAOSTAT	Food and Agriculture Organization Corporate Statistical Database
FFDCA	Federal Food, Drug, and Cosmetic Act
F	fumigants
FIFRA	Federal Insecticide, Fungicide, and Rodenticide Act
FYM	farm yard manure
G	granules
GAP	good agricultural practice
GEMs	genetically engineered microorganisms
GMOs	genetically modified organisms
GMPs	genetically modified plants
GPX	guaiacol peroxidase

GR	glutathione reductase
GSH	glutathione
HCB	hexachlorobenzene
HCBD	hexachlorobutadiene
HEPT	heptachlor
HIV/AIDS	human immunodeficiency virus/acquired immunodeficiency syndrome
HMmNs	heavy metal micronutrients
HMNs	heavy metal nutrients
HMPs	heavy metal pollutants
IAA	indole-3-acetic acid
IPPC	International Plant Protection Convention
IPM	integrated pest management
IRS	indoor residual spraying
LC_{50}	lethal concentration required to kill 50% of the population
LD_{50}	lethal dose at which 50% of the population is killed in a given period of time
MDHAR	monodehydroascorbate reductase
MITC	methylisocyanate
NO	nitric oxide
NRC	National Research Council
OCPs	organochlorine pesticides
OECD	Organization for Economic Cooperation Development
OPCs	organophosphorus compounds
PAHO	Pan American Health Organization
PAHs	polycyclic aromatic hydrocarbons
PBDEs	polybrominated diphenyl ethers
PCBs	polychlorinated biphenyls
PCDE	polychlorinated diphenyl ethers
PCN	polychlorinated naphthalene
PFOS	perfluorooctane sulfonate
PGPR	plant growth promoting rhizospheric
POPs	persistent organic pollutants
ppb	part per billion
ppm	part per million
ppq	part per quadrillion
ppt	part per trillion
RBOHs	respiratory burst oxidase homologues
RLEM	redlegged earth mite

RONS	reactive oxygen and nitrogen species
ROS	reactive oxygen species
SOD	superoxide dismutase
SRB	sulphur-reducing bacteria
SS	sewage sludge
SUH	substituted urea herbicides
SWSR	Status of the World's Soil Resources Report
TCE	trichloroethylene
TFM	3-trifluoromethyl-4-nitrophenol
TNT	trinitrotoluene
TPs	transformation products
UNICEF	United Nations International Children's Emergency Fund
USDA/ERS	United States Department of Agriculture/Economic Research Service
USFWS	US Fish and Wildlife Service
USGS	US Geological Survey
WEC	World Energy Congress
WHO	World Health Organization
WP	wettable powder
2,4-D	2,4-dichlorophenoxyacetic acid

Preface

Pollution is the introduction of impurities into the environment that triggers harm or distress to other living creatures and damage the ecosystem in the form of organic ingredients or heat, light, or noise. Pollutants can be naturally arising materials and are considered contaminants when they exceed the normal levels. Nonbiodegradable pollutants cannot be disintegrated by the living organisms and, therefore, persevere in the environment for extremely long phases of time. They contain plastics, metals, pesticides, glass, and radioactive isotopes.

Soils contaminated with heavy metals have become a major challenge worldwide due to an increase in anthropogenic and geologic activities. Reduction in growth as a result of alterations in biochemical and physiological processes in plants cultivating on heavy-metal-polluted soils has been noted. A few direct noxious effects triggered by elevated metal concentration comprise inhibition of cytoplasmic enzymes and impairment to cell assemblies due to oxidative damage. The substitution of necessary nutrients at cation exchange sites in plants is an example of unintended toxic effect due to heavy metal stress. Further, the deleterious impact heavy metals have on the development and activities of soil microorganisms may also ultimately affect the plant growth. Pesticides have been an indispensable part of agriculture to safeguard crops and livestock from pest invasions and yield decline for many years. Despite their effectiveness, pesticides could carry probable risks to food security, the ecosystem, and to the plants. The significance of agricultural pesticides for developing countries is unquestionable. However, with the growing quantities of their usage, apprehension about their unfavorable effects on plants has also developed. When exposed to airborne contaminants, most plants experience biochemical changes before displaying noticeable impairment to leaves. The atmospheric SO_2 badly affect various structural and biochemical physiognomies of plants. High soil moistness and high relative moisture intensify SO_2 injury in plants. Air pollutants can instigate leaf damage, stomatal impairment, early senescence, decrease in photosynthetic efficiency, interrupt membrane perviousness and decrease growth and yield in sensitive plant types.

This book includes 10 chapters emphasizing varied issues by debating the impact and management methods of effects of pollutants in plants. We are optimistic that this book will deliver the prerequisite for all researchers who are working or have interest in this particular field. Undoubtedly, this book will be helpful for those who have interest in plant abiotic stress physiology or for general use of research students, teachers, and others.

We are extremely thankful to all our contributors for not only sharing their research and knowledge, but also for venerably assimilating their knowledge in dispersed information from various fields in constituting the chapters and enduring editorial suggestions to finally produce this endeavor. We also thank Apple Academic Press (CRC Press, Taylor & Francis Group) team for their substantial assistance at every stage of the book's production.

Lastly, thanks are also due to well-wishers, research students, and authors' family members for their moral support, blessings, and inspiration in the compilation of this book.

—Tariq Aftab
(*Aligarh Muslim University, India*)

—Khalid Rehman Hakeem
(*King Abdulaziz University, Saudi Arabia*)

CHAPTER 1

Effect of Heavy Metal Polluted Soil on Physiology and Biochemistry of Plants

MUHAMMAD ASHAR AYUB[1,2*], MUHAMMAD ZIA UR REHMAN[1], MUHAMMAD ZOHAIB ASLAM[1], WAJID UMAR[3], MUHAMMAD ADNAN[4], and HAMAAD RAZA AHMAD[1]

[1]*Institute of Soil and Environmental Sciences, University of Agriculture Faisalabad, Pakistan*

[2]*Plant Root Biology Lab, Indian River Research Centre, Institute of Food and Agriculture Sciences, University of Florida, Fort Pierce, FL, USA*

[3]*Doctoral School of Environmental Sciences, Szent Istvan University, Godollo, Hungary*

[4]*College of Agriculture, University of Sargodha, Sargodha 40100, Pakistan*

[*]*Corresponding author. E-mail: muhammadasharayub@gmail.com*

ABSTRACT

Heavy metal pollution in the soil is an alarming issue worldwide, which is compromising the quality of food as well as human health. Among heavy metals, some plant nutrients (Cu, Mn, Zn, Ni, and Mo), as well as pollutants (Cd, Hg, Pb, and Cr), are classified. A small amount of heavy metal pollutants (HMPs) is toxic to plants while higher concentrations of heavy metal nutrients (HMNs) show toxicity in plants. The HMs' stress disrupts normal plant physiology (cellular functioning, carbon assimilation, growth, and development) as well as plant biochemistry (disrupting oxidative

Environmental Pollution Impact on Plants: Survival Strategies under Challenging Conditions. Tariq Aftab, PhD & Khalid Rehman Hakeem, PhD (Eds.)

balance, biomolecule synthesis, antioxidants production, and phytohormones homeostasis) resulting in extreme stress to plants.

1.1 INTRODUCTION

In the modern era, due to nonstop industrialization and many other anthropogenic events, the soils are being polluted with toxic components such as heavy metal pollutants (HMPs) worldwide.[1] The HMPs are commonly known for their toxicity, bioaccumulation, persistence, and geo/biotransformation in ecosystems.[2] The major sources of these HMPs can be the metal processing and mining sector,[3] which can be an alarming threat to water bodies' health, aquatic life,[4] and plant health as they are highly bio-accumulable, thereby making them a live threat for humans.[5] The accumulation of these HMPs in plants encourages genotoxic and cytotoxic damage, which makes morphological, physiological, and cellular disruptions.[6] Major toxicity effects of HMPs on plant physiology are due to their catalytic activity for reactive oxygen species production[7] and denaturation of cellular nucleic acid.[8] The HMPs-related generic damage via biomarkers is a well-known phenomenon reporting HMPs-initiated cell and tissue toxicity.[8] Besides cellular toxicity, plants' physiology and growth are also severely damaged with HMP's contact with the living body.[9]

The HMP accumulation in edible crop plants is an important issue due to its negative effects on crop health, micro-morphological components (gaseous exchange and indices),[10] and photosynthetic components.[11] The attributes of plants are usually followed biomarkers for screening the HMP's adverse effects on plant physiology, growth, and development.[11] To cope with the physiological and biochemical effects of crop plants, a wide range of remedial approaches are being followed ranging from bioremediation to the use of various organic and inorganic amendments for effective locking of these pollutants. This chapter discusses HM types and classification, the negative effects of HMP's pollutants on plants, and related approaches to control it.

1.2 HEAVY METAL SOURCES, CLASSIFICATION, AND EFFECTS ON PLANTS

The term *heavy metal* is more of a physical term used for metal having a specific gravity in between 4 and 5 g/cm^3. Among this category of HMs, a list of elements have been reported to have negative effects on plants, and thus are termed as HMPs. The main source of HMs in agroecosystem are weathering of parent material, excessive use of agrochemicals, and other anthropogenic activities including smelting, mining, fossil fuel burning, untreated sewage sludge, e-waste, and water disposal.[12–15] Worldwide major source of HMPs emissions are smelting, atmospheric depositions, waste inspiration, as well as vehicular exhausts (in urban area) contributing significant parts.[16] In industrial economy-based countries like China, industrial operations, waster water irrigation, and sludge application to agricultural lands are major sources of HMPs.[17] Geological and anthropogenic activities are major contributors in industrial countries where the excessive release of HMs into agroecosystem is evident.[18,19] Among industrial processes, wood/paper processing, plastic, and textile industries also have a significant use and disposal of HMPs into the environment.[6] Among domestic sources, household effluent including washouts and detergents contribute toward HMPs, while in agricultural lands, use of pesticides are major, which is enhanced by soil and water erosion.[20] Major heavy metal pollutants, their chemistry, and their fate in the soil are given below in Table 1.1.

1.2.1 EFFECT OF HM POLLUTED SOIL ON PLANT PHYSIOLOGICAL AND BIOCHEMICAL PARAMETERS

The extreme levels of HMPs in the soil make its higher absorption in plants, which entering the cell, cause oxidative stress and initiate reactive oxygen species (ROS) production. ROS production is a sign of plant response toward stress, but its excessive production is damaging for cellular structures. The initiation of ROS causes reversible DNA as well as cellular protein damage and is well reported in numerous published works.

Among HMPs, lead is an important one and Pb toxicity was observed in *S. procumbens*[44] and maize,[45] which can be associated with oxidative stress and photosynthetic pigment disturbance.[46] The higher accumulation of HMPs has numerous effects on plant physiology and biochemical attributes causing chlorosis, enzymatic deregulation, and metabolic/photosystem disturbance.[47] Figure 1.1 represents the general responses of plants under HMs stress.

It is reported that HMPs accumulation in toxic concentrations causes numerous negative effects on crop physiology like chlorosis, enzymes inactivation, and metabolical blocking of essential components like chlorophyll;[47] this may destabilize cells and cause modifications in metabolism, influencing photosynthetic action,[48] due to damage of thylakoid

TABLE 1.1 Heavy Metals in Soil–Plant System.

	Heavy metal mollutants (HMPs)
HMPs name	**Source and chemistry in soil**
Cadmium (Cd)	Cadmium is a potentially toxic HMP with widespread sources and effects on various crop plants. Its higher solubility in the soil makes it readily available to crop plants for uptake and can translocate and accumulate in edible plant portions in significant amounts making a health hazard for humans.[14,21] Major sources of Cd in agroecosystem are wastewater application, excessive use of contaminated phosphatic fertilizers, e-waste improper dumping, industrial activities, and atmospheric depositions.[14,21]
Lead (Pb)	Another pollutant in the HM category is lead, which also has higher persistence in soil and is not degradable due to its inorganic nature.[6] Upon reaching toxic limits, Pb proves to be a poison for plants disrupting normal functioning and physiological/biochemical processes affecting plant growth and carbon assimilation.[22] Its sources in the environment are widespread ranging from natural (erosion, volcanic eruption, and weathering) as well as anthropogenic activities (industrial use and effluents) and use of agrochemicals.[23]
Chromium (Cr)	Related to leather industry effluents, Cr toxicity in agricultural soils is also being reported in developing countries. The oxidation state of Cr in soil control its fate and toxicity for plants as well as humans.[24]
Mercury (Hg)	Mercury (Hg) being the only liquid metal element at room temperature has also a significant place in the HMPs list. Although rare, it is still found in some rocks as well as near the mining sites and its higher built-up concentration in the soil can disrupt the normal physiological functioning of crop plants.[25]
Arsenic (As) metalloid*	Arsenic, being a non-metal should not be considered in this category, but it falls in the metalloid category and has potentially very toxic effects upon the plants as well as on human health.[26] The variable oxidation states of As in soil environment are primarily in control of soil REDOX potential and can affect its availability to plants.[27]
	Heavy metal micronutrients (HMmNs)
Zinc (Zn)	Zinc is an important plant micronutrient, but its higher concentration can be toxic for plants in which condition it tends to escape it via changing Zn acquisition behavior, accumulation, and assimilation into tissues.[28] Under Zn deficiency, the plant tends to increase its absorption via active transport but, under toxic concertation limits, a plant tends to limit its uptake in plant tissues.[29]

TABLE 1.1 *(Continued)*

HMPs name	Source and chemistry in soil
Copper (Cu)	Cu is another much-needed plant micronutrient and the plant has special transporters to use under its deficiency and toxicity conditions Ctrs and COPT1–6.[30] For the proper functioning of plastocyanin, Cu is much needed, and transporters like Cu/Zn SOD play a vital role under this[31] by helping plants in the bioproduction of biomolecules. If excessive Cu is present, its adverse effects are observed on plant physiology, growth, and metabolism[32] and make plant experience severe oxidative stress via ROS production.[33]
Iron (Fe)	Fe is an important micronutrient involved in various plant biochemical pathways and plants tend to avoid its deficiency; but, on the other hand, its toxicity is also an issue for plants. To cope with its deficiency, plants tend to uptake it more or make it more mobile in the rhizosphere.[34] To cope with it, toxicity plant has specialized mechanism for active exclusion of Fe from their cell as well as modulate and manage the resultant oxidative damage.[35]
Nickel (Ni)	Nickel is another HM micronutrient and proved its essentiality for legumes in the early 80s and since then its been reported to be involved in enzyme activation of the plant as well as in N-metabolism.[36] On contrary, Ni toxicity can also disrupt the normal Ca-binding capability of photosystem–Oxygen–evolution complex and initiate ROS production due to lower antioxidants production as well as plant growth.[37]
Molybdenum (Mo)	The bioavailability of Mo largely depends upon soil physico-chemical properties especially soil pH.[38] If Mo deficiency is experienced, plant metabolic enzymes like nitrate-reductase, sulfite oxidase, xanthine dehydrogenase, aldehyde oxidase, and mitochondrial amidoxime reductase are severely influenced and the plant gets into shock[39] and can be reversed via exogenous application.[39] As Mo behaves differently than other HMNs, so the determination of its toxicity is yet to be determined.[40]
Manganese (Mn)	The soil bioavailable concentration of Mn is also under the control of soil physicochemical properties[41] and involved in enzymatic activation and carbon assimilation in plants[42] as well as help plants encounter biotic and abiotic stresses.[43] That's why it's very much needed for proper growth and functioning of plants but Mn toxicity is also an issue that is dependent upon its concentration, crop type, and growth stage.[28]

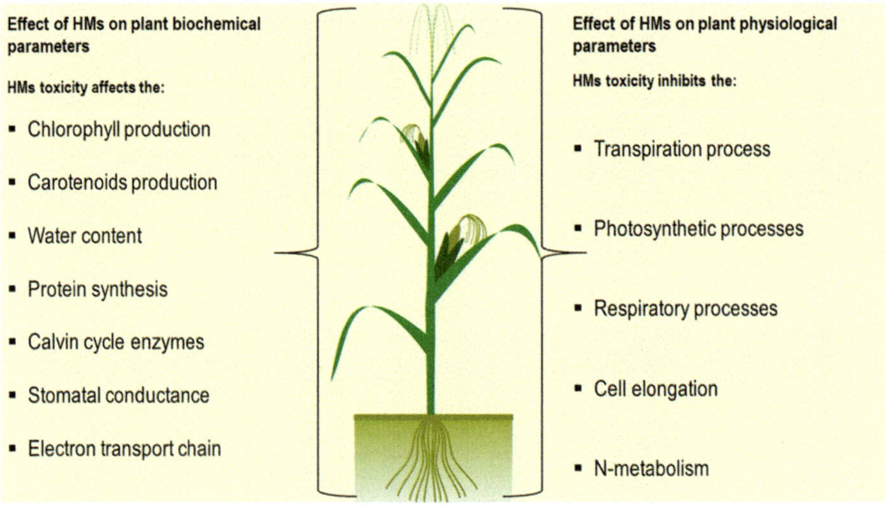

FIGURE 1.1 Effect of HMs toxicity on plants.

and Calvin-cycle machinery, as well as the alteration in pigmentation that causes cell necrosis and chlorosis.[49] For instance, it is documented that Pb relocates Mg in the chlorophyll influencing electron transport and photosynthesis.[30] The HMPs influence many functions of plant photosynthesis and biosynthesis of chlorophyll,[50] reducing chlorophyll a and b total contents, which may cause a reduction in the rate of photosynthesis.[51] The response of plants toward these stresses varies with plant time like in an investigation on the tree. Gomes and colleagues[37] reported that when chlorophyll contents in adult trees grown in tailings and control site were measured, no significant differences were recorded among treatments or in the coverage of stomata. This indicates that over time, in adult trees, adaptations of physiological mechanisms were generated in response to various environmental situations, which permits them to grow and tolerate the negative effects of HM. Among all HMPs, Cd is one of the most prominent pollutants and is responsible for many physiological and biochemical disturbances. In a workshop conducted on mung bean, it was reported that Cd has shown severe physiological and biochemical toxicity effects on the plant. The Cd significantly increased the Cd-increased catalase (CAT), ascorbate peroxidase (APX) activities in stems and roots, and peroxidase activities suggesting the onset of oxidate stress and plant's response to counter it. The higher oxidative stress was also responsible for net decrease disturbance in chlorophyll contents as well as other biomolecules like proline. The exogenous application of ABA has shown a significant effect in tackling Cd toxicity. Among summarized results, it was also seen that Cd toxicity was first observed on plant roots and more absorption of Cd is observed there helping plant to cope with upcoming pollutant.[52]

The cellular damage being experienced by plants under influence of HMPs toxicity is due to higher oxidative stress, which can be observed in the form of higher MDA contents in plants.[53] The ROS produced inside the plant cell are harvested by plants via various mechanisms and pathways. The nonenzymatic antioxidants like AsA, proline, and GSH and enzymatic anti-oxidants like GPX, POD, CAT, SOD, and APX are important components of ROS-scavenging machinery of plant cell helping in net homeostasis and ROS detoxification.[54] Being an inorganic pollutant, the HMPs can not be disintegrated but only their effect can be minimized via adopting such escape measures.[55] The total metal contents, absorption of these metals, and their accumulation in cells/tissue are the main points to consider while working on HMPs-related toxicity in plants.

Among metalloids, the As is considered to be the most toxic pollutant for plants and humans and is responsible for biochemical toxicity in plants. Higher availability of As in soil can disrupt normal growth and biochemical functioning of plants by water homeostasis disruption, and disruption in ATP biosynthesis and photosynthesis.[56] As pollution has reported showing a significant fall in cellular amino acids such as Pro, Met, Thr, Lys, Val, and Trp in *Cicer arietinum* L resulting in necrosis, curling of leaves and leaf blades wilting,[57] which can result in a reduction in leaf area and crop yield.[58] Another general response of As contamination in plants is that ROS production in plants[59] decreased antioxidants production, genotoxicity, biochemical, and physiological mechanisms of plants.[26] Lead is an another HMP being toxic for plants and human health. The fate of Pb in the soil–plant system is primarily controlled by soil physicochemical properties. The Pb toxicity severely affects plant physiology and biochemical attributes negatively influencing plant growth and development. Overall, Pb toxicity can check plant photosynthesis, water balance, and mineral nutrition, while also can disrupt phytohormones, which make membrane permeability questionable.[60] Among this category of HMPs, chromium is another toxic metal responsible for direct disruption of electron transport chain in plants also involved in oxidative damage to biomolecules as well as disturbed quenching of ROS in plant facing Cr toxicity.[61]

Among plant micronutrients, zinc (Zn) stress can also be an issue if its concentration in the soil gets higher than the permissible limit. The higher concentration of Zn is responsible for a net decrease of chlorophyll a, b, and carotenoids production while higher electrolyte leakage can be observed. Furthermore, it was observed that plant's defense system sometimes is not capable to cope with excessive Zn concentrations, which can be observed with an electron microscope via observing ultrastructure alteration.[62] The HMP's pollution tends to decrease antioxidants activity and, thus, increases net oxidative stress in plants.[63] Like, for an essential micronutrient, Ni, the critical limit of toxicity can range between 10 and 50 μg/g dry weight depending upon crop type and metal accumulation capabilities. In lower concentrations, Ni is essential for plants, but as soon as concentrations start to grow up, it becomes toxic. Though Ni itself is not a redox-active metal, so can not produce ROS directly, but can interfere with enzymes responsible for ROS production and thus can cause oxidate stress in plants.[64] Generally, the Ni and its compounds are not reported to be poisoning but have been recognized and suspected to be a carcinogen; that's why special care is required.[65] The Ni overdose in soil and it's over uptake from polluted soils can cause a net disturbance in plants' dietary value, as well as numerous biochemical processes, which are also directly disturbed by its toxicity.[66] Copper (Cu), which is also a potential plant micronutrient, can also behave as a pollutant if its concentration gets higher than the permissible limit in plant range. The most common physiological toxicity symptoms under Cu stress are decreased plant growth and disrupted carbon assimilation while antioxidants disturbance and biomolecule homeostasis dissolution are prominent biochemical toxicity symptoms.[67] A work conducted by Islam et al.[68] reported that Cu toxicity in lentil plants caused an upsurge of oxidative stress, decreased stomatal conductance, and the photosynthetic rate, which were reversed by inoculation of Cu-resistance bacteria. Table 1.2 represents usual biochemical and physiological disturbance in plants experienced under HM's toxicity.

The HMPs pollution is an environmental issue and so also is it's damaging effects on the environment and human health and there is a demand to reform the contaminated soil but it is a very tough task because metals are not biodegradable.[84]

1.2.2 STRATEGIES TO DECREASE THE NEGATIVE INFLUENCE OF HM-POLLUTED SOILS ON PLANT PHYSIOLOGICAL AND BIOCHEMICAL PARAMETERS

The excessive content of HMPs on the plant can only be limited by either their complete exclusion from the soil system or their deactivation. In the

TABLE 1.2 Effect of Heavy Metals on Crop Physiology and Biochemical Attributes.

Heavy metal	Crop	Effect on physiology	Effect on biochemistry	References
Cd, Pb	*Spinacia oleracea* L.	Reduce the photosynthesis and transpiration rate, also lowers the stomatal conductance and membrane stability index	Lowers the carotenoids and chlorophyll contents	[69]
Cu, Pb, Cd	*Triticum aestivum* L.	Lower the root activity, chlorophyll content	Increases the MDA and proline content	[70]
Ni, Cd, Pb	*Lycopersicum esculentum* L.		Significant reduction in carotene and chlorophyll *a* and *b*	[71]
Pb, Zn, Cd	*Sida hermaphrodita* L.	Reduced evapotranspiration	Reduction in chlorophyll	[72]
Cd	*Centella asiatica* L.		Reduces the chlorophyll and carotenoids while increases the flavonoids and phenolics	[73]
Cu, Cd	*Arabidopsis thaliana*	Lower the photosynthetic activity	Reduces the accumulation of chlorophyll	[74]
As	*Cicer arietinum*		Significantly reduces the chlorophyll and increases the MDA contents	[75]
Cr	*Brassica napus* L.	Reduced the stomatal conductance and net photosynthesis	Increases the MDA and H_2O_2, and decrease in chlorophyll and carotenoid was observed	[76]
Pb	*Helianthus annuus* L.		Higher proline and MDA contents and lower chlorophyll and carotenoids were observed	[77]
Pb, Zn, Cd	*Cynara cardunculus* L.	Reduces the transpiration rate, shoot water content, stomatal conductance		[78]
Cr	*Zea mays* L.	Reduces the chlorophyll, carotene, photosynthetic and transpiration rate, and stomatal conductance	Increases the soluble sugars and proline, while decreases the protein and starch contents	[79]

TABLE 1.2 *(Continued)*

Heavy metal	Crop	Effect on physiology	Effect on biochemistry	References
Cu, Cd	*Pisum sativum* L.	Photosynthetic rate and chlorophyll content were reduced significantly		[80]
Cu	*Cucumis sativus* L.	Chlorophyll contents increased with increased Cu conc.		[81]
		Root volume and area was reduced		
Pb, Cd	*Morus alba* L.	The activity of PSI and PSII was reduced	H_2O_2 conc. increased	[82]
Cr	*Helianthus annuus* L.	Reduces the carotenoids and chlorophyll	Reduces the soluble proteins	[24]
As	*Glysine max* L.	Reduces the stomatal conductance	Increases the proline and soluble sugars	[83]

first case, phytoextraction plays a vital role while, in the second case, use of organic and inorganic amendments are well-tested approaches.

1.2.2.1 USING PLANTS FOR REMEDIATION OF HM-POLLUTED SOILS

Bioremediation is the utilization of living organisms for polluted soil treatment. This is a globally accepted way of remediation because it involves natural mechanisms. HH bioremediation can be attained using plants or microorganisms. Phyto-remediation is a process that uses plants for polluted soil treatment. It is appropriate when the contaminants cover a large area or come under the plant root zone.[85] HH phytoremediation includes phytoextraction, phytovolatilization, and phytostabilization. In phytoextraction, the hyperaccumulator plants are used to accumulate HMs in their tissues, which later can be discarded or can be used as a fuel. It is highly dependent upon the nature of the plant, its metal extraction, and assimilation capabilities, as well as its potential to produce under such contaminated environments.[32,86] Phytostabilization, on the other hand, is an approach in which plant roots tend to stabilize metals in their rhizosphere via adopting complexation, sorption, and metal-valance reduction techniques. Its efficiency depends upon the type of crop plant, root structure, and exudates chemistry, as well as amendments being added for the same purpose.[18,87]

1.2.2.2 USING ORGANIC AND INORGANIC AMENDMENTS

Stabilization of HMPs in the soil is another suitable approach in which various organic and inorganic amendments are applied to decrease the bioavailable concentration of metals in soil. A lot of research work has been published regarding gypsum,[88] phosphatic fertilizers,[89] nanoparticles,[90] farm manure, compost,[1] and many more.

1.3 CONCLUSION

The term *heavy metal* is very frequently used for metals with specific gravity >4.5–5. The two major categories in HMs are heavy metal nutrients (HMNs) and heavy metal pollutants (HMPs); excess of both is toxic to plant and HMPs are even toxic in low concentrations. Once onset of HMP's toxicity is experienced, the plant develops severe physiological and biochemical disturbances, like retardation of photosynthesis, chlorophyll biosynthesis, photosystem deactivation, ROS production, antioxidant production disruption, and disturbance in biomolecules biosynthesis. All of these are responsible for deteriorated health of crop plant, which can ultimately affect plant production. The hyperaccumulator class of plants can accumulate significant concentrations of these HMPs and is considered to be a good alternative to be used for remediation of contaminated soil and so does the use of various available organic and inorganic amendments.

KEYWORDS

- **heavy metal nutrients**
- **carbon assimilation**
- **oxidative balance**
- **phytohormones homeostasis**

REFERENCES

1. Lwin, C. S.; Seo, B. H.; Kim, H. U.; Owens, G.; Kim, K. R. Application of Soil Amendments to Contaminated Soils for Heavy Metal Immobilization and Improved Soil Quality—A Critical Review. *Soil Sci. Plant Nutr.* **2018,** *64*, 156–167.
2. Rajput, V.; Minkina, T.; Sushkova, S.; Behal, A.; Maksimov, A.; Blicharska, E.; Ghazaryan, K.; Movsesyan, H.; Barsova, N. ZnO and CuO Nanoparticles: A Threat to Soil Organisms, Plants, and Human Health. *Environ. Geochem. Health* **2020,** *42*, 147–158.
3. Andrey, G.; Rajput, V.; Tatiana, M.; Saglara, M.; Svetlana, S.; Igor, K.; Grigoryeva, T. V.; Vasily, C.; Iraida, A.; Vladislav, Z. The Role of Biochar-Microbe Interaction in Alleviating Heavy Metal Toxicity in *Hordeum vulgare* L. Grown in Highly Polluted Soils. *Appl. Geochem.* **2019,** *104*, 93–101.
4. Onakpa, M. M.; Njan, A. A.; Kalu, O. C. A Review of Heavy Metal Contamination of Food Crops in Nigeria. *Ann. Glob. Health* **2018,** *84*, 488–497.
5. González-Ortegón, E.; Amaral, V.; Baldó, F.; Sánchez-Leal, R. F.; Bellanco, M. J.; Jiménez, M. P.; Forja, J.; Vilas, C.; Tovar-Sanchez, A. Sources and Coastal Distribution of Dissolved Organic Matter in the Gulf of Cadiz. *Sci. Total Environ.* **2018,** *630*, 1583–1595.
6. Nagajyoti, P. C.; Lee, K. D.; Sreekanth, T. Heavy Metals, Occurrence and Toxicity for Plants: A Review. *Environ. Chem. Lett.* **2010,** *8*, 199–216.
7. Ercal, N.; Gurer-Orhan, H.; Aykin-Burns, N. Toxic Metals and Oxidative Stress. Part I: Mechanisms Involved in Metal-Induced Oxidative Damage. *Curr. Top. Med. Chem.* **2001,** *1*, 529–539.

8. Tovar-Sánchez, E.; Cervantes-Ramírez, T.; Castañeda-Bautista, J.; Gómez-Arroyo, S.; Ortiz-Hernández, L.; Sánchez-Salinas, E.; Mussali-Galante, P. Response of *Zea mays* to Multimetal Contaminated Soils: A Multibiomarker Approach. *Ecotoxicology* **2018,** *27*, 1161–1177.
9. Ghazaryan, K. A.; Movsesyan, H. S.; Khachatryan, H. E.; Ghazaryan, N. P.; Minkina, T. M.; Sushkova, S. N.; Mandzhieva, S. S.; Rajput, V. D. Copper Phytoextraction and Phytostabilization Potential of Wild Plant Species Growing in the Mine Polluted Areas of Armenia. *Geochem. Explor. Environ. Anal.* **2019,** *19*, 155–163.
10. Santoyo-Martínez, M.; Mussali-Galante, P.; Hernández-Plata, I.; Valencia-Cuevas, L.; Flores-Morales, A.; Ortiz-Hernández, L.; Flores-Trujillo, K.; Ramos-Quintana, F.; Tovar-Sánchez, E. Heavy Metal Bioaccumulation and Morphological Changes in *Vachellia campechiana* (Fabaceae) Reveal Its Potential for Phytoextraction of Cr, Cu, and Pb in Mine Tailings. *Environ. Sci. Pollut. Res.* **2020,** *27*, 11260–11276.
11. Muro-González, D. A.; Mussali-Galante, P.; Valencia-Cuevas, L.; Flores-Trujillo, K.; Tovar-Sánchez, E. Morphological, Physiological, and Genotoxic Effects of Heavy Metal Bioaccumulation in *Prosopis laevigata* Reveal Its Potential for Phytoremediation. *Environ. Sci. Pollut. Res.* **2020,** *27*, 40187–40204.
12. Antoniadis, V.; Robinson, J; Alloway; B. Effects of Short-Term pH Fluctuations on Cadmium, Nickel, Lead, and Zinc Availability to Ryegrass in a Sewage Sludge-Amended Field. *Chemosphere* **2008,** *71*, 759–764.
13. Ali, H.; Khan, E.; Ilahi, I. Environmental Chemistry and Ecotoxicology of Hazardous Heavy Metals: Environmental Persistence, Toxicity, and Bioaccumulation. *J. Chem.* **2019,** 1–14.
14. Ayub, M. A.; Usman, M.; Faiz, T.; Umair, M., ul Haq, M. A.; Rizwan, M.; Ali, S.; ur Rehman, M. Z. Restoration of Degraded Soil for Sustainable Agriculture. *Soil Health Restoration and Management*; Springer: Singapore, 2020; pp 31–81.
15. Farooqi, Z. U. R.; Ahmad, Z.; Ayub, M. A.; Umar, W.; Nadeem, M.; Fatima, H.; Shoukat, A.; Ali, T.; Ashraf, M. I. Threats to Arable Land of the World: Current and Future Perspectives of Land Use. *Examining International Land Use Policies, Changes, and Conflicts*; IGI Global: Hershey, PA, 2021; pp 186–209.
16. Mostert, M. M.; Ayoko, G. A.; Kokot, S. Application of Chemometrics to Analysis of Soil Pollutants. *TrAC—Trend Anal. Chem.* **2010,** *29*, 430–445.
17. Cheng, S. Heavy Metal Pollution in China: Origin, Pattern and Control. *Environ. Sci. Pollut. Res.* **2003,** *10*, 192–198.
18. Jadia, C. D.; Fulekar, M. Phytoremediation of Heavy Metals: Recent Techniques. *Afr. J. Biotechnol.* **2009,** *8*, 921–928.
19. Knittel, C. R. Automobiles on Steroids: Product Attribute Trade-Offs and Technological Progress in the Automobile Sector. *Am. Econ. Rev.* **2011,** *101*, 3368–3399.
20. Sardar, K.; Ali, S.; Hameed, S.; Afzal, S.; Fatima, S.; Shakoor, M. B.; Bharwana, S. A.; Tauqeer, H. M. Heavy Metals Contamination and What Are the Impacts on Living Organisms. *GJEMPS* **2013,** *2*, 172–179.
21. Naeem, A.; Zafar, M.; Khalid, H.; Rehman, M. Z.; Ahmad, Z.; Ayub, M. A; and Qayyum, M. F. Cadmium-Induced Imbalance in Nutrient and Water Uptake by Plants. *Cadmium Toxicity and Tolerance in Plants*; Academic Press: Cambridge, MA, 2019; pp 299–326.
22. Zulfiqar, U.; Farooq, M.; Hussain, S.; Maqsood, M.; Hussain, M.; Ishfaq, M.; Ahmad, M.; Anjum, M. Z. Lead Toxicity in Plants: Impacts and Remediation. *J. Environ. Manage.* **2019,** *250*, 109557.

23. Gottesfeld, P.; Were, F. H.; Adogame, L.; Gharbi, S.; San, D.; Nota, M. M.; Kuepouo, G. Soil Contamination from Lead Battery Manufacturing and Recycling in Seven African Countries. *Environ. Res.* **2018,** *161*, 609–614.
24. Farid, M.; Ali, S.; Rizwan, M.; Ali, Q.; Abbas, F.; Bukhari, S. A. H.; Wu, L. Citric Acid Assisted Phytoextraction of Chromium by Sunflower: Morpho-physiological and Biochemical Alterations in Plants. *Ecotoxicol. Environ. Saf.* **2017,** *145*, 90–102.
25. Beckers, F.; Rinklebe, J. Cycling of Mercury in the Environment: Sources, Fate, and Human Health Implications: A Review. *Crit. Rev. Environ. Sci. Technol.* **2017,** *47*, 693–794.
26. Abbas, G.; Murtaza, B.; Bibi, I.; Shahid, M.; Niazi, N. K.; Khan, M. I.; Amjad, M.; Hussain, M. Arsenic Uptake, Toxicity, Detoxification, and Speciation in Plants: Physiological, Biochemical, and Molecular Aspects. *Int. J. Environ. Res. Public Health* **2018,** *15*, 59.
27. Fendorf, S.; Nico, P. S.; Kocar, B. D.; Masue, Y; Tufano, K. J. Arsenic Chemistry in Soils and Sediments. *Developments in Soil Science*; Elsevier: Amsterdam, 2010; pp 357–378.
28. Broadley, M.; Brown, P.; Cakmak, I. Function of nutrients. *Marschner's Mineral Nutrition of Higher Plants*; Academic Press: London, 2012; pp 191–248.
29. Waters, B.; Sankaran, R. Moving Micronutrients from the Soil to the Seeds: Genes and Physiological Processes from a Biofortification Perspective. *Plant Sci.* **2011,** *180*, 562–574.
30. Singh, R.; Tripathi, R.; Dwivedi, S.; Kumar, A.; Trivedi, P.; Chakrabarty, D. Lead Bioaccumulation Potential of an Aquatic Macrophyte *Najas indica* Are Related to Antioxidant System. *Bioresour. Technol.* **2010,** *101*, 3025–3032.
31. Ravet, K.; Pilon, M. Copper and Iron Homeostasis in Plants: The Challenges of Oxidative Stress. *Antioxid. Redox Signal.* **2013,** *19*, 919–932.
32. Marques, A. P.; Rangel, A. O.; Castro, P. M. Remediation of Heavy Metal Contaminated Soils: Phytoremediation as a Potentially Promising Clean-Up Technology. *Crit. Rev. Environ. Sci. Technol.* **2009,** *39*, 622–654.
33. Silva, J. C.; Echeveste, P.; Lombardi, A. T. Higher Biomolecules Yield in Phytoplankton under Copper Exposure. *Ecotoxicol. Environ. Saf.* **2018,** *161*, 57–63.
34. Nozoye, T.; Nagasaka, S.; Kobayashi, T.; Sato, Y.; Uozumi, N.; Nakanishi, H. The Phytosiderophore Efflux Transporter TOM2 Is Involved in Metal Transport in Rice. *J. Biol. Chem.* **2011,** *290*, 27688–27699.
35. Wang, M.; Kawakami, Y.; Bhullar, N. K. Molecular Analysis of Iron Deficiency Response in Hexaploid Wheat. *Front. Sustain. Food Syst.* **2019,** *3*, 67.
36. Nkrumah, P. N.; Echevarria, G.; Erskine, P. D.; Chaney, R. L.; Sumail, S.; Van der Ent, A. Effect of Nickel Concentration and Soil pH on Metal Accumulation and Growth in Tropical Agromining 'Metal Crops'. *Plant Soil* **2019,** *443*, 27–39.
37. Gomes, M. P.; Nogueira, M. O. G.; Castro, E. M. D.; Soares, A. M. Ecophysiological and Anatomical Changes Due to Uptake and Accumulation of Heavy Metal in *Brachiaria decumbens*. *Sci. Agric.* **2011,** *68*, 566–573.
38. Mendel, R. R.; Kruse, T. Cell Biology of Molybdenum in Plants and Humans. *Biochim. Biophys. Acta Mol. Cell Res.* **2012,** *1823*, 1568–1579.
39. Rana, M.; Bhantana, P.; Sun, X. C.; Imran, M.; Shaaban, M.; Moussa, M.; Afzal, J. Molybdenum as an Essential Element for Crops: An Overview. *Int. J. Scien. Res. Growth* **2020,** *24*, 18535.

40. McGrath, S. P.; Zhao, F. J. Phytoextraction of Metals and Metalloids from Contaminated Soils. *Curr. Opin. Biotechnol.* **2003,** *14*, 277–282.
41. Kluwer, R.; Millaleo, R.; Reyes-Dı´az, M.; Ivanov, A. G.; Mora, M. L.; Alberdi, M. Manganese as Essential and Toxic Element for Plants: Transport, Accumulation and Resistance Mechanisms. *J. Soil Sci. Plant Nutr.* **2010,** *10*, 470–448.
42. Andresen, E.; Peiter, E.; Küpper, H. Trace Metal Metabolism in Plants. *J. Exp. Bot.* **2018,** *69*, 909–954.
43. Yano, J.; Yachandra, V. Mn4Ca Cluster in Photosynthesis: Where and How Water Is Oxidized to Dioxygen. *Chem. Rev.* **2014,** *114*, 4175–4205.
44. Rosas-Ramírez, M. Relación entre la bioacumulación de metales pesados y la concentración de clorofila en sanvitalia procumbens, Tesis de licenciatura. Facultad de Ciencias Biológicas, Universidad Autónoma, 2018.
45. Yllanes, P.; Vélez-Azañero, A.; Lozano, S. Efectos fitotóxicos del plomo en maíz híbrido dekalb (*Zea mays* L.) en suelo arenoso y limoso. *Biologist* **2014,** *12*, 337–348.
46. DalCorso, G. Heavy Metal Toxicity in Plants. *Plants and Heavy Metals*; Springer: Berlin, 2012; pp 1–25.
47. Manara, A. Plant Responses to Heavy Metal Toxicity. *Plants and Heavy Metals*; Springer: Berlin, 2012; pp 27–53.
48. Ruiz Huerta, E. A.; Armienta Hernández, M. A. Acumulación de arsénico y metales pesados en maíz en suelos cercanos a jales o residuos mineros. *Rev. Int. Contam. Ambient.* **2012,** *28*, 103–117.
49. Yadav, K.; Singh, N. Effects of Benzoic Acid and Cadmium Toxicity on Wheat Seedlings. *Chil. J. Agric. Res.* **2013,** *73*, 168–174.
50. Aggarwal, A.; Sharma, I.; Tripathi, B.; Munjal, A.; Baunthiyal, M.; Sharma, V. Metal Toxicity and Photosynthesis. *Photosynthesis: Overviews on Recent Progress and Future Perspectives*; I.K. International Publishing House Pvt. Ltd.: New Delhi, 2012, pp 229–236.
51. Pourrut, B.; Shahid, M.; Dumat, C.; Winterto, P.; Pinelli, E. Lead Uptake, Toxicity, and Detoxification in Plants. *Rev. Environ. Contam. T* **2011,** *213*, 113–136.
52. Leng, Y.; Li, Y.; Ma, Y. H.; He, L. F.; Li, S. W. Abscisic Acid Modulates Differential Physiological and Biochemical Responses of Roots, Stems, and Leaves in Mung Bean Seedlings to Cadmium Stress. *Environ. Sci. Pollut. Res.* **2021,** *28*, 6030–6043.
53. Farooq, M. A.; Niazi, A. K.; Akhtar, J.; Farooq, M.; Souri, Z.; Karimi, N.; Rengel, Z. Acquiring Control: The Evolution of ROS-Induced Oxidative Stress and Redox Signaling Pathways in Plant Stress Responses. *Plant Physiol. Biochem.* **2019,** *141*, 353–369.
54. Suzuki, N.; Koussevitzky, S.; Mittler, R.; Miller, G. Ros and Redox Signalling in the Response of Plants to Abiotic Stress. *Plant Cell Environ.* **2012,** *35*, 259–270.
55. Abd-Alla, M. H.; Bagy, M. K.; El-enany, A.; Bashandy, S. R. Activation of *Rhizobium tibeticum* with Flavonoids Enhances Nodulation, Nitrogen Fixation, and Growth of Fenugreek (*Trigonella foenum-graecum* L.) Grown in Cobalt-Polluted Soil. *Arch. Environ. Con. Toxicol.* **2014,** *66*, 303–315.
56. Khalid, S.; Shahid, M.; Niazi, N. K.; Rafiq, M.; Bakhat, H. F.; Imran, M.; Abbas, T.; Bibi, I.; Dumat, C. Arsenic Behaviour in Soil-Plant System: Biogeochemical Reactions and Chemical Speciation Influences. *Enhancing Cleanup of Environmental Pollutants*; Springer: Berlin, 2017; pp 97–140.

57. Finnegan, P.; Chen, W. Arsenic Toxicity: The Effects on Plant Metabolism. *Front. Physiol.* **2012,** *3*, 182.
58. Duman, F.; Ozturk, F.; Aydin, Z. Biological Responses of Duckweed (*Lemna minor* L.) Exposed to the Inorganic Arsenic Species as (III) and as (V): Effects of Concentration and Duration of Exposure. *Ecotoxicology* **2010,** *19*, 983–993.
59. Rafiq, M.; Shahid, M.; Shamshad, S.; Khalid, S.; Niazi, N. K.; Abbas, G.; Saeed, M. F.; Ali, M.; Murtaza, B. A Comparative Study to Evaluate Efficiency of EDTA and Calcium in Alleviating Arsenic Toxicity to Germinating and Young *Vicia faba* L. Seedlings. *J. Soil. Sediment.* **2018,** *18*, 2271–2281.
60. Nas, F. S.; Ali, M. The Effect of Lead on Plants in Terms of Growing and Biochemical Parameters: A Review. *MOJ Ecol. Environ. Sci.* **2018,** *3*, 265–268.
61. Vajravel, S.; Saravanan, P. Accumulation of Chromium and Its Effects on Physiological and Biochemical Parameters of *Alternanthera philoxeroides* Seedlings. *J. Pharm. Res.* **2013,** *7*, 633–639.
62. Subba, P.; Mukhopadhyay, M.; Mahato, S. K.; Bhutia, K. D.; Mondal, T. K.; Ghosh, S. K. Zinc Stress Induces Physiological, Ultra-structural and Biochemical Changes in Mandarin Orange (*Citrus reticulata blanco*) Seedlings. *Physiol. Mol. Biol. Plants* **2014,** *20*, 461–473.
63. Verma, S.; Dubey, R. Lead Toxicity Induces Lipid Peroxidation and Alters the Activities of Antioxidant Enzymes in Growing Rice Plants. *Plant Sci.* **2003,** *164*, 645–655.
64. Lin, Y. C.; Kao, C. H. Nickel Toxicity of Rice Seedlings: The Inductive Responses of Antioxidant Enzymes by $NiSO_4$ in Rice Roots. *Crop Environ. Biol.* **2005,** *2*, 330–335.
65. Khan, M. A.; Ahmad, I.; Rahman, I. U. Effect of Environmental Pollution on Heavy Metals Content of *Withania somnifera. J. Chin. Chem. Soc.* **2007,** *54*, 339–343.
66. Khan, M. A.; Saeed, S.; Ullah, N.; Rukh, S.; Javed, M. S.; Amjad, A.; Jamil, M. U.; Shah, M. Effect of Nickel on the Germination and Biochemical Parameters of Two Rice Varieties. *Fresenius Environ. Bull.* **2020,** *29*, 956–963.
67. Azooz, M. M.; Abou-Elhamd, M. F.; Al-Fredan, M. A. Biphasic Effect of Copper on Growth, Proline, Lipid Peroxidation and Antioxidant Enzyme Activities of Wheat ('*Triticum aestivum*' cv. *Hasaawi*) at Early Growing Stage. *Aust. J. Crop Sci.* **2012,** *6*, 688–694.
68. Islam, F.; Yasmeen, T.; Ali, Q.; Mubin, M.; Ali, S.; Arif, M. S.; Hussain, S.; Riaz, M.; Abbas, F. Copper-Resistant Bacteria Reduces Oxidative Stress and Uptake of Copper in Lentil Plants: Potential for Bacterial Bioremediation. *Environ. Sci. Pollut. Res.* **2016,** *23*, 220–233.
69. Desoky, E. S. M.; Merwad, A. R. M.; Semida, W. M.; Ibrahim, S. A.; El-Saadony, M. T.; Rady, M. M. Heavy Metals-Resistant Bacteria (HM-RB): Potential Bioremediators of Heavy Metals-Stressed *Spinacia oleracea* Plant. *Ecotoxicol. Environ. Saf.* **2020,** *198*, 110685.
70. Liang, P. P.; Chen, Z.; Yuan, L.; Geng, J. J.; Yuan, C.; Chen, D. H.; Zhang, X. Effects of Sodium Benzoate on Growth and Physiological Characteristics of Wheat Seedlings under Compound Heavy Metal Stress. *J. Integr. Agric.* **2020,** *19*, 1010–1018.
71. Zeeshan, M.; Ahmad, W.; Hussain, F.; Ahamd, W.; Numan, M.; Shah, M.; Ahmad, I. Phytostabalization of the Heavy Metals in the Soil with Biochar Applications, the Impact on Chlorophyll, Carotene, Soil Fertility and Tomato Crop Yield. *J. Clean. Prod.* **2020,** *255*, 120318.

72. Bury, M.; Rusinowski, S.; Sitko, K.; Krzyżak, J.; Kitczak, T.; Możdżer, E.; Pogrzeba, M. Physiological Status and Biomass Yield of *Sida hermaphrodita* (L.) Rusby Cultivated on Two Distinct Marginal Lands in Southern and Northern Poland. *Ind. Crops Prod.* **2021,** *167*, 113502.
73. Biswas, T.; Parveen, O.; Pandey, V. P.; Mathur, A.; Dwivedi, U. N. Heavy Metal Accumulation Efficiency, Growth and Centelloside Production in the Medicinal Herb *Centella asiatica* (L.) Urban under Different Soil Concentrations of Cadmium and Lead. *Ind. Crops Prod.* **2020,** *157*, 112948.
74. Maksymiec, W.; Krupa, Z. Jasmonic Acid and Heavy Metals in *Arabidopsis* plants—A Similar Physiological Response to Both Stressors? *J. Plant Physiol.* **2002,** *159* (5), 509–515.
75. Adhikary, A.; Kumar, R.; Pandir, R.; Bhardwaj, P.; Wusirika, R.; Kumar, S. *Pseudomonas citronellolis*: A Multi-metal Resistant and Potential Plant Growth Promoter against Arsenic (V) Stress in Chickpea. *Plant Physiol. Biochem.* **2019,** *142*, 179–192.
76. Zaheer, I. E.; Ali, S.; Saleem, M. H.; Imran, M.; Alnusairi, G. S.; Alharbi, B. M.; Soliman, M. H. Role of Iron–Lysine on Morpho-physiological Traits and Combating Chromium Toxicity in Rapeseed (*Brassica napus* L.) Plants Irrigated with Different Levels of Tannery Wastewater. *Plant Physiol. Biochem.* **2020,** *155*, 70–84.
77. Saleem, M.; Asghar, H. N.; Zahir, Z. A.; Shahid, M. Impact of Lead Tolerant Plant Growth Promoting Rhizobacteria on Growth, Physiology, Antioxidant Activities, Yield and Lead Content in Sunflower in Lead Contaminated Soil. *Chemosphere* **2018,** *195*, 606–614.
78. Hernández-Allica, J.; Garbisu, C.; Barrutia, O.; Becerril, J. M. EDTA-Induced Heavy Metal Accumulation and Phytotoxicity in Cardoon Plants. *Environ. Exp. Bot.* **2007,** *60*, 26–32.
79. Bashir, M. A.; Naveed, M.; Ahmad, Z.; Gao, B.; Mustafa, A.; Núñez-Delgado, A. Combined Application of Biochar and Sulfur Regulated Growth, Physiological, Antioxidant Responses and Cr Removal Capacity of Maize (*Zea mays* L.) in Tannery Polluted Soils. *J. Environ. Manage.* **2020,** *259*, 110051.
80. Hattab, S.; Dridi, B.; Chouba, L.; Kheder, M. B.; Bousetta, H. Photosynthesis and Growth Responses of Pea *Pisum sativum* L. under Heavy Metals Stress. *J. Environ. Sci.* **2009,** *21*, 1552–1556.
81. Feil, S. B.; Pii, Y.; Valentinuzzi, F.; Tiziani, R.; Mimmo, T.; Cesco, S. Copper Toxicity Affects Phosphorus Uptake Mechanisms at Molecular and Physiological Levels in *Cucumis sativus* Plants. *Plant Physiol. Biochem.* **2020,** *157*, 138–147.
82. Huihui, Z.; Xin, L.; Zisong, X.; Yue, W.; Zhiyuan, T.; Meijun, A.; Guangyu, S. Toxic Effects of Heavy Metals Pb and Cd on Mulberry (*Morus alba* L.) Seedling Leaves: Photosynthetic Function and Reactive Oxygen Species (ROS) Metabolism Responses. *Ecotoxicol. Environ. Saf.* **2020,** *195*, 110469.
83. Vezza, M. E.; Llanes, A.; Travaglia, C.; Agostini, E.; Talano, M. A. Arsenic Stress Effects on Root Water Absorption in Soybean Plants: Physiological and Morphological Aspects. *Plant Physiol. Biochem.* **2018,** *123*, 8–17.
84. Rai, P. K.; Lee, S. S.; Zhang, M.; Tsang, Y. F.; Kim, K. H. Heavy Metals in Food Crops: Health Risks, Fate, Mechanisms, and Management. *Environ. Int.* **2019,** *125*, 365–385.
85. Garbisu, C.; Alkorta, I. Basic Concepts on Heavy Metal Soil Bioremediation. *Eur. J. Min. Process. Environ. Protect.* **2003,** *3*, 58–66.

86. Jabeen, R.; Ahmad, A; Iqbal, M. Phytoremediation of Heavy Metals: Physiological and Molecular Mechanisms. *Bot. Rev.* **2009,** *75*, 339–364.
87. Chibuike, G. U.; Obiora, S. C. Heavy Metal Polluted Soils: Effect on Plants and Bioremediation Methods. *Appl. Environ. Soil. Sci.* **2014,** *2014*, 1–13.
88. Dubrovina, T. A.; Losev, A. A.; Karpukhin, M. M.; Vorobeichik, E. L.; Dovletyarova, E. A.; Brykov, V. A.; Brykova, R. A.; Ginocchio, R.; Yáñez, C.; Neaman, A. Gypsum Soil Amendment in Metal-Polluted Soils, an Added Environmental Hazard. *Chemosphere* **2021,** *281*, 130889.
89. Gupta, D. K.; Chatterjee, S.; Datta, S.; Veer, V.; Walther, C. Role of Phosphate Fertilizers in Heavy Metal Uptake and Detoxification of Toxic Metals. *Chemosphere* **2014,** *108*, 134–144.
90. Baragaño, D.; Forján, R.; Welte, L.; Gallego, J. L. R. Nanoremediation of As and Metals Polluted Soils by Means of Graphene Oxide Nanoparticles. *Sci. Rep.* **2020,** *10*, 1–10.

CHAPTER 2

Pesticides and Their Impacts: Benefits and Hazards

LUBNA NAJAM[1*] and TANVEER ALAM[2]

[1]*DAV(PG) College Muzaffarnagar, CCS University, Meerut, UP, India*

[2]*Natural and Medical Sciences Research Center, University of Nizwa, Sultanate of Oman*

[*]*Corresponding author. E-mail: najam.lubna@gmail.com*

ABSTRACT

Pesticides are chemical compounds used to kill pests, such as insects, rodents, fungus, and weeds (unwanted plants). More than 1000 types of pesticides (insecticides, herbicides, and fungicides) are used to kill pests all over the world. Pesticides are a necessary component of human life since they are used to prevent the growth of undesirable organisms in crop areas. According to studies, just 0.1% of pesticides reach the target pest and the remaining amount of the pesticides is absorbed by the surrounding environment. In fact, there are two aspects to evaluate the application of pesticides. The first one is to evaluate the risk assessment (hazards) and the second one is the economic and agriculture benefits. Pesticides, despite their advantages, may be harmful to both humans and the environment. Therefore, in this chapter, the author aims to provide a scientifically based assessment of the benefits and hazards of pesticides used in human life.

Environmental Pollution Impact on Plants: Survival Strategies under Challenging Conditions. Tariq Aftab, PhD & Khalid Rehman Hakeem, PhD (Eds.)

2.1 INTRODUCTION

2.1.1 PESTICIDES

Pesticides are chemical compounds used to kill, prevent, repel any pests, for example, insects, rodents, plant pathogens, molluscs, birds, mammals, fungi, and unwanted plants (weeds).[73] More than 1000 types of pesticides (insecticides, herbicides, and fungicides) are used to kill pests globally.

> Pesticides are chemicals that are used to prevent, kill, repel, or minimize pests, such as insects, animals, and plants, as well as microorganisms.
>
> —Grube et al.[105]

According to Food and Agriculture Organization,[83]

> A pesticide is any substance or mixture of substances intended for preventing, destroying or controlling any pest, including vectors of human or animal disease, unwanted species of plants or animals, causing harm during or otherwise interfering with the production, processing, storage, transport, or marketing of food, agricultural commodities, wood and wood products or animal feedstuffs, or substances that may be administered to animals for the control of insects, arachnids, or other pests in or on their bodies.

In general, pesticides should be toxic to the pests that are being targeted, but not to nontarget species, including human beings. It is said to say that this is not the case. A major debate about pesticide usage and abuse has erupted recently. The uncontrolled applications of these pesticides, under the proverb, "if little is good, a lot more will be better" has developed chaos in human and other life forms.

Chronic pesticide risk has been enhanced by applying modern technology. In the 1950s, tiny amounts of pesticides could be detected at one part per million (ppm). Scientists discovered a trace quantity of pesticide in one part per billion (ppb) by 1965. By 1975, part per trillion (ppt) was used to detect the traces of pesticides, and presently we are using one part per quadrillion (ppq) or picograms per liter (pg/L) for detection. After decades of the use of pesticides in the United States, it is found that pesticide residues are present almost everywhere. Pesticide residues found in our food, air, or water may be quantified using advanced detection technology. Knowing the importance of sensitive pesticides residue detection abilities, the maximum allowable residue levels have been established by the EPA, which is called "tolerances." The presence of pesticide residues according to the EPA's range of "acceptable" risks is permissible in food commodities. The EPA has gone to extra mile to make the tolerances conventional. Pesticide usage in India is drastically different from that in the rest of the globe. It can be presented in Figure 2.1. In India, 76% of insecticide is used as a pesticide while it is 44% globally.[154]

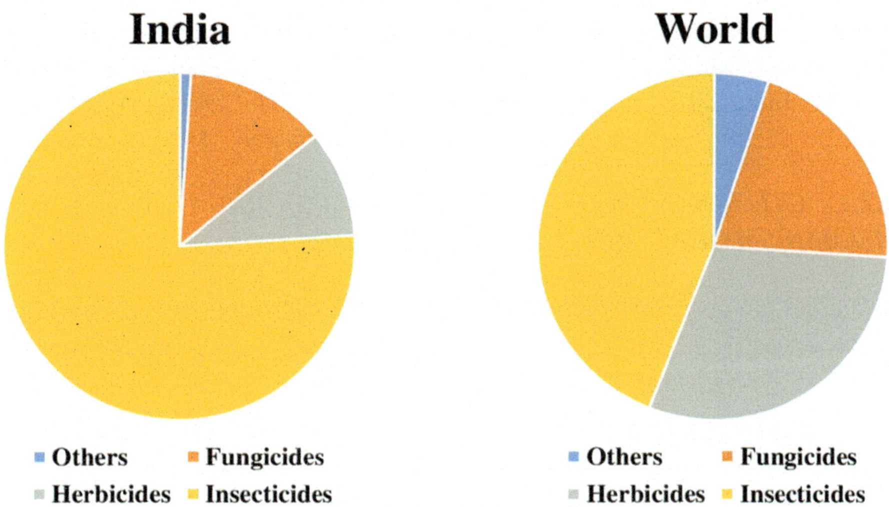

FIGURE 2.1 Consumption pattern of pesticides.

2.1.2 FORMULATIONS OF PESTICIDES

In the markets, pesticides are found in three different forms, namely, liquid, solid, and gaseous forms.

Emulsifiable Concentrates (E or EC): Emulsifiable Concentrates are tiny suspensions of oil droplets in water that have a milky appearance. Emulsifiable Concentrates do not need to be mixed before use.

Wettable Powders (WP): Wettable Powders are water-based suspensions of tiny particles. Before each use, the suspensions must be shaken.

Granules (G): Granules are produced by blending the active component with clay for outside uses.

Baits: Baits are produced by blending the main components with the food base.

Dust (D): Dust applied directly with the help of common carriers, namely, clay, talc, silica gel.

Fumigants (F): Fumigants are found in gaseous forms and generally have been packed under pressure and kept as liquids. Some fumigants are available in tablet forms that release gas after mixing with water.

2.2 CLASSIFICATION OF PESTICIDES

A pesticide can be differentiated by its physical and chemical properties. However, pesticides can be broadly classified in different ways depending based on their mechanism of action (Table 2.1), their origin (Table 2.2), targeted pest species they kill (Table 2.3), and the pesticide's chemical content (Table 2.4).

2.2.1 CATEGORIZATION OF PESTICIDES ACCORDING TO THEIR MECHANISM OF ACTION

Pesticides are grouped according to their work to accomplish the objectives (Table 2.1).

TABLE 2.1 Categorization of Pesticides According to Their Mechanism of Action.

S. No.	Mode of Action	Sources	Examples
1.	Contact (nonsystemic)	Pesticides cannot penetrate the plant tissues and not translocate via phloem or xylem	Pirymethanil, chlorpyrifos
2.	Systemic	Pesticides that invade the vascular system adequately and translocate via phloem or xylem	Acetamiprid, boscalid or difenoconazole[43]

2.2.2 CATEGORIZATION OF PESTICIDES BASED ON ORIGIN FROM NATURAL SOURCES

In this type of classification, pesticides are differentiated based on their occurrence (Table 2.2).

TABLE 2.2 Classification of Pesticides Based on Origin.

S. No.	Sources	Examples
1.	Organic sources	Phytoconstituents (volatile oils, herbal extracts), synthetically produced by chemical reactions, e.g., DDT, BHC
2.	Inorganic sources	Ferrous sulfate, copper sulphate, sulfur, and copper
3.	Biological sources	Bacteria, viruses, and fungi

2.2.3 CLASSIFICATIONS OF PESTICIDES BASED ON TARGET PEST SPECIES

In this sort of classification, pesticides are classified based on the respective pest which they kill as given in Table 2.3.

TABLE 2.3 Classifications of Pesticides Based on Target Pest Species.

S. No.	Pesticide Class	Purpose	Example
1.	Acaricides	To kill mites	Bifonazole
2.	Algaecides	To kill algae	Benzalkonium chloride
3.	Avicides	To repel birds	Strychnine
4.	Bactericides	To kill bacteria	Agri-Mycin 17 (streptomycin sulfate)
5.	Fungicides	To kill fungi	**Actinovate AG (*streptomyces lydicus*)**
6.	Herbicides	To kill weeds	2,4-dichlorophenoxyacetic acid (2,4-D)
7.	Insecticides	To kill or repel insects	Aldicarb (Temik)
8.	Larvicides	To kill larvae	Methoprene
9.	Molluscicides	To kill or repel snail	Metaldehyde
10.	Nematicides	To kill nematodes	Aldicarb
11.	Ovicides	An agent that used to kill eggs	Benzoxazine
12.	Piscicides	To kill fishes	TFM
13.	Repellents	To repel insects	Anthraquinone
14.	Rodenticides	To kill rodents	Bromadiolone
15.	Termiticides	To kill termites	Imidacloprid
16.	Viricides	An agent that kills viruses	Cyanovirin-N

2.2.4 CATEGORIZATION OF PESTICIDES BASED ON THEIR CHEMICAL CONSTITUTION

Pesticides are categorized into six groups based on their chemical constitution, namely, organochlorine, organophosphorus, carbamates, pyrethrin and pyrethroids, and sulfonylurea.[43] These groups of pesticides are presented in Table 2.4.

TABLE 2.4 Categorization of Pesticides Based on the Chemical Constitution.

S. No.	Group of Pesticides	Function	Examples
1.	Organochlorines	Organochlorine pesticides are organic compounds that contain five or more chlorine atoms in their structure	Heptachlor, toxaphene, and chlordane
2.	Organophosphate	One phosphate group is attached to the basic structure of organophosphorus insecticides	Fenthion, dichlorvos, chlorpyrifos, and ethion
3.	Carbamate	Pesticides derived from carbamic acid come under the category of carbamates pesticides	Carbaryl, aldicarb, propoxur, terbucarb, and oxamyl
4.	Pyrethrin and Pyrethroids	Natural identical pyrethrins are called pyrethroids	Permethrin, cypermethrin, and deltamethrin
5.	Sulfonylurea	Sulfonylurea is a substituted urea herbicide (SUHs)	Chlorsulfuron, rimsulfuron, terbacil, and foramsulfuron

2.3 ENTRY OF PESTICIDES INTO ANIMALS AND PLANTS

The selection of pesticide and its form (solid, liquid, granule, or aerosol) depends on the target insect's habits, for example, adult mosquitoes (flying pests) are killed by aerosol sprays or fogs and crawling insects are killed by surface powders, sprays, or granules. When pesticides invade the bodies of insects and kill them, they function as poison. The entry of pesticides into the body of pests may be possible in three different ways.

2.3.1 DERMAL ENTRY

Insects absorb pesticides into their bodies through their skin. The skin of an insect is called the *cuticle*. These types of pesticides are called contact poisons.

Dermal entry can happen when:

- Droplets of aerosol spray hit the insect.
- Insects walk over and thereby come into contact with powder or granule forms of insecticide.

2.3.2 ORAL ENTRY

Pesticide enters into the body of insects with the help of their mouth when the insect eats it. Pesticides that are swallowed are known as ingested poisons. So, when insect swallows the insecticides, oral entry may occur in the form of poisonous bait.

- When it cleans itself after the poison comes into contact with its body.

2.3.3 RESPIRATORY ENTRY

This type of pesticide enters the body of insects when the insect breathes in the insecticide. These types of pesticides are called *inhaled poisons*.

Mostly, the animals breathe through the mouth but insects do not breathe through their mouth. Insects breathe through their ***spiracles*** (tiny miniature along the side of the abdomen).

2.4 IMPACTS OF PESTICIDES: BENEFITS AND HAZARDS

The application of pesticides (herbicides, insecticides, and fungicides) increased agricultural yields over the last five decades. Pesticides contribute to increased yields of field crops and improved product quality by eliminating or controlling the growth of weeds, insects, nematodes, and plant diseases when used properly. Apart from this, herbicides can minimize labor cost, machinery, and fuel consumption for mechanical weed control.

If the credits are there for pesticides in food production and vector-borne disease control, then the debits are for pesticides in the form of serious health issues in humans and livestock and impacts on the environment. There is so much overcome evidence that pesticides can represent a risk to humans and livestock, as well as have unintended consequences for the environment. The impacts of pesticides used are summarized in the form of their benefits and hazards.

2.4.1 BENEFITS OF PESTICIDES

The applications of pesticides are very important in human life. Pesticides empower farmers to grow and harvest more food in less area by protecting crops from pests, diseases, and weeds and increasing production per hectare. Pesticides have contributed to more than quadruple the output of major crops since 1960.[84] Rice, for example, is utilized by over half of the global population, and production has more than doubled. The production of wheat has increased by nearly 160%. Most of the farmers (including organic farmers) use pesticides, either from synthetic or natural sources. Pesticides derived from natural sources are used only by farmers. But both the pesticides (synthetic and natural) possess different levels of toxicity. Pests, weeds, and diseases would damage more than half of the crops if pesticides are not being used. Twenty-six percent to 40% of crops of the world's potential production are lost in a year because of pests, weeds, and diseases. This loss could be double without crop protection.[174]

There are different pests (including 30,000 species of weeds, 3000 species of worms, and 10,000 species of plant-eating insects), which attack food crops and destroy the crops. The life span of food crops can be enhanced and losses can be stopped after harvesting with the help of pesticides. Approximately, 925 million people worldwide are currently unable to obtain sufficient nourishment. We require to increase food production to reduce hunger in the world. Pesticides are the only way to help the farmers in the production of food. With the intervention of insecticides, farmers cultivated safe, high-quality meals at a minimal price. Pesticides also assist farmers in producing a huge supply of nutritious crops throughout the year which is mandatory for the good health of the people. Fruits and vegetables, which are highly nutritious, are more accessible and economical.

The quantity and quality of the food depend on crop protection. A US study found that production of most fruits and vegetable yields would decrease by 50–90% without the use of fungicides.[98] Pesticides are used in everyday life to help protect food supplies, reduction in the price of foods for consumers, and provide excellent quality of food products.

The application of pesticides (herbicides, fungicides, and insecticides) has two levels of benefits, namely, primary and secondary. Direct gains from the application of pesticides come under the category of primary benefits and secondary benefits are less immediate or less obvious effects and are more long-term. These benefits are presented in Figure 2.2.[58]

2.4.1.1 PRIMARY BENEFITS—CONTROLLING PESTS AND PLANT DISEASE VECTORS

2.4.1.1.1 Improved Productivity

Pesticides and fertilizers in combination with upgraded seed varieties help to increase crop production over the last 80 years. Productivity of the crops has increased in most of the countries, for example, In 1998, Austin reported that yields of wheat in the United Kingdom increased from 2.5 t/ha in 1948 to 7.5 t/ha in 1997.[15] Between 1920 and 1980, open-pollinated maize yields in the United States increased from 30 bu/ac (2.02 t/ha) to over a hundred per acre.[137] Over the past 70 years, cotton yields rose nearly fourfold, soybean yields increased more than threefold, and wheat yields climbed more than 2.5-fold.[87]

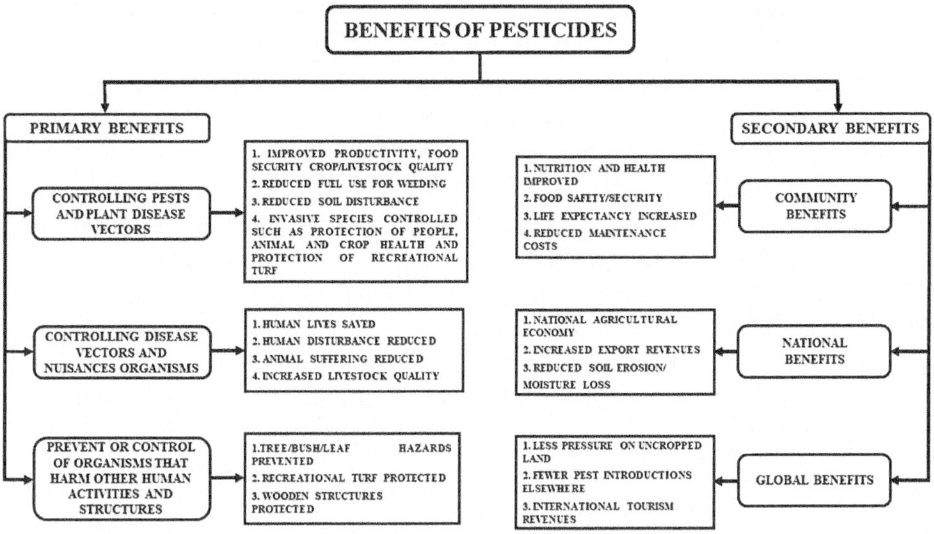

FIGURE 2.2 Benefits of pesticides used.

In 1948–49, India's food grain output was 50 million tons, but by the end of 1996–97, it had nearly quadrupled to 198 million tons. By using high-yield varieties of seeds, advanced and modern technologies of irrigation and fertilizers/pesticides, this result has been achieved.[75]

Warren[250] also found that agricultural production in the United States has increased throughout the 20th century. The average US yields for nine crops for 10 years of periods increase two to seven times. This enhancement can be related to a range of soil and water management, better plant types, and fertilizer and pesticide use. Webster et al.[254] narrated that "considerable economic losses" can be developed without the application of pesticides, and the yield increases in gross margin due to the use of pesticides. Webster and Bowles[253] stated that without the application of pesticides, apple crops would not be commercially economic and the land of the farmers would have to be used for other crops. In Russia, Petrusheva[187] assigned that orchard yield can be increased by 1.5–2 times with the help of pesticides application. On the same continent,[52,130,260] all found that the financial expenses incurred on pesticides were remunerated and boosted yields by four to six times.

2.4.1.1.2 Food Security

The UN projects that by 2050, the world population will have risen to 9 billion people.[234] To fulfill the need of this increased population, food production should be more as well. Nevertheless, new agricultural land is limited, as a result, achieving global food security requires realistic production and boosting the productivity of current agricultural land.[189]

World Food Summit[90] mentioned that Food security "exists when all people have physical and economic access to adequate, safe, and nutritious food to fulfill their dietary needs and food choices for an active and healthy life at all times." Modern technology, such as the use of pesticides to manage insects, weeds, and disease-inducing chemicals, has enabled food production for the world's 7 billion inhabitants. In Africa, starvation is a terrible reminder that when agricultural systems collapse, the populace cannot live or suffers from severe malnutrition. The American Heritage Dictionary defines malnutrition as "suffering from defective nutrition, especially as a result of an inadequate or poorly balanced diet." Pesticides are required for food production, otherwise, food production would be reduced further, and the number of people suffering from hunger would rise. In 1787, 90% of the people of the United States lived on farms and produced enough food to feed themselves. By 1950, the percentage of individuals had decreased to 16% who lived on farms. In 1990, only 2% of the population in the United States depended on farms, but they produced food for 120 people in addition to themselves. About 95 of the 120 persons are from the United States, while the remaining 25 are from other countries.[225]

In the whole world, the application of herbicides has provided a 10–20% yield increase in bread grains. This is enough for 15 bread loaf for every individual on the planet. In the poorest countries, 95% of the population generates enough food to sustain themselves, while the other 5% remain hungry. In industrialized nations, on the other hand, just 3–5% of the population generates enough resources to feed the population at large.[113]

2.4.1.1.3 Food Quality

Pesticides can improve product quality by preventing mold development, scarring, flaws, and contamination from insects and weeds. In the first world countries, Brown[41] observed that a diet rich in fresh fruits and vegetables is considerably more beneficial than eating crops with very minimal pesticide residues. Dietary Guidelines[69] narrates that consumption of fruit and vegetables daily reduces the risk of many diseases, such as high blood pressure, cancers, diabetes, stroke, and chronic diseases.

Kolbe[133] found that pesticides can also increase the product's quality including its safety. The harmful microorganisms that cause disease in the plants, release acutely toxic chemicals, for example, Ergot (*Claviceps purpurea*) is a cereal disease that creates very poisonous compounds and occasionally deadly alkaloids in the grain. This infection can be protected with the help of a fungicide. Aflatoxin (a kind of mycotoxins) is a powerful carcinogen and immunotoxic substance and can delay the growth when used. Fumonism, a naturally occurring toxin associated with *Fusarium* fungi (molds) that damages the brain and kidney and is also responsible for several health problems in animals and humans.

Bruns[42] found that these mycotoxins can be produced in various crops, such as maize grain in the field or in storage when conditions are favorable for them. Such types of fungal infections can reduce by the application of fungicides.[125] The impact of 10 commercial fungicides and insecticides on *Aspergillus parasiticus* growth and aflatoxin B1 production was studied by Etcheverry et al.[79] Four of the five fungicides tested were shown to suppress growth and toxin generation in laboratory media at doses similar to those used in commercial usage. By applying insecticides, the crop's life span may be extended and massive postharvest losses from pests and illnesses can be avoided.

Grain that has been stored in bags or bins can be protected from insect spoilage by using insecticides.[64] Trials in Tanzania on food grains demonstrated that applying pesticide combinations in modest amounts as protectants of shelled maize can diminish the bigger grain borer *Prostephanus truncatus* and *Sitophilus* species for a minimum of 9 months. Zettler and Arthur[261] also studied fumigants and residual treatments for chemical control of stored product insects and reported that pesticides are the cheapest and most efficient way to control the insects.

Raw ingredients and processed foods can protect from insect contamination by the controlled application of insecticides during processing, manufacturing, and packaging sites. In grocery stores, pesticides are used to manage insects, and food and food leftovers attract rats.

2.4.1.1.4 Protection of Crops Losses

Globally, 14% Postharvest losses are determined and it can be reached up to 50% in some developing countries. Savary et al.[206] reported that the maximum loss of crops is seen especially in the regions where the population is growing very fast and food security is already a concern. On the other hand, locations with a high level of production and surpluses show the minimum loss of crop. Weed management studies were conducted on tomatoes, and it was found that weeds can reduce the yield of tomatoes by 47%. Fluchloralin and metolachlor controlled *Celosia argentea* and *Cyperus iria* both effectively resulted in the yield of tomato fruits.[26]

A second study of GRC emphasized only the differences of a fungicide ban. These fungicides control plant disease caused by fungi. If there is no use of fungicide, fungi can kill the crop and sometimes produce toxins. A fungicide prohibition in the United States would lower fruit productivity by 32%, vegetables by 21%, peanuts by 68%, and maize and wheat yield by 6% each. These numbers are even more alarming when we realize that eating fruits and vegetables might help prevent heart disease and some malignancies.[104] National Research Council (NRC)[173] suggested that in the absence of fungicides, consumption of these nutritious foods per capita would drop by 24%, indicating detrimental impacts on our country's health. Farmers have almost eliminated the cotton boll weevil in broad parts of the southeast with precisely scheduled pesticide treatments. This insect decimated the cotton-based southern agricultural economy around the turn of the 20th century.[113] Without pesticides, the industrialized world's food production efficiency would be impossible to achieve.

2.4.1.2 PRIMARY BENEFITS-CONTROLLING HUMAN/LIVESTOCK DISEASE VECTORS AND NUISANCE ORGANISMS

2.4.1.2.1 Controlling Disease Vector in Human and Nuisances Organism

Especially in hot climates, insects can spread many tick- and insect-borne diseases in humans, such as dog heartworms, encephalitis, malaria, plague, river blindness, rocky mountain spotted, sleeping sickness, typhoid fever, and yellow fever. Insecticides are now used to keep these diseases at distance. Insecticides can be used to manage fleas, cockroaches, and flies in our houses, increasing sanitation and comfort. In the US, millions of homes protected themselves from termites by using long-lasting soil pesticides.

Vector-borne diseases in the world are practically handled by destroying the vectors. Insecticides are the only way to control insects that transmit fatal infections like malaria, which kills an estimated 5000 people per day.[103,202,231]

In western Kenya, bed nets sprayed with insecticide substantially decreased new-born mortality, with no additional death in older children due to delayed malaria immunity.[146] Bhatia[35] stated that in the developing world, malaria is one of the main reasons for morbidity and mortality and it is a big public health problem around the globe. Yadav and Sampath[257] carried out the field trials and found that in *Anopheles* mosquitoes, bed nets treated with deltamethrin reduced indoor resting density, biting, light trap catches human-sourced engorgement rate and parous incidence of malaria infection. It was discovered that using treated bed nets reduces the number of infective bites per person per year by 75%.[63]

According to the World Health Organization, life will stay in a risky zone for the majority of mankind unless chemical control measures are used.[10] The World Health Organization (WHO) is now recommending that indoor residual spraying (IRS) be used not only in epidemic regions but also in places where malaria transmission is persistent and strong such as throughout Africa (http://www.who.int/mediacentre/news/releases/2006/pr50/en/index.html).

Hoffman et al.[114] studied and found that acute malaria infection increases the chances of HIV load, and this viral load can be controlled by antimalarial medications. This malaria-associated infection could lead to increased transfer of HIV disease in humans.

2.4.1.2.2 Controlling Livestock Disease Vectors

The most significant outcome of controlling the broad spectrum of livestock disease vectors is to reduce suffering from diseases and save lives or else big loss. The controlling disease vectors converted into secondary benefits, namely, livestock revenue generation and veterinary and medicine costs reduction.

By spraying a single dose of insecticide in clover to control red-legged earth mite costs $10/ha. In annual pastures, the application of a Timerite spring spray to control the red-legged earth mite (RLEM), *Halotydeus destructor* (Tucker) (Acari: Penthaleidae) was studied, and it appeared to produce higher RLEM control.[199] Kamuanga[128] found that tsetse control programs utilizing insecticide-impregnated targets and pour-on treatments of all livestock with deltamethrin 1% were used to combat trypanosomiasis in Burkina Faso. The results show a 25% increase in herd size and an increase in the number of oxen per household from 0.1 to 1.1. A reduction in mortality from 63.1% to 7.1%, as well as reductions in abortion and stillbirth rates of 55.9% and 51.3%, respectively; and an increase in the rate of live births of

57.6%, as well as increases in milk yield from 0.2 to 0.4 L/cow/day in the dry season.

In Ethiopia, *Anopheles arabiensis* is the main vector of malaria, when it feeds on animals and human beings. In the studies, it was found that when animals were treated to control tsetse fly, they were healthy and the infection of malaria was reduced.

2.4.1.3 PRIMARY BENEFITS—CONTROLLING OF ORGANISMS THAT HARM OTHER HUMAN ACTIVITIES AND STRUCTURES

Pesticides, especially herbicides, are widely utilized in the transportation industry. The turf on sports (cricket, football, etc.) pitches and golf courses are maintained by the use of pesticides (herbicides and insecticides). Buildings and other wooden materials can be saved from the termites and wood-boring insects' infection by applying insecticides. Similarly, many organisms have an adverse effect on human life, infrastructure, and common goods of everyday life without control. Pesticides (insecticides, herbicides, and fungicides) play an important role to check such types of adverse effects.

2.4.1.3.1 Protection of Transport System

The transport department have widely used the pesticides, mainly herbicides to verify that roads, railway lines, and waterways are free from vegetation. The vegetation might cause a danger or nuisance. For example, if the grasses and weeds are permitted to grow along with roadsides, it disturbs the driver's vision on the road, and big branches of the trees or vegetation onto the road which can produce an obstacle or road become very slippery. Pesticides are used to manage the vegetation on the roadside which gives secondary benefits of safer transport systems with a lower rate of accidents and less tension for users.

The water hyacinth is a nuisance species that can become barricades for waterways. In the 1950s, South America and many other countries use the water hyacinth as an ornament plant that can provide food for livestock, and pollution can be controlled by absorbing heavy metals. It is unfortunate that the water hyacinth multiplied very fast and spread over many countries. Other plants cannot grow due to a lack of space and water in the rivers, lakes, and ponds. Anon[7] found that Africa has spent an estimated USD 60 million yearly in controlling alien weeds (water hyacinth) for their great thirst for water. Every year, the Worldwide Fund for Nature eliminates invasive plants from 200,000 ha of land. Herbicides are employed in this program and have shown to be highly effective in controlling invasive species.[7]

The development of fungi, algae, and weeds can be controlled by the use of pesticides on water craft and produces secondary benefits by reducing costs of cleaning by hands, and increased fuel efficiency of a water plane.

The vegetation shows huge destruction; the growth of vegetation above the ground can enhance corrosion on metal structures. The spreading roots of plants may be seen breaking pipes, opening potholes in the road, and dislodging railway lines.

In metro cities, most people want that footpaths, roads, and gutters remain clean and weed-free, but they do not know that the regular application of herbicides is the reason. Thus, pesticides (herbicides, fungicides, and insecticides) generate primary benefits which are related to preventing these problems and transform into secondary benefits like maintenance costs reduction and more safety for the transport.

2.4.1.3.2 Protection of Recreational Turf

Exercise, games, and entertainment are extremely important for the development of people's physical and mental health during this period of sitting work. Herbicides and insecticides can be used to preserve the grass on sports fields, cricket grounds, and golf courses, resulting in secondary advantages, such as increased health and fitness, decreased stress, and a higher quality of life. Similarly, house owners can apply pesticides to keep their edible or ornamental plants in their domestic gardens healthy and free of pests and diseases. In the United Kingdom, gardening is the most popular recreation and pesticides are used to ease popular hobbies that generate natural air and exercise for a larger group of people in the world. The hard work of UK people reflected in the growth of the economy in different modes, for example, exercise nurture health and minimize medical assistant, and pleasant and beautiful gardens increase the value of their homes.[111,193]

With the help of pesticides, ornamental plants and trees in public gardens may be protected from pests and diseases, and this sort of civic vegetation makes urban landscapes more appealing to live in, improving life quality and reducing stress. Plants, particularly trees, give shade in warmer countries, lowering the cost of cooling energy, while crops in windy areas can be protected by planting trees as windbreaks.[224]

2.4.1.3.3 Protection of Buildings and Wooden Structures

Buildings and other wooden structures can be saved from damage by wood-boring insects (like termites) with the help of insecticides. Thus, insecticides can decrease the maintenance costs and increase the longevity and safety of buildings.

Methylisocyanate (MITC) is used as a fumigant to control an existing internal decay and insect attack of wood.[230] Borates are having a toxic effect on all wood-damaging insects and fungi. Borates are water-soluble and penetrate the wood by diffusion methods and protect them from infection.[256] The growth of harmful microorganisms (fungi, bacteria, and viruses) can be retarded by using antimicrobial pesticides (also called biocides). These microorganisms can decay spoilage, deteriorate, or foul of materials, such as chilling plants, heat exchangers, turbines, reverse osmosis membranes, garments, and paper products and lead to secondary benefits include increased shelf life and product longevity, as well as lower maintenance expenses.[58]

Most all families (97.8%) used pesticides at least one time a year and two-thirds of families used pesticides more than five times a year.[66] About 80% of the families use pesticides in their home only one-time in a year. About 57% of the families use herbicide to control yard weeds and 50% of the families use insecticide to control fleas and ticks on pets. A substantial number of families (33%) also used pesticides in their garden or orchard to control the pests. It is concluded that proper application of pesticides improves and enhances the quality of life, protects our property, and develops a better environment.

2.4.1.4 SECONDARY BENEFITS—COMMUNITY BENEFITS

2.4.1.4.1 Nutrition and Health Improved

The flavor, nutrition, and health advantages of apples and blueberries in the US diet were detailed by Lewis and Ruud.[142,143] They found that they are the richest source of anthocyanin which has the highest antioxidant activity. Antioxidants act as protectants against oxidative stress which produces aging, cancer, heart disease, and urinary tract infections. Lewis assigned the double production of wild blueberry crop and consequent increases in consumption by using herbicide which helps in weed control.

Gianessi[98] attributed inexpensive foods and good quality fresh fruit and vegetables to be available throughout for US consumers. The appearance and quality of the food products are due to the contribution of chemical pesticides.

2.4.1.4.2 Reduced Stress and Life Expectancy Increased

Better nutrition in the communities can come if food is more, which proceeds into healthy lives. Healthy people in the communities are happy people and are more productive and give better to their community.

We can ensure food security and stable lives from farming with the help of pesticides. Using pesticides can interrupt the loop that affects personal livelihoods and the quality of life. In rural communities, the quality of life can be changed by improved nutrition and reduced drudgery.

Nowadays, people expect and enjoy a long and healthy life in comparison to the past. In 1900, the average life expectancy in the United States was just 47 years, but it has already risen to 78 years.[9]

According to Atreya,[14] life expectancies in France have increased by 3 months per year over the previous 50 years. In many other regions of the world, better medical care and medicinal treatments, as well as better living circumstances and sanitation, have all contributed to increased life expectancies. The value of nutritious, safe, and affordable food can also be a health promoter that extends life expectancy.[14,80,236] Gattuso[94] wrote that fruits and vegetables play a vital role in protecting against cancer. To ban some pesticides would limit the availability, affordability, and total intake of vegetables and fruits.

2.4.1.4.3 Reduced Drudgery of Weeding and Maintenance Costs

Weeds are the main obstruction for the good yield in many crops and they may be controlled by pesticides, notably herbicides. Herbicides, which account for almost half of all pesticides used worldwide, are the most common crop protection agents. Insecticides and fungicides each account for approximately 17%.[62] According to Anon,[11] there would be a $13.3 billion loss in agriculture revenue in the United States if herbicides were not used. Yancy and Cecil[258] estimated the yearly benefits of herbicide usage to be $21 billion, compared with a cost of $6.6 billion for excellent crops that decreased weed losses by 23% and avoided an $8 billion loss in agricultural revenue. The US losses $4 billion due to weeds in place of $20 billion with the use of herbicides.[40] Miller[160] examined the primary and secondary productivity and labor implications of better weed control obtained with herbicides on farmers and rural communities and he summarized that greater agricultural production provides immediate economic advantages to farming households in the form of increased revenue.

2.4.1.4.4 Reduced Fuel Use for Weeding

Behera and Singh[26] found that weeds lower the production of dry land crops by 37–79%. Weeds attack early in the crop's life cycle, resulting in a 40% loss in production. Economic and labor benefits can be achieved with the help of herbicides. Herbicides can be used to replace manual weeding and can minimize the use of fossil fuels in mechanized farming. In Sub-Saharan Africa, where HIV/AIDS is more prevalent, the reduction in the requirement for physical weeding is extremely beneficial, as it results in a labor shortage and many people being sick and unable to work.[107]

Pesticides (insecticides, herbicides, and fungicides) are used to limit the growth of pests, such as weeds, insects, and plant diseases, reducing the amount of labor, fuel, and machinery required for pest management.[93,176,179] US farmers spend approximately $12 billion on pesticides in 2008 by their own will.[235] Many consumers from all over the world can be benefited from abundant, and relatively inexpensive, unblemished foods.[176]

The most transparent and easiest benefits are economic benefits for the farmers which come from the reduction of other costly inputs, such as labor and fuel and the protection of commodity yield and quality.

2.4.1.5 SECONDARY BENEFITS-NATIONAL BENEFITS

2.4.1.5.1 National Agriculture Economy

Pesticides are regulated by The Federal Insecticide, Fungicide, and Rodenticide Act (FIFRA) and the Federal Food, Drug, and Cosmetic Act (FFDCA).[88,178] The application of pesticides gives economic benefits to the producers which leads to consumers. Farm revenue growth is entirely dependent on crop yields and quality, which means the wealthier farmers are having more extra income to boost up the national economy. Higher yields put less pressure on barrel land to cultivate and lead to a wider benefit to biodiversity and the environment.[157]

Knutson et al.[132] estimated the potential impacts on US society after removing some pesticides from the agricultural market. In the absence of pesticides, the production of foods would fall and the prices of the food would be more in the US. Because of decreased output and higher costs, American farmers would be less competitive in global markets for key cereals, cotton, and peanuts. Oerke et al.[175] found that the global losses from pests for eight crops in some regions were more than 50% of attainable crop output.

2.4.1.5.2 Increased Export Revenues

As a result, regional and national agricultural economies become more fluid, and profits from high-quality product exports bring in foreign currency. This is especially significant for some poor nations that export fruits and vegetables to the United States and Europe, where the presence of unwanted flora and fauna in the goods can be a substantial trade obstacle.[121]

Rehman et al.[196] studied cotton crops and analyzed that by the use of pesticides, crop protection *increased and gave good yield which leads to* earns the country's largest *export revenues.*

2.4.1.6 SECONDARY BENEFITS—GLOBAL BENEFITS

2.4.1.6.1 Biodiversity Conservation

All animals, plants, and microorganisms constitute biodiversity. Pesticides (insecticides, herbicides, and fungicides) play an important role in the ecological consequences of the increasing loss of biodiversity.[99]

Pesticides can sometimes help to restore ecosystems that have been damaged by alien species. In the Laurentian Great Lakes, for example, the Atlantic sea lamprey (*Petromyzon marinus*) is an invasive species that acts as a predator. They attacked and killed native species of fishes, namely, lake trout (*Salvelinus namaycush*) and lake whitefish (*Coregonus clupeaformis*), and the populations of lake trout were destroyed. Lamprey numbers in the Great Lakes have been reduced through the use of lampricide chemicals (usually TFM) and the lake trout population is now rebounding.[243]

Furthermore, pesticides are an effective weapon for combating invasive species, which pose a significant danger to native ecosystems. For example, when Pacific rat (*Rattus exulans*), a small Southeast Asian species, Norway rat (*R. norvegicus*), and ship (roof) rat (*R. rattus*) are introduced onto islands, they showed a devastating effect on local fauna in particular, on insects, amphibians, land birds, and reptiles. Pesticides can play a significant role in biodiversity conservation in this way.[229]

2.4.1.6.2 Reduce Soil Erosion, Moisture Loss, and Global Warming

Herbicides have the potential to minimize the usage of mechanical cultivation in large-scale agriculture which show the clear national and international benefits in the form of reduced production of greenhouse gases emissions (GHG), reduced soil erosion on sloping terrain, and reduced moisture loss from soil surfaces.[8,22] Organic farming reduces soil losses and has the potential to diminish soil erosion.[76,209]

2.4.1.6.3 Reduced International Spread of Diseases

The risk of bringing new plant pests and diseases has grown as trading has expanded. These pests can spread and damage native plant species, resulting in economic losses.

Effective pest management on export commodities can help to prevent pest introductions in other nations.

The neotropical parasitoid *Apoanagyrus* (*Epidinocarsis*) *lopezi* was the most successful natural enemy introduced to attack the cassava mealybug, *Phenacoccus manihoti*. Biocides (natural enemies) are introduced in 26 African countries, resulting in the reduction of *P. manihoti* population in most farmer's lands.[171]

2.4.2 HAZARDS OF PESTICIDES

Pesticides are toxic chemicals that pose risks to living things and its surrounding. The substance's toxicity and the amount of exposure are the main functions of risk factors. As an ancient precept, "the dose makes the poison." With the help of skin, mouth, eyes, or lungs, pesticides can enter the body. The risks of pesticides used are presented in Figure 2.3.

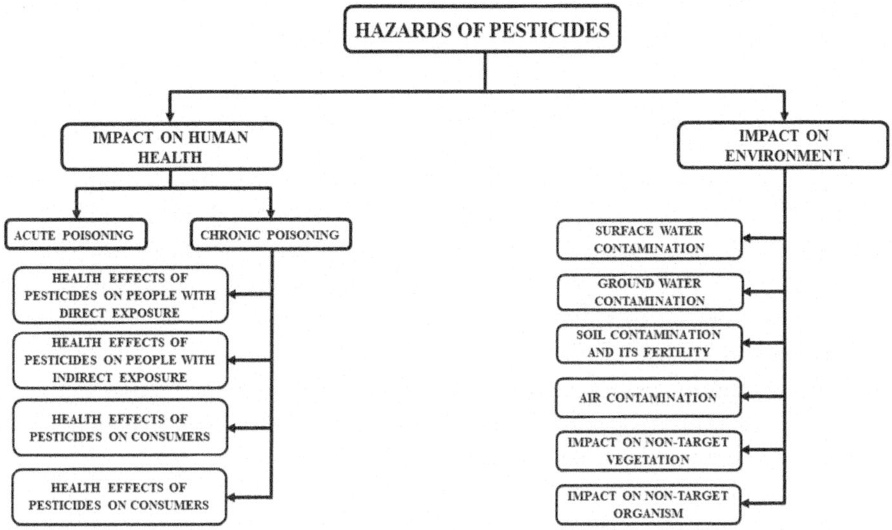

FIGURE 2.3 Hazards of pesticides used.

2.4.2.1 IMPACT OF PESTICIDES ON HUMAN HEALTH

Pesticide toxicity and persistence in the environment have negative impacts on human health, and their potential to penetrate the food chain has dangerous implications.

Pesticides can penetrate the human body by direct touch, food, contaminated water, or polluted air. The main objective of guidelines across the globe is to be healthy humans and the environment. This part of the chapter describes the correlation between health risk and pesticide exposure. The relevant scientific researches show that pesticides have an adverse effect on the human being. Beyond the literature, the outcomes also highlight the requirement to incorporate these findings for development guidelines and programs for the community. There should be control in production, use, storage, packaging, and disposal of packaging after use processes of pesticides, these are high-risk factors for surroundings and humans. There should be awareness programs for farmers and exposed workers about the misuse of pesticides, application errors, such as protective clothing use during the application, rules of personal hygiene, overdose and unnecessary duplication, exposure, and contact to pesticides. Pesticides should be used with caution, and persistent organic pollutants (PoPs) should be avoided. Minimization of application of pesticides can be encouraged by the guidelines such as good agricultural practices and pest management. Two types of toxicity have been observed in human health.

1. Acute toxicity
2. Chronic toxicity

Scientists apply a measurement to describe the acute toxicity of a substance termed the LD_{50} (Lethal Dose), which is required to kill 50% of tested animals. It is measured in milligrams per kilogram body weight. As the value of the LD_{50} is low, the more threatening will be the poison. The lethal dose is determined for many pests and test animals. Some substances can be hazardous at a sufficiently high dose for instance, LD_{50} of ordinary table salt is 3 grams per kilogram body weight; a lethal dose of table salt for a small child is about two tablespoons. Acute and chronic diseases occur from pesticide exposure which is summarized below.

2.4.2.1.1 Acute Toxicity

Pesticides are responsible for <4% of mortality from intentional, occupational, or accidental poisoning as per reports from poison control centers because earlier worldwide data of APP are not available. Between 1992 and 2000, the number of APP cases in Central America increased from 6.3 to 19.5 per 100,000 people, with a corresponding increase in the mortality rate from 0.3 to 2.1 per 100,000. In recent years, there has been considerable improvement in the recording of statistics[180] and extensive application of pesticides results from the increasing numbers of acute pesticide poisoning cases are highlighting, but these data are unable to distinguish between occupational, accidental, and intentional poisoning. Furthermore no information about qualitative and the severity of poisoning. According to available recordings, suicide attempts are the most common cause of acute poisoning in poor nations, accounting for 44–91% and 26–60% of acute pesticide poisonings in Southeast Asia and Central America, respectively.[31,180] There is usually involvement of organophosphorus compounds in a suicide attempt but not entirely.[31] Poisoning severity varies according to a nation with death rates ranging from 1% to 24%. The general population is not much aware of the proper way of application or handling of pesticides intended for amateur uses or pest management in a home which are the occupational causes of mostly unintentional acute pesticide poisoning.[124] In Southeast Asia, 5–32% of occupational poisoning cases occur, and in Central American Countries, it is about 36%.[31,180] Pesticide laws are being implemented, such as restrictions on the use of the most hazardous substances, and a reduction in public access to agrochemicals in general[131] results in a remarkable reduction of acute pesticides poisoning including suicides. It has been proven that simple availability to pesticides is a major factor in the occurrence of acute suicide poisoning.[51]

Acute toxic effects resulting from a short contact or exposure to the pesticides. Pesticide drift from agricultural fields, pesticide exposure during the application, and intentional or unintentional poisoning are the most common causes of acute disease in people.[67,141] Pesticide poisoning can induce headaches, skin rashes, body pains, nausea, poor focus, blurred vision, dizziness, cramping, panic attacks, and in severe instances, coma and death.[182] The degree of hazards is determined by the toxicity and quantity of pesticides used, as well as the form of application, mode of action, frequency and length of contact with pesticides, and the person.[198] There are about 3 million cases of acute pesticides poisoning outlined worldwide yearly, of which two-thirds of cases are suicide attempts and the rest are incidents of professional or unintentional poisoning.[213] The main reason for suicide attempts in rural regions is mostly due to the wide availability of pesticides.[67,198] The only notable case of pesticide-related human poisoning happened in Bhopal, India, in 1984, when a large amount of toxic methyl isocyanate vapor (chemical intermediate) was released during the manufacturing of an agricultural pesticide. It killed 2800 people and more than 20,000 people were badly injured.[233] Agricultural and occupational workers who handle pesticides face acute poisoning as a persistent problem and trail them into their homes, so family members get exposed. People can be exposed by drift from aerial spraying who live near agricultural fields.

In Sri Lanka and many rural regions of Asia, a major clinical and public health problem is pesticide self-poisoning which continues to be a significant cause of death in Sri Lanka, particularly in men. Pesticides were widely used in agricultural applications during the Agricultural Revolution of the 1950s and 1960s, which was immediately followed by an exponential increase in the number of cases and mortalities. Most of the mortalities in 30 years occurred because of OP insecticides, locally used in agriculture, many deaths occur before patients get to the hospital. Due to poor medical management, patient survival is sometimes impossible in the best hospitals. Each year, 250–350,000 deaths are estimated due to pesticide poisoning. The contribution of these pesticides is gradually decreasing by regulatory intervention to restraint the application of the pesticide. The highest case fatality among all pesticides is paraquat, so carbamates and the herbicide paraquat are becoming more important. Ongoing regulatory interventions will prevent deaths from acute pesticides poisoning in near future by educating health care workers, improving medical management, making storage of pesticides in communities safer, and developing new antidotes and diagnostic methods. Nevertheless, Sri Lanka has proclaimed that the death rate can be substantially decreased by the regulation of pesticides without affecting agricultural output.

2.4.2.1.2 Chronic Toxicity

The continuous and frequent application of pesticides has also markedly risen the concentration of pesticides in food and surroundings which impacts human health badly. People coming into contact with pesticides as daily lives activity can face more dangerous consequences. Nevertheless, farmers are developing more interest in the frequent application of pesticides due to pest challenges on crop production. Moreover, the majority of the farmers are unlearned about the safe handling of pesticides. Farmers and workers on commercial and on small-hold farms can have serious health issues because of unsafe handling and interaction of these chemicals.[153] Pesticide poisoning cases are dozen of million worldwide yearly.[198] Additionally, continuous pesticide exposure causes chronic health consequences, such as neurological problems, cancer, diabetes, respiratory illnesses, fatal diseases, and genetic abnormalities. Depending on the degree and kind of exposure, these health consequences vary, there can be variation in health impact on four types of people exposed up to several years.

1. Pesticides' health impacts on people with direct exposure.
2. Pesticides' health impacts on people with indirect exposure.
3. Pesticides' health impacts on consumers.
4. Pesticides' impact on children's health.

2.4.2.1.3 Pesticides' Health Impacts on People with Direct Exposure

Industrial workers, farm workers, and farmers are the perilous community exhibited to pesticides who prepare the pesticides mixture for application, pesticide seed dressings, sprayed crops, washing, and discarding of pesticide containers. The probability of hazards may be highest during the production of pesticides as procedures are not safe at that time to the exposed person. Workers at industrial plants handle a variety of hazardous substances, such as insecticides and raw materials, and toxic solvents are at elevated risk, for instance, a group of farmworkers was deadly affected by pesticides while working in an onion field in Caldwell, Idaho. July 2005. At night time, three pesticides were applied to the farmland without giving the farm owner any information. In the morning, a group of 29 workers started weeding the farm without any precautionary measures. Suddenly, they felt that their clothes became wet but they did not take it seriously and thought it was just moisture. after some time, they were puking, headaches, nausea, and diarrhea. Here some chronic diseases are discussed on long-term exposure to pesticides in Table 2.5.

TABLE 2.5 Chronic Diseases Which Arise Due to Long Time Exposure of Pesticides.

S. No.	Health Hazards	References
1.	Cancer	[2,23,91,150,151]
2.	Depression and neurological deficits	[33,106,126,127,172,203]
3.	Diabetes	[60,161]
4.	Respiratory diseases	[116–119,214]
5.	Reproductive disorders and gene mutation	[85,86,139]
6.	Dermatitis	[166,216]
7.	Cardiovascular diseases	[68]
8.	General health, multiple diseases, and others	[17,61,102,169;186,210]

2.4.2.1.4 Pesticides' Health Impact on People with Indirect Exposure

The impact of pesticides on the health of people with indirect exposure is also needed to concern and aware of its consequences. Mostly farmers' family persons or people living in areas where the extensive application of pesticides is done are categorized in that group. Another means of exposure to pesticides of this group is the "Take-Home pathway" in which workers with direct exposure can take them back into their homes and contaminate family members.[18] In Table 2.6, the authors summarized the serious illness of the indirectly exposed group.

Recently published article in Exposed and Ignored, after knowing the consequences of health issues of pesticides, many US consumers are more concerned about the protection of family's health, for this, they decreased their utilization of pesticides in agricultural practices. Traditional growing practices can be a potential method for agriculture production while no pesticides are in demand. The farmworkers can be exposed at greater levels than others to pesticides, so some efforts have done to protect them from poisonous pesticides used in the fields.

TABLE 2.6 Chronic Diseases Which Arise Due to Indirect Exposure of Pesticide.

S. No.	Health Hazards	References
1.	Cancer	[89]
2.	Depression and neurological deficits	[32,38,77,78,109,259]
3.	Diabetes	[81,215]
4.	Lungs-related diseases	[19]
5.	Endocrine disorders and fatal growth	[48,255]
6.	Cardiovascular diseases	[32]
7.	General health problems and others	[5,170]

2.4.2.1.5 Pesticides' Health Impact on Consumers

One of the major sources of pesticide exposure to humans is through the food chain, mainly top-level consumers are at high risk. Sea foods are the most contaminated source of pesticide.[1,25] Industrial and agricultural activities have released pesticides to the surroundings which contaminate crops, domestic animals, groundwater, soil, seafood, and drinking water. For instance, DDT and Dieldrin extensively applied in agricultural fields to rise the production of crops and to kill unwanted pests which cause contamination into crops, water, and environment. Researchers find out that the main residues of contaminations in food are pesticides.[208]

There are several steps involved in the food preparation process which may be the probable source of pesticide contamination, at a very early stage during crop production, pesticides can be entered from the surroundings into raw material. The persistent nature of pesticides facilitates their existence in the environment for a long time. Released pesticides become a potential source of contamination of food and feed supplies. Plants uptake pesticides from air to plant also and eventually to food. Additionally, processed food steps, processing, packaging, transportation, and storage could be a source of contamination of pesticides. Some findings of pesticides contamination in food are given in Table 2.7.

Pesticide bioaccumulation and persistent properties in fatty tissues make pesticides a life-threatening factor to human health. Various serious health issues are associated with pesticide exposure and it is worth studying the possible health effects to protect human health. Possible human health hazards due to exposure to pesticide-contaminated food are given in Table 2.8.

TABLE 2.7 Summary of Contaminated Food with Pesticide.

Pesticides	Edible Items	References
PCBs, PAHs, furans/Dioxins and OCPs	Milk Products	[12,29,53,123,181,188,190, 227]
PCBs, OCPs, HCBD, PCN, Dioxins/furans	Meat-based products	[29,59,70,184,190,217,262]
Polyaromatic hydrocarbon (PAH)	Bran and Gram flour	[227]
Organochlorine pesticides	Honey	[227]
Poly aromatic hydrocarbon	Oyster	[227]
OCPs, PAHs, PCBs, and PFOS	Water	[44,74,223]
PBDEs, PCBs, OCPs, Dioxins/ furans PFOS and HBCDs	Fish	[46,50,53,70,97,123,145,158, 190,212,223,241,252]
PCBs, OCPs, PBDEs	Mussel	[162,218,221,223]
HBCDs, OCPs, PCBs, and Dioxan/Furans	Vegetable oil	[59,129,184]
PCBs, OCPs, and PAHs	Fruits, Vegetables, and Rice	[34,74,188,247,248]

2.4.2.1.6 Pesticides' Health Impact on Children

Due to age factors and small body mass, pesticide exposure risk is higher in children who are the most susceptible group. At the age of 6–12 months, their metabolic processes are not mature and unable to detoxify hazardous substances which are major reasons for the sensitivity of children to pesticides. The possible increased susceptibility of health issues in children because of pesticides exposure is highlighted in Table 2.9.

Routes are correlated with different settings and media of exposure. Ten micrometers of particulate material is more threatening because it is breathable and easily passes through alveoli. Spraying areas are full of such particulate matter, children near such areas will have a high risk of pesticides exposure. Soil is also one of the main routes for pesticides exposure, crawling and hand-to-mouth activities enhance vulnerability. There are some routes and media of exposure discussed in Table 2.10.

TABLE 2.8 Health Hazards Associated with POPs in Food.

Pesticide	Health Issues	References
OCPs	Endocrine—disruption, Neurological Disorders, Diabetes, Obesity, Heart-Related Problems, Infertility and Birth Defects in Fetal, Blood Pressure, Cancer and Reproductive Problems	[3,70,144,168, 263]
PAHs	DNA Damage, Carcinogenicity and Mutagenicity, Respiratory Diseases Oxidative Stress, Impaired Male Fertility	[3,70,112,227]
PCBs	Carcinogenic, Endocrine Interferences, Hepatic Dysfunction, Neurological Problem, Obesity, Sugar Intolerances, and Cardiovascular Diseases	[3,70,249]
Dioxins/furans	Cognitive Dysfunction among children, Carcinogenic, Diabetes, Endocrine Disruption, High Blood Pressure, and Cardiovascular Diseases	[3,70,249]
PFOS and PFOA	Breast Cancer	[3]
PBDE	Reproductive Problems, Diabetes, Testicular Cancer, Obesity and Heart Diseases	[3]
HBCD	Reproductive Disorders Endocrine Disruption, and Behavioral Effects	[149]
PCN	Cancers	[70]
PCDE	Cancers	[70]

TABLE 2.9 Health Hazards Associated with Pesticides in Children.

S. No.	Possible Hazards	References
1.	Carcinogenicity Epidemiological risk of cancer	[65,120,242]
2.	Birth defects in an unborn child	[27,78]
3.	Asthma	[204]
4.	Developmental toxicity, multigeneration reproductive toxicity, and acute and short-term neurotoxicity	[191]
5.	Cognitive dysfunction among children	[192]

2.4.2.2 IMPACT ON ENVIRONMENT

Surface and groundwater, soil, turf, and other vegetation can be polluted by the direct and indirect usage of pesticides. Besides destroying weeds or insects, nontarget plants and organisms can be adversely affected because of pesticides application. Pesticides' persistence and biodegradation processes controlled the pesticide fate into surroundings, extensive application of pesticides in agricultural purposes is the major source of contamination of other matrices of the environment, such as water, air, and land which badly affect the human and wildlife.[36,101,115]

TABLE 2.10 Routes of Exposure in Children.

S. No.	Route of Exposure of Pesticides	
1.	Ingestion	By breastfeeding
		By accidental ingestion
		By residual contents in food
		Mouthing
2.	Inhalation	Pesticides spraying in indoor and outdoor
		Occupational exposure to pesticides
3.	Dermal absorption	Residual contents on surfaces
		Contaminated garments
		Medical use: scabies, head lice
		Occupational exposure
4.	Transplacental	Mother's intake and is transferred to the children through the placenta

2.4.2.2.1 Surface Water Contamination

Pesticide contamination in surface water creates a considerable issue that needs more attention at community, territory, national, and international levels. Pesticide residues reach the aquatic environment through agricultural discharges, leaching, careless discharge of vacant containers, tools washing.[183]

Pesticides can enter surface water along with discharges from industrial plants and through the soil, pesticide impurity in the surface water is worldwide. The US Geological Survey (USGS) studied throughout the country's major river basins and found that one or several pesticides were present in most of the fish and water samples collected from water bodies.[135] Samples were collected from major rivers and metro cities that receive an agricultural and nonagricultural discharge and showed the presence of pesticides. The USGS found that the levels of insecticides in the rivers of metropolitan cities are generally higher than the recommended levels for marine life preservation.[37,239] The point sources, transformation, and transportation processes are needed for expected fate and distribution in sediments and aquatic environments.[95]

To be a better world, unintended pesticide contamination in water should be shunned. In the United States, there are about 800 pounds of pesticides used every year. In agriculture, the most often used pesticides include herbicides, fungicides, and insecticides. Nowadays 85% herbicides, 13% insecticides, and just 2% of fungicides are utilized in the United States. Even the presence of low concentrations of toxic pesticides can create a hazardous environment for the aquatic organism. Pesticide toxicity index (LC_{50}) is accessible for the toxicity of different pesticides in freshwater fish. Sucahyo et al.[220] did a comparative study of the LC_{50} of the pesticides tested for different aquatic organisms.

Similarly, research on fresh water bodies in the UK found that 50% of water samples were highly polluted and 88% of samples of them have a high level of pesticide contamination. Research on Blue-tailed damselflies finds out that neonicotinoids are the main culprit for their diminished numbers. Scientists studied in Japan on the reason for the decline of dragonflies is the application of pesticides in rice fields. In 2019, Japan researchers have findings on fish species that the high toxicity and durability of neonicotinoid pesticides pose a hazard to the water system and effects flowing along with the ecosystem which changes the structure of the food chain.

2.4.2.2.2 Groundwater Contamination

The absence of oxygen and the presence of microorganisms create the worst environment to break down the pesticides in groundwater and became a highly concerning aspect for water contamination. In groundwater, an immensely low rate of breakdown and dilution of pesticides show the presence of residues in the future.[16]

Pesticide contamination underwater is a big issue across the world because groundwater is used for drinking water by most of the rural population of the globe directly. Pesticides can be entered in groundwater through aquifers beneath the ground from agricultural practices, the outpouring of polluted surface water, inappropriate discarding, unintended discharges.

In one study of Indian drinking water, water samples were taken from a variety of hand pumps and wells in the Bhopal area. It was found that organochlorine pesticides were found more than the EPA limits in 58% of the test samples.[134] If the groundwater is polluted with pesticides, it is a time-taking process to clean up the contamination, and it is a very costly and tedious process.[177,240,251]

2.4.2.2.3 Soil Contamination and Its Fertility

Pesticides and transformation products can be categorized based onthe nature: (a) pesticides that are hydrophobic, persistent, and bioaccumulable show high affinity to the soil, these include organochlorine pesticides DDT, endosulfan, endrin, heptachlor, and lindane, as well as their toxicants, even though these pesticides are prohibited in agricultural practices, presence of their residues require remedial processes. (b) herbicides, carbamates, fungicides, and certain organophosphorus insecticides, as well as their TPs are polar pesticides that may be transported from the soil via runoff and leaching, these pesticides constitute an issue for the supply of drinking water. Herbicides are the most studied pesticides in soil because of their transformation to produce toxic phenolic compounds through methylation, hydrolysis, and ring cleavage. Some pesticide TPs have been measured and documented in soil.[20,200,201] For ionizable herbicides, soil pH is particularly important since adsorption rises as soil pH decreases (e.g., 2,4-D,2,4,5-T, picloram, and atrazine).[6]

The number of beneficial soil microorganisms can be reduced if we use more pesticides for the treatment of soil. Dr. Elaine Ingham said that the soil declines if both bacteria and fungus are lost. The use of chemical fertilizers and pesticides in excess degrades soil quality. "If we lose both bacteria and fungi, then the soil degrades." The use of chemical fertilizers and pesticides in excess degrades soil quality and will kill organisms that are enough to hold the micronutrients.[207] Plants, for example, require a variety of soil microbes to convert atmospheric nitrogen into nitrates. Pell et al.[185] concluded that herbicides triclopyr inhibits soil bacteria that help to transform ammonia into nitrite. Triclopyr has also been revealed to be poisonous to a variety of mycorrhizal fungi and their spores[49] and oxadiazon applications.[163] Santos and Flores[205] found that Glyphosate inhibits the growth and activity of soil-dwelling nitrogen-fixing bacteria, and 2,4-D decreases the nitrogen fixation process by the bacteria which exist on the root systems of bean plants.[13,82] Some studies showed that the use of pesticides decreases microbiological biomass in the soil.[159]

2.4.2.2.4 Air Contamination

The atmospheric air can be contaminated by spraying pesticides. Unintentional diffusion of every application of pesticides, even from ground equipment, results in the release of chemicals.[100] Within a few days of application, almost 80–90% of a pesticide can be dispersed and vaporized.[152] The USGS has reported that pesticides can be found in the atmosphere in all the tested locations, that is, the sampled areas of the USA.[207]

US Geological Survey[239] studies found that several pesticides can be found in the air at more than half of the sites tested across the country. Herbicides are used to kill plants and simultaneously if used directly, they can damage or kill the beneficial species, or if they are diffuse or vaporized on them. Herbicides, particularly ester formulations, have been shown to vaporize treated plants with vapors which are enough to damage the other plants.[219]

Pesticides, such as dacthal, chlorothalonil, chlorpyrifos, metolachlor, terbufos, and trifluralin have been found in environmental samples from the Arctic.[92,197] Several studies have been carried out and some of these chemicals have been discovered to be capable of undergoing short-range atmospheric transmission to environmentally sensitive areas in investigations, such as the Chesapeake Bay and the Sierra Nevada mountains.[110,140,156,167,228]

Belzer et al.[28] reported that from February 1996 to March 1997, one long-term research of pesticides in the atmosphere of British Columbia (BC) was conducted, this has revealed that 57 chemicals were studied at two sample locations (Agassiz and Abbotsford) in the Fraser Valley. Verrin et al.[244] found that highly hazardous compounds classified as high-priority pesticides, such as atrazine, malathion, and diazinon were discovered from the end of February (72 pg/m^3) to mid-October (253 pg/m^3), with a peak concentration of 42.7 ng/m^3 in mid-June.

Another organophosphate insecticide, Naled (Dibrom) produces a product, dichlorvos.[108] Dosman and Cockcraft[71] reported that at these two locations, captan and 2,4-D had the greatest concentrations and deposition rates, followed by dichlorvos and diazinon. In Alberta, air concentrations of presently used pesticides were studied in 1999 at four different sample locations selected based on the data available on pesticide sales and geography.[138] The major pesticides found at the four locations were triallate and trifluralin. At the end of June in 1989 and 1990, the levels of 2,4-D in South of Regina, Saskatchewan, Canada were 3.9 and 3.6 ng/m^3, respectively.[246]

Waite et al.[245] studied two agricultural sites Lake, Bratt's 35 km southwest of Regina and Hafford to the north and a background site at Waskesiu for spatial variations of chosen herbicides along a 500-km transect. In 1995, atrazine, a non-toxic pesticide, was also studied.[194] Even though the pesticide dacthal was not used in this region, throughout the sample periods in 1994, 1995, and 1996, dacthal was detected (<20–300 pg/m^3).[195]

2.4.2.2.5 Impact on Nontarget Vegetation

The continuous application of pesticides can destroy a variety of species that are not pests. Pesticides having a broad range of toxicity (i.e., not simply poisonous to the insect) are broadcast sprayed across vast regions, such as entire agricultural fields or woods for this reason. In addition to the targeted pests, many non-pest organisms come under direct exposure to these sorts of treatments. This exposure can result in toxicity and inevitable nontarget death, depending on the pesticide and the sensitivity of the non-pest species. For example, in agricultural land or forest, there are just a few plant species present which interrupt the growth of crop plants. These non-crop species of plants are called weeds that might be killed by the use of herbicide. However, several other plant species present in the same ecosystem do not interfere with the growth of the crop plants. The herbicide treatment affects these nontarget plants as well, but they serve no use in terms of pest management. Nontarget plants may serve a positive function in their environment by preventing erosion and nutrient leaching, as well as providing food and habitat for animal species.

Pesticide exposure can cause toxic effects on plants in addition to killing nontarget plants. Phenoxy herbicides, including 2,4-D, can damage the trees and shrubs which are near to targeted pests if they spray or vaporize on the leaves.[72] Locke et al.[147] found that exposure to the herbicide glyphosate can severely reduce the quality of seed, and it can also make certain plants more susceptible to disease.[39] The US Fish and Wildlife Service (USFWS) has recognized 74 endangered plants which can be threatened by only glyphosate.[238] Yields in potato plants can be reduced by direct exposure to the herbicide clopyralid.[148] EPA examined that vaporization of 1% of applied clopyralid is sufficient to kill nontarget plants.[237] Dreistadt et al.[72] found that plants can also be damaged by pesticides and fungicides.

2.4.2.2.6 Impact on Nontarget Organism

The abundance of pesticides in agricultural fields, farms, parks, ways, and water bodies shows that organisms have been exposed in many ways. They are affected indirectly as Crop sprays that accumulate outside the intentional spraying area. The quality of soils is affected by the pesticide spray badly and it will be highly toxic to living organisms that exist in the soil. When agricultural fields are ploughed, pesticide residues run through landscapes and over the field. Pesticides transfer within soils and then rivers and sea, furthermore induce the marine life worse. Pesticide residues go up via the food cycle as birds eat annelida and insects. The number and diversity of insects diminish due to the application of insecticides targeting crop-damaging pests in an ecosystem.[232]

Consequences of organochlorine pesticides in water birds and marine mammals are recorded in the region of Europe and North America.[21,57,136] As a result of India's position as a top producer and consumer of pesticides, wild birds in the nation are exposed to high levels of OC pesticides.[222] Scientists study that organochlorine pesticides exposure is correlated with a reduction in the numbers of bats around the world.[4,54–56,96,122,226] Insects are the main source of diet for mostly bats which further facilitate the accumulation of organo pesticides in their body.[155] Pesticide residues were discovered in the livers and eggs of birds in several investigations conducted in industrialized nations.[24,30,45,47,164,165] Similarly, a multitude of research studies showed organochlorine pesticides are found in humans and wildlife in India.[211]

In Netherlands and France, several studies have related bird decline to pesticide exposure. Similarly, a study in the UK for the Health and Safety Executive found that there is a reduction in farmland birds because of farming practices. A study of gray partridge, for example, discovered strong evidence that herbicides have had a substantial influence on their decrease. Long-term research (1985–2012) reported a reduction in 15 species of butterfly due to extensive usage of pesticides in those regions. Similarly, a study across 19 European nations reported a huge reduction in butterflies since 1990. The study team investigated that pesticides are particularly problematic in Western Europe's heavy agricultural sectors. The lack of food in sterile fields where many pesticides and chemicals are used, there are also larger scale farms, so there are fewer hedgerows for hedgehogs to use, according to the British Hedgehog Preservation Society. Amphibians, such as frogs, toads, and reptiles are also at risk as a result of laboratory and field experiments linking pesticides passed as safe to frog and toad death.

2.5 CONCLUSION

This chapter deliberated an overview of the benefits and hazards of pesticides used in human life. Consumption of food is increasing day by day and agriculture industries are extensively using pesticides (insecticides, herbicides, and fungicides) to grow food production. The use of pesticides leads to many problems for the environment and humanity. It is evident from the different studies that pesticides possess some benefits and risks in humans and the environment. The best probability to increase the benefits and decrease the risk factors of pesticides used to invest time, funds, and good efforts into the development of new strategies to control the pests. It will improve viable production, quality of the environment, and health of human beings. In this regard, further clinical studies need to be carried out on pesticides to explore the side effects of pesticides application, which can be helpful for the welfare of mankind.

KEYWORDS

- **agriculture**
- **benefits**
- **DDT**
- **hazards**
- **pesticides**
- **plants and animals**

REFERENCES

1. Ábalos, M.; Barceló, D.; Parera, J.; Farré, M. L.; Llorca, M.; Eljarrat, E.; Giulivo, M.; Capri, E.; Paunovic, M.; Milacic, R.; Abad, E. Levels of Regulated POPs in Fish Samples from the Sava River Basin. Comparison to Legislated Quality Standard Values. *Sci. Total Environ.* **2019,** *647*, 20–28.
2. Alavanja, M.; Samanic, C., Dosemeci, M.; Lubin, J.; Tarone, R.; Lynch, C.; Knott, C.; Thomas, K.; Hoppin, J.; Barker, J.; J, Coble, J.; Sandler, D.; Blair, A. Use of Agricultural Pesticides and Prostate Cancer Risk in the Agricultural Health Study Cohort. *Am. J. Epidemiol.* **2003,** *157* (9), 800–814.
3. Alharbi, O. M. L.; Basheer, A. A.; Khattab, R. A.; Ali, I. Health and Environmental Effects of Persistent Organic Pollutants. *J. Mol. Liq.* **2018,** *263*, 442–453.
4. Altenbach, J. S.; Geluso, K. N.; Wilson, DE. Population Size of *Tadaria brasiliensis* at Carlsbad Caverns in 1973. In *Biological investigations in Guadelupe Mountains National Park*; Genoways, H. H., Baker, R. J., Eds.; Texas Natl Park Service Proc Trans Ser No. 4, Washington, DC, 1979; vol 341, pp 341–348.
5. Andersen, H. R.; Wohlfahrt-Veje, C.; Dalgård, C.; Christiansen, L.; Main, K. M.; Nellemann, C.; Murata, K.; Jensen, T. K.; Skakkebaek, N. E.; Grandjean, P. Paraoxonase 1 Polymorphism and Prenatal Pesticide Exposure Associated with Adverse Cardiovascular Risk Profiles at School Age. *PLoS ONE* **2012,** *7* (5), e36830.

6. Andreu, P. V.; Pico, Y. Determination of Pesticides and Their Degradation Products in Soil: Critical Review and Comparison of Methods. *Trends Anal. Chem.* **2004,** *23* (10–11), 772–789.
7. Anon. *Water Matters for Sustainable Agriculture—A Collection of Case Studies*; CropLife International: Brussels, 2004. www.croplife.org/pub/watermatters.
8. Anon. *The World Health Organization*, 2004. http://www.who.int/mediacentre/news/releases/2006/pr50/en/index.html.
9. Anon. *Population and Vital Statistics Report, Series: A, No. 236/237*, **2005,** *LVIII*, (1).
10. Anon. *Conservation Technologies and the Plant Science Industry: Managing Natural Resources Sustainably*; CropLife International: Belgium, 2005; 42 pp.
11. Anon. Herbicide Use Essential to Crop Production. *Chem. Mark. Rep.* **2003,** *263* (18), 4.
12. Archer, J. C.; Jenkins, R. G. Automated Milk Fat Extraction for the Analyses of Persistent Organic Pollutants. *J. Chromatogr. B Anal. Technol. Biomed. Life Sci.* **2017,** *1041*, 70–76.
13. Arias, R. N.; Fabra, P. A. Effects of 2,4-Dichlorophenoxyacetic Acid on *Rhizobium* sp. Growth and Characterization of Its Transport. *Toxicol. Lett.* **1993,** *68*, 267–273.
14. Atreya, N. Chemophobia-Pesticide Residues in Food. *Outlooks Pest Manage.* **2006,** *17*, 242.
15. Austin, R. B. Yield of Wheat in the UK: Recent Advances and Prospects. In *Annual Meeting of the Crop Science Society of America, Symposium—1998 ASA Meeting*, Baltimore, 1998.
16. Aydinalp, C.; Porca, M. M. The Effects of Pesticides in Water Resources. *J. Cent. Eur. Agric.* **2004,** *5*, 5–12.
17. Azmi, A.; Naqvi, S.; Azmi, A.; Aslam, M. Effect of Pesticide Residues on Health and Different Enzyme Levels in the Blood of Farm Workers from Gadap (Rural Area) Karachi, Pakistan. *Chemosphere* **2006,** *64* (1), 1739–1744.
18. Ballantyne, B.; Marrs, T. C. Pesticides: An Overview of Fundamentals. *Pesticide Toxicology and International Regulation*, 2004; pp 1–23. doi:10.1002/0470091673.ch1.
19. Balluz, L.; Philen, R.; Brock, J.; Falter, K.; Kiefer, M.; Hart, R.; Hill, R. Health Complaints Related to Pesticide Stored at a Public Health Clinic. *Environ. Res.* **2000,** *82* (1), 1–6.
20. Barcelo, D.; Hennion, M. C. *Trace Determination of Pesticides and Their Degradation Products in Water*; Elsevier: Amsterdam, The Netherlands, 1997.
21. Barron, M. G.; Galbraith, H.; Beltman, D. Comparative Reproduction and Developmental Toxicology of Birds. *Comp. Biochem. Physiol.* **1995,** *112c*, 1–14.
22. Bates, G.; Denton, H. P. *No-Till Establishment of Forage Crops Agricultural Extension Service*. The University of Tennessee: Knoxville, TN, 2007. http://www.utextension.utk.edu/publications/spfiles/sp435c.pdfS.
23. Beane, L.; Bonner, M.; Blair, A.; Hoppin, J.; Sandler, D.; Lubin, J.; Dosemeci, M.; Lynch, C.; Knott, C.; Alavanja, M. Cancer Incidence among Male Pesticide Applicators in the Agricultural Health Study Cohort Exposed to Diazinon. *Am. J. Epidemiol.* **2005,** *162* (11), 1070–1079.
24. Becker, P. H. Seabirds as Monitor Organisms of Contaminants along the German North Sea Coast. *Hel Merr.* **1989,** *43*, 395–403.
25. Bedi, J. S.; Singh, V.; Gupta, A.; Gill, J. P. S.; Aulakh, R. S. Persistent Organic Pollutants (POPs) in Fresh Water Farm Fish Species from Punjab (India) and Evaluation of Their Dietary Intake for Human Risk Assessment. *Hum. Ecol. Risk Assess.* **2018,** *24*, 1659–1672.
26. Behera, B.; Singh, S. G. Studies on Weed Management in Monsoon Season Crop of Tomato. *Indian J. Weed Sci.* **1999,** *31* (1–2), 67.

27. Bell, E. M.; Hertz-Picciotto, I.; Beaumont, J. J. A Case-Control Study of Pesticides and Fetal Death Due to Congenital Anomalies. *Epidemiology* **2001,** *12* (2), 148–156.
28. Belzer, W.; Evans, C.; Poon, A. Atmospheric Concentrations of Agricultural Chemicals in the Lower Fraser Valley. *FRAP Study Report*; Aquatic and Atmospheric Science Division, Environment Canada: Vancouver, BC, 1998.
29. Bernard, A.; Hermans, C.; Broeckaert, F.; De Poorter, G.; De Cock, A.; Houins, G. Food Contamination by PCBs and Dioxins. *Nature* **1999,** *401*, 231–232.
30. Bernardz, J. C.; Klem, D.; Goodrich, L. J.; Senner, S. E. Migration Counts of Raptors at Hawk Mountain, Pennsylvania, as Indicators of Population Trends, 1934–1986. *Auk* **1990,** *107*, 96–109.
31. Besbelli, N. Harmonized Collection of Data on Human Pesticide Exposures. In *Seventh GINC Tokyo Meeting for Information Exchange and Collaboration in Asia on Chemical Management and Pesticide Poisoning*, 2001. http://www.nihs.go.jp/GINC/meeting/7th/meet-rep.html.
32. Beseler, C.; Stallones, L.; Hoppin, J.; Alavanja, M.; A. Blair, A.; Keefe, T.; Kamel, F. Depression and Pesticide Exposures in Female Spouses of Licensed Pesticide Applicators in the Agricultural Health Study Cohort. *J. Occup. Environ. Med.* **2006,** *48* (10), 1005–1013.
33. Beseler, C.; Stallones, L.; Hoppin, J.; Alavanja, M.; Blair, A.; Keefe, T.; Kamel, F. Depression and Pesticide Exposures among Private Pesticide Applicators Enrolled in the Agricultural Health Study. *Environ. Health Perspect.* **2008,** *116* (12), 1713–1719.
34. Bhanti, M.; Taneja, A. Contamination of Vegetables of Different Seasons with Organophosphorous Pesticides and Related Health Risk Assessment in Northern India. *Chemosphere* **2007,** *69* (1), 63–68.
35. Bhatia, R.; Ortega, L.; Dash, A. P.; Mohamed, A. J. Vector-Borne Disease in South-East Asia: Burdens and Key Challenges to Be Addressed. *WHO South-East Asia J. Public Health* **2014,** *3* (1), 2–4.
36. Bilal, M.; Iqbal, H. M.; Barcelo, D. Persistence of Pesticides-Based Contaminants in the Environment and Their Effective Degradation Using Laccase-Assisted Bio-catalytic Systems. *Sci. Total. Environ.* **2019,** *695*, 133896.
37. Bortleson, G. C; Davis, D. A. *Pesticides in Selected Small Streams in the Puget Sound Basin, 1987–1995*; US Geological Survey and Washington State Department of Ecology: Lacey, WA, 1997; pp 1–4.
38. Bouchard, M.; Chevrier, J.; Harley, K., Kogut, K.; Vedar, M.; Calderon, N.; Trujillo, C., Johnson, C.; Bradman, A.; Boyd Barr, D.; Eskenazi, B. Prenatal Exposure to Organophosphate Pesticides and IQ in 7-Year-Old Children. *Environ. Health Perspect.* **2011,** *119* (8), 1189–1195.
39. Brammall, R. A.; Higgins, V. J. The Effect of Glyphosate on Resistance of Tomato to Fusarium Crown and Root Rot Disease and on the Formation of Host Structural Defensive Barriers. *Can. J. Bot.* **1988,** *66*, 1547–1555.
40. Bridges, D., Ed. *Crop Losses Due to Weeds in the United States*. Weed Sci. Soc. Am.: Champaign, IL, 1992.
41. Brown, I. *UK Pesticides Residue Committee Report*, 2004. http://www.pesticides.gov.uk/uploadedfiles/Web_Assets/PRC/PRCannual-report2004.pdf.
42. Bruns, A. Controlling Aflatoxin and Fumonisin in Maize by Crop Management. *J. Toxicol. Toxin Rev.* **2003,** *22* (2), 153–173.
43. Buchel, K. H. *Chemistry of Pesticides*; John Wiley and Sons, Inc.: New York, NY, 1983.

44. Buczynska, A.; Szadkowska, I. Identification of Health Hazards to Rural Population Living near Pesticide Dump Sites in Poland. *Int. J. Occup. Med. Environ. Health* **2005,** *18* (4), 331–339.
45. Cade, T. J.; Lincer, J. L.; White, C. M.; Rosenau, D. G.; Swartz, L. G. DDE Residues and Eggshell Changes in Alaskan Falcons and Hawks. *Science* **1989,** *172*, 955–957.
46. Cajka, T.; Hajslova, J.; Kazda, R.; Poustka, J. Challenges of Gas Chromatography-High-Resolution Time-of-Flight Mass Spectrometry for Simultaneous Analysis of Polybrominated Diphenyl Ethers and Other Halogenated Persistent Organic Pollutants in Environmental Samples. *J. Sep. Sci.* **2005,** *28*, 601–611.
47. Castillo, L.; Thybaud, E.; Caquet, T.; Ramade, F. Organochlorine Contaminants in Common Tern (*Sterna hirundo*) Eggs and Young from the Rhine River Area (France). *Bull. Environ. Contam. Toxicol.* **1994,** *53*, 759–764.
48. Cecchi, A.; Rovedatti, M.; Sabino, G.; Magnarelli, G. Environmental Exposure to Organophosphate Pesticides: Assessment of Endocrine Disruption and Hepatotoxicity in Pregnant Women. *Ecotoxicol. Environ. Saf.* **2012,** *80*, 280–287.
49. Chakravarty, P.; Sidhu, S. S. Effects of Glyphosate, Hexazinone and Triclopyron In Vitro Growth of Five Species of Ectomycorrhizal Fungi. *Eur. J. Forest Pathol.* **1987,** *17*, 204–210.
50. Chan, H. M.; Khoury, M. E.; Sedgemore, M.; Sedgemore, S.; Kuhnlein, H. V. Organochlorine Pesticides and Polychlorinated Biphenyl Congeners in Ooligan Grease: A Traditional Food Fat of British Columbia First Nations. *J. Food Compos. Anal.* **1996,** *9*, 32–42.
51. Chang, S. S.; Lu, T. H.; Sterne, J. A.; Eddleston, M.; Lin, J. J.; Gunnell, D. The Impact of Pesticide Suicide on the Geographic Distribution of Suicide in Taiwan: A Spatial Analysis. *BMC Public Health* **2012,** *12*, 260. doi:10.1186/1471-2458-12-260.
52. Chenkin, A. F. Economic Effects of Plant Protection in the Russian Federated Republic. In *Papers at Sessions, VIII International Congress of Plat Protection, Moscow, USSR,* 1975, vol 2; p 27.
53. Chung, S. W. C.; Chen, B. L. S. Determination of Organochlorine Pesticide Residues in Fatty Foods: A Critical Review on the Analytical Methods and Their Testing Capabilities. *J. Chromatogr. A* **2011,** *1218*, 5555–5567.
54. Clark, D. R. Death of Bats from DDE, DDT or Dieldrin Diagnosis Via Residues in Carcass fat. *Bull. Environ. Contam. Toxicol.* **1981,** *26*, 367–371.
55. Clark, D. R.; Krynitsky, A. J. DDT: Recent Contamination in New Mexico and Arizona. *Environment* **1983,** *25*, 27–31.
56. Clark, D. R.; Lamont, T. G. Organochlorine Residues in Females and Nursing Young of the Big Brown Bats (*Eptesicus fuscus*). *Bull. Environ. Contam. Toxicol.* **1976,** *15*, 1–8.
57. Cooke, A. S. Egg Shell Characteristic of Gannets *Sula bassana*, Shags *Phalacrocorax aristotelis* and Great Backed Gulls Larusmarianus Exposed to DDE and Other Environmental Pollutants. *Environ. Pollut.* **1979,** *19*, 47–65.
58. Cooper, J.; Dobson, H. The Benefits of Pesticides to Mankind and the Environment. *Crop Protect.* **2007,** *26*, 1337–1348.
59. Costopoulou, D.; Vassiliadou, I.; Leondiadis, L. Infant Dietary Exposure to Dioxins and Dioxin-Like Compounds in Greece. *Food Chem. Toxicol.* **2013,** *59*, 316–324.
60. Cox, S.; Niskar, A.; Narayan, V.; Marcus, M. Prevalence of Self-reported Diabetes and Exposure to Organochlorine Pesticides among Mexican Americans Hispanic Health and Nutrition Examination Survey, 1982–1984. *Environ. Health Perspect.* **2007,** *115* (12), 1747–1752.

61. Crawford, J.; Hoppin, J.; Alavanja, M.; Blair, A.; Sandler, D.; Kamel, F. Hearing Loss among Licensed Pesticide Applicators in the Agricultural Health Study. *J. Occup. Environ. Med.* **2008,** *50* (7), 817–826.
62. CropLife. *Guidelines of Personal Protection When Using Crop Protection Products in Hot Climates*, 2004. https://croplife.org/wp-content/uploads/2014/04/Guidelines-for-personal-protection-when-using-crop-protection-products-in-hot-climates.pdf.
63. Curtis, C.; Jana-Kara, B.; Maxwell, C. A. Insecticide Treated Nets: Impact on Vector Populations and Relevance of Initial Intensity of Transmission and Pyrethroid Resistance. *J. Vector Borne Dis.* **2003,** *40* (1–2), 1–8.
64. Dales, M. J.; Golob, P. The Protection of Maize against *Prostephanus truncatus* (Horn), Using Insecticide Sprays in Tanzania. *Int. J. Pest Manage.* **1997,** *43* (1), 39–43.
65. Davis, J. R.; Brownson, R. C.; Garcia, R.; Bentz, B. J.; Turner, A. Family Pesticide Use and Childhood Brain Cancer. *Arch. Environ. Contam. Toxicol.* **1993,** *24* (1), 87–92.
66. Davis, J. R.; Brownson, R. C.; Garcia, R. Family Pesticide Use in the Home, Garden, Orchard, and Yard. *Arch. Environ. Contam. Toxicol.* **1992,** *22*, 260–266.
67. Dawson, A. H.; Eddleston, M.; Senarathna, L.; Mohamed, F.; Gawarammana, I.; Bowe, S. J.; Manuweera, G.; Buckley, N. A. Acute Human Lethal Toxicity of Agricultural Pesticides: A Prospective Cohort Study. *PLoS Med.* **2010,** *7*, e1000357.
68. Dayton, S.; Sandler, D.; Blair, A.; Alavanja, M.; Beane, L.; Hoppin, J. Pesticide Use and Myocardial Infarction Incidence among Farm Women in the Agricultural Health Study. *J. Occup. Environ. Med.* **2010,** *52* (7), 693–697.
69. US Department of Health and Human Services. *Dietary Guidelines for Americans*; US Department of Agriculture, 2005. www.healthierus.gov/dietaryguidelines.
70. Domingo, J. L. Concentrations of Environmental Organic Contaminants in Meat and Meat Products and Human Dietary Exposure: A Review. *Food Chem. Toxicol.* **2017,** *107*, 20–26.
71. Dosman, J. A.; Cockcraft, D. W. *Principles of Health and Safety in Agriculture*; CRC Press: Boca Raton, FL, 1989.
72. Dreistadt, S. H.; Clark, J. K.; Flint, M. L. *Pests of Landscape Trees and Shrubs. An Integrated Pest Management Guide*; University of California, Division of Agriculture and Natural Resources: Los Angeles, CA, 1994. Publication No. 3359.
73. Eldridge, B. F. *Pesticide Application and Safety Training for Applicators of Public Health Pesticides*; California Department of Public Health, Vector-Borne Disease Section: Sacramento, CA, 2008.
74. El-Shahawi, M. S.; Hamza, A.; Bashammakh, A. S.; Al-Saggaf, W. T. An Overview on the Accumulation, Distribution, Transformations, Toxicity and Analytical Methods for the Monitoring of Persistent Organic Pollutants. *Talanta* **2010,** *80* (5), 1587–1597.
75. Labour Bureau, Ministry of Labour. *Employment Information: Indian Labour Statistics*. Labour Bureau, Ministry of Labour: Chandigarh, 1994.
76. Erhart, E.; Hartl, W. Soil Protection through Organic Farming: A Review. In *Organic Farming, Pest Control and Remediation of Soil Pollutants*; Lichtfouse, E., Ed.; Springer: Dordrecht, 2009; pp 203–226.
77. Eskenazi, B.; Huen, K.; Marks, A.; Harley, K.; Bradman, A.; Boyd Barr, D.; Holland, N. PON1 and Neurodevelopment in Children from the CHAMACOS Study Exposed to Organophosphate Pesticides *in Utero*. *Environ. Health Perspect.* **2010,** *118* (12), 1775–1781.
78. Eskenazi, B.; Marks, A. M.; Bradman, A.; Harley, K.; Barr, D. B.; Johnson, C.; Morga, N.; Nicholas, P.; Jewell, N. P. Organophosphate Pesticide Exposure and Neurodevelopment in

Young Mexican-American Children. *Environ. Health Perspect.* **2007,** *115* (5), 792–798. doi:10.1289/ehp.9828.

79. Etcheverry, M.; Chulze, S.; Dalcero, A.; Varsavsky, E.; Rodriguez, M.; Moschetti, E.; Ferrero, S. Aflatoxins in Sunflower Seeds: Influence of Fungicides and Insecticides. *Int. J. Food Microbiol.* **1987,** *5*, 103–109.
80. European Food Information Council. 2006. http://www.eufic.org/page/en/health-lifestyle/healthy-eating/.
81. Everett, C. J.; Matheson, E. M. Biomarkers of Pesticide Exposure and Diabetes in the 1999–2004 National Health and Nutrition Examination Survey. *Environ. Int.* **2010,** *36* (1), 398–401.
82. Fabra, A.; Duffard, R.; Evangelista, D. D. A. Toxicity of 2,4-Dichlorophenoxyacetic Acid in Pure Culture. *Bull. Environ. Contam. Toxicol.* **1997,** *59*, 645–652.
83. FAO. *International Code of Conduct on the Distribution and Use of Pesticides*; Chief Publishing Policy and Support Branch, Office of Knowledge Exchange, Research and Extension, FAO: Rome, Italy, 1989.
84. FAOSTAT. *Food and Agriculture Organization of the United Nations*, n.d. http://www.fao.org/faostat/en/#data/QC.
85. Farr, S.; Cai, J.; Savitz, D.; Sandler, D.; Hoppin, J.; Cooper, G. Pesticide Exposure and Timing of Menopause. *Am. J. Epidemiol.* **2006,** *163* (8), 731–742.
86. Farr, S.; Cooper, G.; Cai, J.; Savitz, D.; Sandler, D. Pesticide Use and Menstrual Cycle Characteristics among Premenopausal Women in the Agricultural Health Study. *Am. J. Epidemiol.* **2004,** *160* (12), 1194–1204.
87. Fernandez-Cornejo, J. *The Seed Industry in US Agriculture: An Exploration of Data and Information on Crop Seed Markets, Regulation, Industry Structure, and Research and Development, Agricultural Information Bulletins 33671*; United States Department of Agriculture, Economic Research Service, 2004.
88. Fernandez-Cornejo, J.; Jans, S.; Smith, M. Issues in the Economics of Pesticide Use in Agriculture: A Review of the Empirical Evidence. *Rev. Agric. Econ.* **1998,** *20* (2), 462–488.
89. Ferreira, J. D.; Couto, A. C.; Pombo-de-Oliveira, M. S.; Koifman, S. Brazilian Collaborative Study Group of Infant Acute Leukemia: In Utero Pesticide Exposure and Leukemia in Brazilian Children <2 Years of Age. *Environ. Health Perspect.* **2013,** *121* (2), 269.
90. Food and Agriculture Organization. *World Food Summit Plan of Action 1996*; Food and Agriculture Organization: Rome, Italy, 1996. http://www.fao.org/wfs/index_en.htm.
91. Frumkin, H. Agent Orange and Cancer: An Overview for Clinicians. *CA Cancer J. Clin.* **2003,** *53*, 245.
92. Garbarino, J. R.; Snyder-Conn, E.; Leiker, T. J.; Hoffman, G. L. Contaminants in Arctic Snow Collected over Northwest Alaskan Sea Ice. *Water, Air, Soil Pollut.* **2002,** *139*, 183–214.
93. Gardner, J.; Nehring, R.; Nelson, C. Genetically s Crops and Household Labor Savings in US Crop Production. *AgBioForum* **2009,** *12* (3–4), 303–312.
94. Gattuso, D. Understanding the Benefits of Pesticides. *Consum. Res. Mag.* **2000,** *83* (2), 34.
95. Gavrilescu, M. Fate of Pesticides in the Environment and Its Bioremediation. *Eng. Life Sci.* **2005,** *5*, 497–526.
96. Geluso, K. N.; Altenbach, J. S.; Wilson, D. E. Bat Mortality: Pesticide Poisoning and Migratory Stress. *Science* **1976,** *194*, 184–186.

97. Gerig, B. S.; Chaloner, D. T.; Janetski, D. J.; Rediske, R. R.; O'Keefe, J. P.; Moerke, A. H.; Lamberti, G. A. Congener Patterns of Persistent Organic Pollutants Establish the Extent of Contaminant Biotransport by Pacific Salmon in the Great Lakes. *Environ. Sci. Technol.* **2016,** *50*, 554–563.
98. Gianessi, L. Beneficial Impacts of Pesticide Use for Consumers. In *Pesticides: Managing Risks and Optimizing Benefits. American Chemical Society Symposium Series #734*; Ragsdale, N., Seiber, J., Eds.; American Chemical Society: Washington, DC, 1999; p 207.
99. Gianessi, L. P. *Pesticide Use and Biodiversity Conservation on Farms*; CropLife Foundation: Washington, DC, 2010. www.croplifefoundation.org.
100. Glotfelty, D. E.; Sch6omburg, C. J. Volatilization of Pesticides from Soil. In *Reactions and Movements of Organic Chemicals in Soil*; Sawhney, B. L., Brown, K., Eds.; Soil Science Society of America Special Pub.: Madison, WI, 1989; Vol. 22.
101. Goh, K.; Bret, B.; Potter, T. L.; Gan, J. *Pesticide Mitigation Strategies for Surface Water Quality*; ARS, 2011, vol 1075. http://pubs.acs.org/isbn/9780841226579.
102. Goldner, W.; Sandler, D.; Yu, F.; Hoppin, J.; Kamel, F.; Tricia, L. Pesticide Use and Thyroid Disease among Women in the Agricultural Health Study. *Am. J. Epidemiol.* **2010,** *171* (4), 455–464.
103. Gratz, N. G.; Jany, W. C. What Role for Insecticides in Vector Control Programs? *Am. J. Trop. Med. Hyg.* **1994,** *50* (6), 11–20.
104. GRC Economics. *The Value of Fungicides to the Availability of a Healthy and Affordable Food Supply*; Hill and Knowlton, Inc. Economics Group, Government Research Corporation: Washington, DC, 1989.
105. Grube, A.; Donaldson, D.; Kiely, T.; Wu, L. *Pesticide Industry Sales and Usage. 2006 and 2007 Market Estimates. Biological and Economic Analysis*; Office of Chemical Safety and Pollution Prevention, United States Environmental Protection Agency: Washington, DC, 2011.
106. Gupta, S. K.; Jani, J. P.; Saiyed, H. N.; Kashyap, S. K. Health Hazards in Pesticide Formulators Exposed to a Combination of Pesticides. *Indian J Med Res.* **1984,** *79*, 666.
107. Hainsworth, S. D.; Eden-Green, S. J., Eds. In *Sustaining Change: Proceedings of a Workshop on the Factors Affecting Uptake and Adoption of Department for International Development (DFID) Crop Protection Programme (CPP) Research Outputs*; Imperial College at Wye: Kent, UK, 2000.
108. Hall, G. L.; Mourer, C. R.; Shibamoto, T.; Fitzell, D. Development and Validation of an Analytical Method for Naled and Dichlorvos in Air. *J. Agric. Food Chem.* **1997,** *45* (1), 145–148.
109. Harari, R.; Julvez, J.; Murata, K.; Barr, D.; Bellinger, D.; Debes, F.; P. Grandjean, P. Neurobehavioral Deficits and Increased Blood Pressure in School-Age Children Prenatally Exposed to Pesticides. *Environ. Health Perspect.* **2010,** *118* (6), 890–896.
110. Harman-Fetcho, J. A.; McConnell, L. L.; Rice, C. P.; Baker, J. E. Wet Deposition and Air-Water Gas Exchange of Currently Used Pesticides to a Sub-estuary of the Chesapeake Bay. *Environ. Sci. Technol.* **2000,** *34*, 1462–1468.
111. Henry, M. S. J. *J. Environ. Health.* **1994,** *12*, 65–70.
112. Hernández, Á. R.; Boada, L. D.; Mendoza, Z.; Ruiz-Suárez, N.; Valerón, P. F.; Camacho, M.; Zumbado, M.; Almeida-González, M.; Henríquez-Hernández, L. A.; Luzardo, O. P. Consumption of Organic Meat Does Not Diminish the Carcinogenic Potential Associated with the Intake of Persistent Organic Pollutants (POPs). *Environ. Sci. Pollut. Res. Int.* **2017,** *24*, 4261–4273.

113. Hodgson, E. Pesticides: Past, Present and Future. In *Pesticides and the Future: Toxicological Studies of Risks and Benefits*; Hodgson, E., Roe, R. M., Motoyama, N., Eds. North Carolina State University: Raleigh, NC, 1991.
114. Hoffman, I. F.; Jere, C. S.; Taylor, T. E.; Munthali, P.; Dyer, J. R.; Wirima, J. J.; Rogerson, S. J.; Kumwenda, N.; Eron, J. J.; Fiscus, S. A.; Chakraborty, H.; Taha, T. E.; Cohen, M. S.; Molyneux, M. E. The Effect of *Plasmodium falciparum* Malaria on HIV-1 RNA Blood Plasma Concentration. *AIDS* **1999,** *13* (4), 487–494. doi:10.1097/00002030-199903110-00007.
115. Honeycutt, R. C.; Schabacker, D. J. *Mechanisms of Pesticide Movement into Ground Water*; CRC Press: Boca Raton, FL, 2018.
116. Hoppin, J.; Umbach, D.; Kullman, G.; Henneberger, P.; London, S.; Alavanja, M.; Sandler, D Pesticides and Other Agricultural Factors Associated with Self-Reported Farmer's Lung among Farm Residents in the Agricultural Health Study. *Occup. Environ. Med.* **2007,** 64 (1), 334–342.
117. Hoppin, J.; Umbach, D.; London, S.; Alavanja, M.; Sandler, D. Chemical Predictors of Wheeze among Farmer Pesticide Applicators in the Agricultural Health Study. *Am. J. Respir. Crit. Care Med.* **2002,** *165* (1), 683–689.
118. Hoppin, J.; Umbach, D.; London, S.; Henneberger, P.; Kullman, G.; Coble, J.; Alavanja, M.; Beane, L.; Sandler, D. Pesticide Use and Adult-Onset Asthma among Male Farmers in the Agricultural Health Study. *Eur. Respir. J.* **2009,** *34* (6), 1296.
119. Hoppin, J.; Valcin, M.; Henneberger, P.; Kullman, G.; Umbach, D.; London, S.; Alavanja, M.; Sandler, D. Pesticide Use and Chronic Bronchitis among Farmers in the Agricultural Health Study. *Am. J. Ind. Med.* **2007,** 50 (12), 969–979.
120. Infante-Rivard, C.; Weichenthal, S. Pesticides and Childhood Cancer: An Update of Zahm and Ward's 1998 Review. *J. Toxicol. Environ. Health B Crit. Rev.* **2007,** *10*, 81–99.
121. IPPC. *International Plant Protection Convention—New Revised Text Approved by the FAO Conference at Its 29th Session*, November 1997, 1997.
122. Jefferies, D. J. Organochlorine Insecticide Residue in British Bats and Their Significance. *J Zool.* **1976,** *166*, 245–251.
123. Jensen, E.; Bolger, P. M. Exposure Assessment of Dioxins/Furans Consumed in Dairy Foods and Fish. *Food Addit. Contam.* **2001,** *18*, 395–403.
124. Jeyaratnam, J. Acute Pesticide Poisoning: A Major Global Health Problem. *World Health Stat. Q.* **1990,** *43* (3), 139–144.
125. Joshi, M. *The Future Role of Pesticides in US Agriculture. Committee on the Future Role of Pesticides in US Agriculture, Board of Agriculture and Natural Resources and Board on Environmental Studies and Toxicology, Commission on Life Sciences*; National Academy Press: Washington, DC, 2001; p 31.
126. Kamel, F.; Engel, L.; Gladen, B.; Hoppin, J.; Alavanja, M.; Sandler, D. Neurologic Symptoms in Licensed Private Pesticide Applicators in the Agricultural Health Study. *Environ. Health Perspect.* **2005,** *113* (7), 877–882.
127. Kamel, F.; Rowland, A.; Park, L.; Anger, K.; Baird, D.; Gladen, B.; Moreno, T.; Stallone, L.; Sandler, D. Neurobehavioral Performance and Work Experience in Florida Farmworkers. *Environ. Health Perspect.* **2003,** *111* (14), 1765–1772.
128. Kamuanga, M.; Sigué, H.; Swallow, B.; Bauer, B.; d'Ieteren, G. Farmers' Perceptions of the Impacts of Tsetse and Trypanosomosis Control on Livestock Production: Evidence from Southern Burkina Faso. *Trop. Anim. Health Prod.* **2001,** *33* (2), 141–153. doi:10.1023/a:1005286631470.

129. Kedikoglou, K.; Costopoulou, D.; Vassiliadou, I.; Bakeas, E.; Leondiadis, L. An Effective and Low Cost Carbon Based Clean-Up Method for PCDD/Fs and PCBs Analysis in Food. *Chemosphere* **2018,** *206*, 531–538.
130. Keiserukhsky, M. G.; Kashirsky, O. P. Economics of Plant Protection in the USSR. In *VIII International Congress of Plant Protection*, Moscow, USSR, 1975, vol 2; p 12.
131. Knipe, D. W.; Metcalfe, C.; Fernando, R.; Pearson, M.; Konradsen, F.; Eddleston, M.; Gunnell, D. Suicide in Sri Lanka 1975–2012: Age, Period and Cohort Analysis of Police and Hospital Data. *BMC Public Health* **2014,** *14*, 839. doi:10.1186/1471-2458-14-839.
132. Knutson, R. D.; Taylor, C. R.; Penson, J. B.; Smith, E. G. *Economic Impacts of Reduced Chemical Use*; Knutson and Associates: College Station, TX, 1990.
133. Kolbe, W. Effect of Different Crop Protection Programmes on Yield and Quality of Apples II (1967–1981). *AGRIS* **2013,** *35* (2), 189–204.
134. Kole, R. K.; Bagchi, M. M. Pesticide Residues in the Aquatic Environment and Their Possible Ecological Hazards. *J. Inland Fish Soc. India* **1995,** *27* (2), 79–89.
135. Kole, R. K.; Banerjee, H.; Bhattacharyya, A. Monitoring of Market Fish Samples for Endosulfan and Hexachlorocyclohexane Residues in and around Calcutta. *Bull. Environ. Contam. Toxicol.* **2001,** *67*, 554–559.
136. Kubiak, T. J.; Harris, H. J.; Smith, L. M.; Schwartz, T. R.; Stalling, P. L.; Trick, L.; Sielo, D. E.; Pocherty, P. D.; Erdman, T. C. Microcontaminants and Reproductive Impairment of the Forster's Tern in Green Bay, Lake Michigan. *Arch. Environ. Contam. Toxicol.* **1989,** *18*, 706–727.
137. Kucharik, C. J.; Ramankutty, N. Trends and Variability in US Corn Yields over the 20th Century. *Earth Interact.* **2005,** *9*, 1–29.
138. Kumar, Y. *Pesticides in Ambient Air in Alberta*; Air Research Users Group, Alberta Environment: Edmonton, AB, 2001. ISBN 0-7785-1889-4.
139. Lacasaña, M.; López-Flores, I.; Rodríguez-Barranco, M.; Aguilar-Garduño, C.; Blanco-Muñoz, J.; Pérez-Méndez, O.; Gamboa, R.; Gonzalez-Alzaga, B.; Bassol, S.; Cebrian, M. E. Interaction between Organophosphate Pesticide Exposure and PON1 Activity on Thyroid Function. *Toxicol. Appl. Pharmacol.* **2010,** *249* (1), 16–24.
140. Le Noir, J. S.; McConnell, L. L.; Fellers, G. M.; Cahill, T. M.; Seiber, J. N. Summertime Transport of Current Use Pesticides from California's Central Valley to the Sierra Nevada Mountain Range, USA. *Environ. Toxicol. Chem.* **1999,** *18*, 2715–2722.
141. Lee, S. J.; Mehler, L.; Beckman, J.; Diebolt-Brown, B.; Prado, J.; Lackovic, M.; Waltz, J.; Mulay, P.; Schwartz, A.; Mitchell, Y.; Moraga-McHaley, S.; Gergely, R.; Calvert, G. M. Acute Pesticide Illnesses Associated with Off-Target Pesticide Drift from Agricultural Applications: 11 States, 1998–2006. *Environ. Health Perspect.* **2011,** *119*, 1162.
142. Lewis, N.; Ruud, J. Apples in the American Diet. *Nutr. Clin. Care* **2004,** *7* (2), 82.
143. Lewis, N.; Ruud, J. Blueberries in American Diet. *Nutr. Today* **2005,** *40* (2), 92.
144. Li, Q.; Loganath, A.; Seng Chong, Y.; Tan, J.; Philip Obbard, J. Persistent Organic Pollutants and Adverse Health Effects in Humans. *J. Toxicol. Environ. Health A* **2006,** *69*, 1987–2005.
145. Li, X.; Gan, Y.; Yang, X.; Zhou, J.; Dai, J.; M. Xu, M. Human Health Risk of Organochlorine Pesticides (OCPs) and Polychlorinated Biphenyls (PCBs) in Edible Fish from Huairou Reservoir and Gaobeidian Lake in Beijing, China. *Food Chem.* **2008,** *109* (1), 348–354.
146. Lindblade, K. A.; Eisele, T. P.; Gimnig, J. E.; Alaii, J. A.; Odhiambo, F.; ter Kuile, F. O.; Hawley, W. A.; Wannemuehler, K. A.; Phillips-Howard, P. A.; Rosen, D. H.; Nahlen, B.

L.; Terlouw, D. J.; Adazu, K.; Vulule, J. M.; Slutsker, L. Sustainability of Reductions in Malaria Transmission and Infant Mortality in Western Kenya with Use of Insecticide-Treated Bednets: 4 to 6 Years of Follow-Up. *JAMA* **2004,** *291* (21), 2571–2580. doi:10.1001/jama.291.21.2571.

147. Locke, D.; Landivar, J. A.; Moseley, D. The Effects of Rate and Timing of Glyphosate Applications of Defoliation Efficiency, Regrowth Inhibition, Lint Yield, Fiber Quality and Seed Quality. In *Proc. Beltwide Cotton Conf., National Cotton Council of America,* 1995; pp 1088–1090.
148. Lucas, W. J.; Lobb, P. G. Response of Potatoes, Tomatoes and Kumaras to Foliar Applications of MCPA, MCPB, 2,4-D, Clopyralid, and Amitrole. In *Proc. 40th N.Z. Weed and Pest Control Conf.*, 1987; pp 59–63.
149. Lyche, J. L.; Rosseland, C.; Berge, G.; Polder, A. Human Health Risk Associated with Brominated Flame-Retardants (BFRs). *Environ. Int.* **2015,** *74*, 170–180.
150. Lynch, S.; Mahajan, R.; Beane, L.; Hoppin, J.; Alavanja, M. Cancer Incidence among Pesticide Applicators Exposed to Butylate in the Agricultural Health Study (AHS). *Environ. Res.* **2009,** *109* (1), 860–868.
151. Lynch, S.; Rusiecki, J.; Blair, A.; Dosemeci, M.; Lubin, J.; Sandler, D.; Hoppin, J.; Lynch, C.; Alavanja, M. Cancer Incidence among Pesticide Applicators Exposed to Cyanazine in the Agricultural Health Study. *Environ. Health Perspect.* **2006,** *114* (8), 1248–1252.
152. Majewski, M. S.; Capel, P. D. Pesticides in the Atmosphere: Distribution, Trends, and Governing Factors. *Pesticides in the Hydrologic System*; Ann Arbor Press Inc.: Chelsea, MI, Boca Raton, FL, 1996; vol 1.
153. Marquis, S. Agrochemical Use in South Africa: The Toxicological and Sociological Impacts of Agrochemicals on Susceptible Populations in South Africa's Agricultural Sector. *Final Thesis. Environment.* Cape Town University SA: Cape Town, 2013.
154. Mathur, S. C. Future of Indian Pesticides Industry in Next Millennium. *Pesticide Inf.* **1999,** *24* (4), 9–23.
155. McBee, K.; Bickham, J. Mammals as Bioindicators of Environmental Toxicity. In *Current Mammalogy*; Genoways, H. H., Ed.; Plenum Publ. Corp.: New York, NY, 1992; vol 2, p 37.
156. McConnell, L. L.; Nelson, E.; Rice, C. P.; Baker, J. E.; Johnson, W. E.; Harman, J. A.; Bialek, K. Chlorpyrifos in the Air and Surface Water of Chesapeake Bay: Predictions of Atmospheric Deposition Fluxes. *Environ. Sci. Technol.* **1997,** *31*, 1390–1398.
157. McNeely, J. A.; Scherr, S. J. *Ecoagriculture: Strategies to Feed the World and Save Wild Biodiversity*; Island Press: Washington, DC, 2003.
158. Megson, D.; Reiner, E. J.; Jobst, K. J.; Dorman, F. L.; Robson, M.; Focant, J.-F. A Review of the Determination of Persistent Organic Pollutants for Environmental Forensics Investigations. *Anal. Chim. Acta* **2016,** *941*, 10–25.
159. Merrington, G.; Rogers, S. L.; Van Zwieten, L. The Potential Impact of Long-Term Copper Fungicide Usage on Soil Microbial Biomass and Microbial Activity in an Avocado Orchard. *Soil Res.* **2002,** *40* (5), 749–759.
160. Miller, S. F. The Effects of Weed Control Technology Change on Rural Communities. *Outlook Agric.* **1982,** *11* (4), 172.
161. Montgomery, M.; Kamel, F.; Saldana, T.; Alavanja, M.; Sandler, D. Incident Diabetes and Pesticide Exposure among Licensed Pesticide Applicators: Agricultural Health Study, 1993–2003. *Am. J. Epidemiol.* **2008,** *167* (10), 1235–1246.

162. Moon, H. B.; Kim, H. S.; Choi, M.; Yu, J.; Choi, H. G. Human Health Risk of Polychlorinated Biphenyls and Organochlorine Pesticides Resulting from Seafood Consumption in South Korea, 2005–2007. *Food Chem. Toxicol.* **2009,** *47* (1), 1819–1825.
163. Moorman, T. B. A Review of Pesticide Effects on Microorganisms and Microbial Processes Related to Soil Fertility. *J. Prod. Agric.* **1989,** *2* (1), 14–23.
164. Mora, M. A. Transboundary Pollution: Persistent Organochlorine Pesticides in Migrant Birds of the Southeastern United States and Mexico. *Environ. Toxicol. Chem.* **1997,** *16*, 3–11.
165. Mora, M. A. Conger-Specific Polychlorinated Biphenyl Patterns in Eggs of Aquatic Birds from the Lower Laguna, Madre, Texas. *Environ. Toxicol. Chem.* **1996,** *15*, 1003–1010.
166. Moretto, A. Occupational Aspects of Pesticide Toxicity in Humans. In *Pesticide Toxicology and International Regulation*; Marrs, T., Ballantyne, B., Eds.; Wiley: London, 2002; pp 431.
167. Muir, D. C. G.; Teixeira, C.; Wania, F. Empirical and Modelling Evidence of Regional Atmospheric Transport of Current-Use Pesticides. *Environ. Toxicol. Chem.* **2004,** *23*, 2421–2432.
168. Multigner, L.; Kadhel, P.; Rouget, F.; Blanchet, P.; Cordier, S. Chlordecone Exposure and Adverse Effects in French West Indies Populations. *Environ. Sci. Pollut. Res. Int.* **2016,** *23*, 3–8.
169. Murphy, H.; Sanusi, A.; Dilts, R.; Djajadisastra, M.; Hirschhorn, N.; Yuliantiningsih, S. Health Effects of Pesticide Use Among Indonesian Women Farmers: Part I: Exposure and Acute Health Effects. *J. Agromed.* **1999,** *6* (3), 61–85.
170. Murphy, R.; Haith, D. Inhalation Health Risk to Golfers from Turfgrass Pesticides at Three Northeastern US Sites. *Environ. Sci. Technol.* **2007,** *41* (3), 1038–1043.
171. Neuenschwander, P. Biological Control of the Cassava Mealybug in Africa: A Review. *Biol. Control* **2001,** *21* (3), 214–229.
172. Nigam, S. K.; Karnik, A. B.; Chattopadhyay, P.; Lakkad, B. C.; Venkaiah, K.; Kashyap, S. K. Clinical and Biochemical Investigations to Evolve Early Diagnosis in Workers Involved in the Manufacture of Hexachlorocyclohexane. *Int. Arch. Occup. Environ. Health* **1993,** *65*, S193.
173. NRC. *Diet and Health: Implications for Reducing Chronic Disease Risk*; National Research Council (US) Committee on Diet and Health, National Academies Press: Washington, DC, 1989. doi:10.17226/1222.
174. Organisation of Economic Co-operation and Development and Food and Agriculture Organization of the United Nations. *OECD-FAO Agricultural Outlook*, 2012. https://www.oecd-ilibrary.org/agriculture-and-food/oecd-fao-agricultural-outlook-2012_agr_outlook-2012-en.
175. Oerke, E. C.; Dehne, H. W.; Schonbeck, F.; Weber, A. *Crop Production and Protection: Estimated Losses in Major Food and Cash Crops*, 1st ed. Elsevier Science: Amsterdam, 1994. doi:10.1016/C2009-0-00683-7.
176. Ollinger, M.; Fernandez-Cornejo, J. Innovation and Regulation in the Pesticide Industry. *Agric. Resour. Econ. Rev.* **1998,** *27* (1). doi:10.1017/S1068280500001660.
177. O'Neil, W.; Raucher, R.; Wayzata, M. N. *Groundwater Policy Education Project*, 1998. http://www.dnr.state.wi.us/org/water/dwg/gw/costofgw.htm.
178. Osteen, C. Pesticide Regulation Issues: Living with the Delaney Clause. *J. Agric. Appl. Econ.* **1994,** *26* (1), 60–74. doi:10.1017/S1074070800019167.
179. Osteen, C. D.; Szmedra, P. I. *Agricultural Pesticide Use Trends and Policy Issues. AER-622*. US Department of Agriculture, Economic Research Service, 1989.

180. Pan American Health Organization (PAHO). Epidemiological Situation of Acute Pesticide Poisoning in Central America Isthamus, 1992–2000. *Epidemiol. Bull.* **2002,** *23* (5).
181. Pandit, G.; Sahu, S. Assessment of Risk to Public Health Posed by Persistent Organochlorine Pesticide Residues in Milk and Milk Products in Mumbai, India. *J. Environ. Monit.* **2002,** *4* (1), 182–185.
182. Pan-Germany. *Pesticide and Health Hazards. Facts and Figures*, 2012; pp 1–16. www.pangermany.org/download/Vergift_EN-201112-web.pdf.
183. Parastar, S.; Ebrahimpour, K.; Hashemi, M.; Maracy, M. R.; Ebrahimi, A.; Kelishadi, R.; Poursafa, P. Association of Urinary Concentrations of Four Chlorophenol Pesticides with Cardiometabolic Risk Factors and Obesity in Children and Adolescents. *Environ. Sci. Pollut. Res.* **2018,** *25* (5), 4516–4523.
184. Patel, K.; Fussell, R. J.; Hetmanski, M.; Goodall, D. M.; Keely, B. J. Evaluation of Gas Chromatography-Tandem Quadrupole Mass Spectrometry for the Determination of Organochlorine Pesticides in Fats and Oils. *J. Chromatogr. A* **2005,** *1068*, 289–296.
185. Pell, M.; Stenberg, B.; Torstensson, L. Potential Denitrification and Nitrification Tests for Evaluation of Pesticide Effects in Soil. *Ambio* **1998,** *27*, 24–28.
186. Perry, M. J.; Venners, S. A.; Chen, X.; Liu, X.; Tang, G.; Xing, H.; Barr, D. B.; Xu, X. Organophosphorous Pesticide Exposures and Sperm Quality. *Reprod. Toxicol.* **2011,** *31* (1), 75–79.
187. Petrusheva, N. I. Protection of Horticultural Crops in the USSR and Future Perspectives. In *Papers at Sessions, VIII International Congress of Plant Protection*; USSR: Moscow, 1975; vol 2, p 124.
188. Polder, A.; Savinova, T. N.; Tkachev, A.; Løken, K. B.; Odland, J. O.; Skaare, J. U. Levels and Patterns of Persistent Organic Pollutants (POPS) in Selected Food Items from Northwest Russia (1998–2002) and Implications for Dietary Exposure. *Sci. Total Environ.* **2010,** *408*, 5352–5361.
189. Popp, J.; Peto, K.; Nagy, J. Pesticide Productivity and Food Security. A Review. *Agron. Sustain. Dev.* **2013,** *33*, 243–255.
190. Portoles, T.; Sales, C.; Abalos, M.; Saulo, J.; Abad, E. Evaluation of the Capabilities of Atmospheric Pressure Chemical Ionization Source Coupled to Tandem Mass Spectrometry for the Determination of Dioxin-Like Polychlorobiphenyls in Complex-Matrix Food Samples. *Anal. Chim. Acta* **2016,** *937*, 96–105.
191. Raffaele, K. C.; Rowland, J.; May, B.; Makris, S. L.; Schumacher, K.; Scarano, L. J. The Use of Developmental Neurotoxicity Data in Pesticide Risk Assessments. *Neurotoxicol. Teratol.* **2010,** *32*, 563–572.
192. Rauh, V.; Arunajadai, S.; Horton, M.; Perera, F.; Hoepner, L.; Barr, D.; Whyatt, R. Seven-Year Neurodevelopmental Scores and Prenatal Exposure to Chlorpyrifos, a Common Agricultural Pesticide. *Environ. Health Perspect.* **2011,** *119* (8), 1196–1201. doi:10.1289/ehp.1003160.
193. Ravlin, F. W.; Robinson, W. H. Audience for Residential Turfgrass Pest Management Programs. *Bull. Entomol. Soc. Am.* **1985,** *31*, 45–50.
194. Rawn, D. F. K.; Halldorson, T. H. J.; Muir, D. C. G. Triazines Herbicides Risk Assessment. *ACS Symp. Ser.* **1998,** *683*, 158–176.
195. Rawn, D. F. K.; Muir, D. C. G. Sources of Chlorpyrifos and Dacthal to a Small Canadian Prairie Watershed. *Environ. Sci. Technol.* **1999,** *33*, 19, 3317–3323.
196. Rehman, A.; Jingdong, L.; Chandio, A. A.; Hussain, I.; Wagan, S. A.; Memon, Q. U. A. Economic Perspectives of Cotton Crop in Pakistan: A Time Series Analysis (1970–2015) (*Part 1*). J. Saudi Soc. Agric. Sci. **2019,** *18* (1), 49–54.

197. Rice, C. P.; Cherniak, S. M. Marine Arctic Fog: An Accumulator of Currently Used Pesticide. *Chemosphere* **1997,** *35*, 867–878.
198. Richter, E. D. Acute Human Pesticide Poisonings. *Encyclopaedia of Pest Management*; Routledge: Oxfordshire, 2002; pp 3–6.
199. Ridsill-Smith, J.; Pavri, C. Single Spring Spray Protects Pastures. *Farm. Ahead* **2000,** *103*, 60–63.
200. Roberts, T. R. Metabolic Pathway of Agrochemicals. Part I. *Herbicides and Plant Growth Regulators*; The Royal Society of Chemistry: Cambridge, UK, 1998.
201. Roberts, T. R.; Hutson, D. H. Metabolic Pathway of Agrochemicals. Part II. *Insecticides and Fungicides*; The Royal Society of Chemistry: Cambridge, UK, 1999.
202. Ross, G. Risks and Benefits of DDT. *Lancet* **2005,** *366*, 1771–1772. doi:10.1016/S0140-6736 (05)67722-7.
203. Saiyed, H. N.; Sadhu, H. G.; Bhatnagar, V. K.; Dewan, A.; Venkaiah, K.; Kashyap, S. K. Cardiac Toxicity Following Short-Term Exposure to Methomyl in Spraymen and Rabbits. *Hum. Exp. Toxicol.* **1992,** *11*, 93.
204. Salam, M. T.; Li, Y. F.; Langholz, B.; Gilliland, F. D. Children's Health Study. Early-Life Environmental Risk Factors for Asthma: Findings from the Children's Health Study. *Environ. Health Perspect.* **2004,** *112*, 760–765.
205. Santos, A.; Flores, M. Effects of Glyphosate on Nitrogen Fixation of Free-Living Heterotrophic Bacteria. *Lett. Appl. Microbiol.* **1995,** *20*, 349–352.
206. Savary, S.; Willocquet, L.; Pethybridge, S. J.; Esker, P.; McRoberts, N.; Nelson, A. The Global Burden of Pathogens and Pests on Major Food Crops. *Nat. Ecol. Evol.* **2019,** *3*, 430–439.
207. Savonen, C. Soil Microorganisms Object of New OSU Service. *Good Fruit Grower*, 1997. http://www.goodfruit.com/archive/1995/6other.html.
208. Schafer, K. S.; Kegley, S. E. Persistent Toxic Chemicals in the US Food Supply. *J. Epidemiol. Comm. Health* **2002,** *56*, 813.
209. Seitz, S.; Goebes, P.; Puerta, V. L.; Pereira, E. I. P.; Wittwer, R.; Six, J.; van der Heijden, M. G. A.; Scholten, T. Conservation Tillage and Organic Farming Reduce Soil Erosion. *Agron. Sustain. Dev.* **2019,** *39*, 4.
210. Sekiyama, M.; Tanaka, M.; Gunawan, B.; O. Abdoellah, O.; C. Watanabe, C. Pesticide Usage and Its Association with Health Symptoms among Farmers in Rural Villages in West Java, Indonesia. *Environ. Sci.* **2007,** *14* (1), 23–33.
211. Senthilkumar, K.; Kannan, K.; Subramanian, A.; Tanabe, S. Accumulation of Organochlorine Pesticides and Polychlorinated Biphenyls in Sediments, Aquatic Organisms, Birds, Bird Eggs and Bat Collected from South India. *Environ. Sci. Pollut. Res.* **2000,** *7*, 1–13.
212. Shoiful, A.; Fujita, H.; Watanabe, I.; Honda, K. Concentrations of Organochlorine Pesticides (OCPs) Residues in Foodstuffs Collected from Traditional Markets in Indonesia. *Chemosphere* **2013,** *90*, 1742–1750.
213. Singh, B.; Mandal, K. Environmental Impact of Pesticides Belonging to Newer Chemistry. In *Integrated Pest Management*; Dhawan, A. K., Singh, B., Brar-Bhullar, M., Arora, R., Eds.; Scientific Publishers: Jodhpur, India, 2013; pp 152–190.
214. Slager, R.; Poole, J.; LeVan, T.; Sandler, D.; Alavanja, M.; J, Hoppin, J. Rhinitis Associated with Pesticide Exposure among Commercial Pesticide Applicators in the Agricultural Health Study. *Occup. Environ. Med.* **2009,** *66* (11), 718–724.
215. Son, H.; Kim, S.; Hang, J.; Chang, Y.; Park, S.; Lee, S.; Jacobs, D.; Lee, D. Strong Associations between Low-Dose Organochlorine Pesticides and Type 2 Diabetes in Korea. *Environ. Int.* **2010,** *36* (5), 410–414.

216. Spiewak, R. Pesticides as a cause of occupational skin diseases in farmers. *Ann. Agric. Environ. Med.* **2001,** *8*, 1.
217. Stefanelli, P.; Santilio, A.; Cataldi, L.; Dommarco, R. Multiresidue Analysis of Organochlorine and Pyrethroid Pesticides in Ground Beef Meat by Gas Chromatography-Mass Spectrometry. *J. Environ. Sci. Health B* **2009,** *44*, 350–356.
218. Stephenson, M. D.; Martin, M.; Tjeerdema, R. S. Long-Term Trends in DDT, Polychlorinated Biphenyls, and Chlordane in California Mussels. *Arch. Environ. Contam. Toxicol.* **1995,** *28*, 443–450.
219. Straathoff, H. Investigations on the Phytotoxic Relevance of Volatilization of Herbicides. *Mededelingen van de Faculteit Landbouwweten-schappen, Rijksuniversiteit Gent.* **1986,** *51* (2A), 433–438.
220. Sucahyo, D.; Van Straalen, N. M.; Krave, A.; Van Gestel, C. A. Acute Toxicity of Pesticides to the Tropical Freshwater Shrimp *Caridina laevis*. *Ecotoxicol. Environ. Saf.* **2008,** *69*, 421–427.
221. Tadeo, J. L.; Sanchez-Brunete, C.; Albero, B.; Garcia-Valcarcel, A. I. Application of Ultrasound-Assisted Extraction to the Determination of Contaminants in Food and Soil Samples. *J. Chromatogr. A* **2010,** *1217*, 2415–2440.
222. Tanabe, S.; Senthilkumar, K.; Kannan, K.; Subramanian, A. N. Accumulation Features of Polychlorinated Biphenyls and Organochlorine Pesticides in Resident and Migratory Birds from South India. *Arch. Environ. Contam. Toxicol.* **1998,** *34*, 387–397.
223. Tang, H. P.-O. Recent Development in Analysis of Persistent Organic Pollutants under the Stockholm Convention. *Trends Anal. Chem.* **2013,** *45*, 48–66.
224. Templeton, S. R.; Zilberman, D.; Jick Too, S. An Economic Perspective on Outdoor Residential Pesticide Use. *Policy Anal.* **1998,** *32* (17), 416 A–423 A.
225. The Food Equation. Feeding a Growing Global Family. *Detail Video Produced by Monsanto Company in Cooperation with the National Corn Growers Association, the W. J. Morse Foundation, and the American Soybean Association*; Monsanto Detail Program: Hazelwood, MO, 1991.
226. Thies, M. L.; McBee, K. Cross-Placental Transfer of Organochlorine Pesticides in Mexican Free-Tailed Bats from Oklahoma and New Mexico. *Arch. Environ. Contam. Toxicol.* **1994,** *27* (2), 239–242.
227. Thompson, L. A.; Darwish, W. S. Environmental Chemical Contaminants in Food: Review of a Global Problem. *J. Toxicol.* **2019,** 1–14, 2345283. doi:10.1155/2019/2345283.
228. Thurman, E. M.; Cromwell, A. E. Atmospheric Transport, Deposition and Fate of Triazine Herbicides and Their Metabolites in Pristine Areas at Isle Royale National Park. *Environ. Sci. Technol.* **2000,** *34*, 3079–3085.
229. Towns, D. R.; Atkinson, I. A. E.; Daugherty, C. H. Have the Harmful Effects of Introduced Rats on Islands Been Exaggerated? *Biol. Invasions* **2006,** *8*, 863–891.
230. Townsend, L. *Wood Preservation (Wood Preservative Pesticides)*; University of Kentuckey, Department of Entomology, College of Agriculture: Lexington, KY, 2018.
231. Townson, H.; Nathan, M. B.; Zaim, M.; Guillet, P.; Manga, L.; Bos, R.; Kindhauser, M. Exploiting the Potential of Vector Control for Disease Prevention. *Bull. World Health Organ.* **2005,** *83* (12), 942–947.
232. Turnbull, L.; Hector, A. Applied Ecology: How to Get Even with Pests. *Nature* **2010,** *466* (7302), 36–7. doi:10.1038/466036a.
233. Tyagi, Y. K.; Rosencranz, A. Some International Law Aspects of the Bhopal Disaster. *Soc. Sci. Med.* **1988,** *27* (10), 1105–1112. doi:10.1016/0277-9536(88)90305-X.

234. UN DESA. *World Population Prospects*, 2015. http://www.un.org/en/development/desa/news/population/2015-report.html.
235. US Department of Agriculture (USDA); Economic Research Service (ERS). Pesticide Expenditures by State: 1949–2009. *Farm Income: Data Files*, 2010. http://www.ers.usda.gov/data/FarmIncome/FinfidmuXls.htm.
236. US Department of Health and Human Services, US Department of Agriculture. *US Dietary Guidelines*, 2005. www.healthierus.gov/dietaryguidelines.
237. US Environmental Protection Agency. *Clean Air Act*; US Environmental Protection Agency: Washington, DC, 1990. http://www.epa.gov/oecaagct/lcaa.html.
238. US EPA Office of Pesticides and Toxic Substances. *Guidance for the Reregistration of Pesticide Products Containing Glyphosate as the Active Ingredient*; US EPA Office of Pesticides and Toxic Substances: Washington, DC, 1986.
239. US Geological Survey. *The Quality of Our Nation's Waters—Nutrients and Pesticides. Circular 1225*; US Geological Survey: Reston, VA, 1999. http://water.usgs.gov/pubs/circ/circ1225/.
240. US EPA. Managing Small-Scale Application of Pesticides to Prevent Contamination of Drinking Water. *Water Protection Practices Bulletin*; Office of Water: Washington, DC, 2001. EPA 816-F-01-031.
241. Usydus, Z.; Szlinder-Richert, J.; Polak-Juszczak, L.; Komar, K.; Adamczyk, M.; Malesa-Ciecwierz, M.; Ruczynska, W. Fish Products Available in Polish Market-Assessment of the Nutritive Value and Human Exposure to Dioxins and Other Contaminants. *Chemosphere* **2009,** *74*, 1420–1428.
242. Van Wijngaarden, E.; Stewart, P. A.; Olshan, A. F.; Savitz, D. A.; Bunin, G. R. Parental Occupational Exposure to Pesticides and Childhood Brain Cancer. *Am. J. Epidemiol.* **2003,** *157*, 989–997.
243. Vélez-Espino, L. A.; McLaughlin, R. L.; Pratt, T. C. Management Inferences from a Demographic Analysis of Sea Lamprey (*Petromyzon marinus*) in the Laurentian Great Lakes. *Can. J. Fish. Aquat. Sci.* **2008,** *65* (2), 227–244.
244. Verrin, S. M.; Begg, S. J.; Ross, P. S. Pesticide Use in British Columbia and the Yukon: An Assessment of Types, Applications and Risks to Aquatic Biota. In *Canadian Technical Report on Fisheries and Aquatic Sciences*, 2004; p 2517.
245. Waite, D. T.; Bailey, P.; Sproull, J. F.; Quiring, D. V.; Chau, D. F.; Bailey, J.; Cessna, A. J. Atmospheric Concentrations and Dry and Wet Deposits of Some Herbicides Currently Used on the Canadian Prairies. *Chemosphere* **2005,** *58*, 693–703.
246. Waite, D. T.; Cessna, A. J.; Grover, R.; Kerr, L. A.; Snihura, A. D. Environmental Concentrations of Agricultural Herbicides: 2,4-D and Triallate. *J. Environ. Q.* **2002,** 31, 129–144.
247. Wang, G.; Lu, Y.; Han, J.; Luo, W.; Shi, Y.; Wang, T.; Sun, Y. Hexachlorobenzene Sources, Levels and Human Exposure in the Environment of China. *Environ. Int.* **2010,** *36*, 122–130.
248. Wang, H. S.; Sthiannopkao, S.; Du, J.; Chen, Z. J.; Kim, K. W.; Mohamed Yasin, M. S.; Hashim, J. H.; Wong, C. K. C.; Wong, M. H. Daily Intake and Human Risk Assessment of Organochlorine Pesticides (OCPs) Based on Cambodian Market Basket Data. *J. Hazard. Mater.* **2011,** *192*, 1441–1449.
249. Wang, S. L.; Tsai, P. C.; Yang, C. Y.; Guo, L. Y. Increased Risk of Diabetes and Polychlorinated Biphenyls and Dioxins: A 24-Year Follow-Up Study of the Yucheng Cohort. *Diabetes Care* **2008,** *31* (8), 1574–1579.

250. Warren, G. F. Spectacular Increases in Crop Yields in the United States in the Twentieth Century. *Weed Technol.* **1998,** *12*, 752.
251. Waskom, R. Best Management Practices for Private Well Protection. Colorado State Univ. Cooperative Extension: Fort Collins, CO, 1994. http://hermes.ecn.purdue.edu:8001/cgi/convertwq7488.
252. Weber, R.; Herold, C.; Hollert, H.; Kamphues, J.; Blepp, M.; Ballschmiter, K. Reviewing the Relevance of Dioxin and PCB Sources for Food From Animal Origin and the Need for Their Inventory, Control and Management. *Environ. Sci. Eur.* **2018,** *30*, 42.
253. Webster, J. P. G.; Bowles, R. G. Estimating the Economic Costs and Benefit of Pesticide Use in Applies. In *Proceedings of BCPC Conference*, Brighton, 1996; pp 325–330.
254. Webster, J. P. G.; Bowles, R. G.; Williams, N. T. Estimating the Economic Benefits of Alternative Pesticide Usage Scenarios: Wheat Production in the United Kingdom. *Crop Prot.* **1999,** *18*, 83.
255. Wickerham, E.; Lozoff, B.; Shao, J.; Kaciroti, N.; Xia, Y.; J. Meeker, J. Reduced Birth Weight in Relation to Pesticide Mixtures Detected in Cord Blood of Full-Term Infants. *Environ. Int.* **2012,** *47*, 80–85.
256. Williams, L. H. Borate Wood-Protection Compounds: A Review of Research and Commercial Use. *J. Preservat. Technol.* **1996,** *27* (4), 46–51.
257. Yadav, R. S.; Sampath, R. R. Deltamethrin Treated Bednets for Control of Malaria Transmitted by *Anopheles culicifacies* (Diptera: Culicidae) in India. *J. Med. Entomol.* **2001,** *38* (5), 613–622.
258. Yancy, J.; Cecil, H. *Study Touts Herbicide Benefits*; Southeast Farm Press: Raleigh, NC, 2005; vol 32, no 11, p 16.
259. Yesavage, J. A.; Sheikh, J.; Noda, A.; Murphy, G.; O'Hara, R.; Hierholzer, R.; Battista, M.; Ashford, J. W.; Kraemer, H. C.; Tinklenberg, J. Use of a VA Pharmacy Database to Screen for Areas at High Risk for Disease: Parkinson's Disease and Exposure to Pesticides. *J. Geriatr. Psychiatry Neurol.* **2004,** *17* (1), 36–38.
260. Zakharenko, V. A. Economics and Prospects of Pesticide Application as Related to the Intensification of Farming. In Abstract of the *Eighth International Plant Protection Congress*, Moscow, 1975; vol 2, pp 1–6.
261. Zettler, L. J.; Arthur, H. Chemical Control of Stored Product Insects with Fumigants and Residual Treatments. *Crop Prot.* **2000,** *19*, 577–582.
262. Zhao, R.; Chu, S.; Zhao, R.; Xu, X.; Liu, X. Ultrasonic Extraction Followed by Sulfuric Acid Silica Gel Cleanup for the Determination of α-Hexachlorocyclohexane Enantiomers in Biota Samples. *Anal. Bioanal. Chem.* **2005,** *381*, 1248–1252.
263. Zong, G.; Valvi, D.; Coull, B.; Göen, T.; Hu, F. B.; Nielsen, F.; Grandjean, P.; Sun, Q. Persistent Organic Pollutants and Risk of Type 2 Diabetes: A Prospective Investigation among Middle-Aged Women in Nurses' Health Study II. *Environ. Int.* **2018,** *114*, 334–342.

WEB REFERENCES

https://croplife.org/wp-content/uploads/2014/04/Guidelines-for-personal-protection-when-using-crop-protection-products-in-hot-climates.pdf.

http://www.who.int/mediacentre/news/releases/2006/pr50/en/index.html.

CHAPTER 3

Antioxidant, Photosynthesis, and Growth Characteristics of Plants Grown in Polluted Soils

PARNEET KAUR[1], RUCHIRA GHOSH[2], ZOYA SHAIKH[3], PRITI SOHAL[1], SAURABH KULSHRESHTHA[1], JESSICA PANDOHEE[4*], and AHMAD ALI[3]

[1]*School of Biotechnology, Faculty of Applied Sciences and Biotechnology, Shoolini University, Solan, Himachal Pradesh, India*

[2]*Centre for Sustainable Technologies, Belfast School of Architecture and the Built Environment, Ulster University, Northern Ireland, United Kingdom*

[3]*Department of Life Sciences, University of Mumbai, Vidyanagari, India*

[4]*Centre for Crop and Disease Management, School of Molecular and Life Sciences, Curtin University, Bentley, WA 6102, Australia*

[*]*Corresponding author. E-mail: jessica.pandohee@gmail.com*

ABSTRACT

Environmental pollution is a global threat arising from growing human activities and release of unwanted particles and chemicals in the ecosystem. Since the industrial revolution, rapid industrialization and urbanization have led to uncontrolled release of waste in river streams and lakes. Large-scale and excessive use of fertilizers, pesticides in agricultural pastures, unsustainable manufacture, use and discard of plastic products together with industrial and chemical wastes such as heavy metals dumped into the environment cause many irreversible issues for human and plants. While a large number of research works are being carried out on the remediation of environmental

Environmental Pollution Impact on Plants: Survival Strategies under Challenging Conditions. Tariq Aftab, PhD & Khalid Rehman Hakeem, PhD (Eds.)

pollution and policies such as the Sustainable Development Goal 12 seeking to "ensure sustainable consumption and production patterns," plants have found their own way of living and surviving environmental pollution. This chapter discusses the approaches that plants have matured including antioxidant, photosynthesis, and growth characteristics in order to survive the challenging polluted soils they are growing in.

3.1 INTRODUCTION

Environmental pollution is a global challenge affecting not only the health of humans but also the flora and fauna through the contamination of air, soil, and water bodies.[1] The use of soil is crucial for human survival as soil is an irreplaceable commodity in agriculture, habitat, and leisure. However, this important soil can be rendered unusable and even toxic in some cases due to the exploitation, uncontrolled growth and mismanagement of soil and its resources. Anthropogenic activities, such as intensive farming including stock breeding, pesticide, and fertilizer use and leeching of chemicals into water systems are often to be blamed. Since the industrial revolution, rapid industrialization and urbanization have led to uncontrolled release of waste (heavy metals, oil, and other foreign matters) in river streams and lakes. Rapid growth in population with little waste management including solid waste, sanitary and decontamination is surprisingly still a major problem in many countries around the world. Localized, nuclear testing, and mining are serious sources of pollution leading to air, soil, and radioactive contaminations into the microcosm and natural environment of the soil.[66,77,97]

Soil pollution is the result of unwanted materials that have been released or dumped into the environment. These pollutants are in the form of xenobiotic materials, metal ions, plastic, pharmaceuticals, chemicals, or other foreign particles that either exist at high concentration or do not decompose naturally, therefore altering the natural soil characteristics. Occurrence of such materials creates detrimental effects on land biodiversity, endangering health, and contaminating food chain.[86] Over the past few decades, it has been a cause of concern among scientific community.[17] Heavy metals, in particular, are one of the most dangerous soil pollutants. Although heavy metals, such as copper, iron, zinc, and aluminum exist naturally in the soil and are essential for plant growth, when present in excess such as in high amount in polluted soils, they have the power to alter the photosynthesis process, mineral nutrient uptake, and phytotoxicity.[77] Heavy metals also have the tendency to be transported to long distances, and this leads to their accumulation in the environment. In China for example, more than 16.1% of soil is contaminated including 2.1% with copper.[22] The fact that plants are directly rooted in the soil and are the first component in the food chain, their exposure and accumulation of pollutant make them hazardous for humans.[55]

Plants growing in polluted soils experience abiotic stress and need to adapt through several mechanisms to ensure their survival.[58] For example, environmental stress induces the accumulation of reactive oxygen species (ROS) in plant cells which causes severe oxidative damage to the plants.[20] Moreover, plants living in soil with high level of heavy metal contaminants suffer from reduced photosynthesis, plant growth inhibition, and are in danger of extinction.[99] To overcome these challenging situations, plants adapt various strategies. This chapter discusses the various ways by which plants living in polluted soils have adapted and developed for their survival with a strong focus on the characteristics regarding antioxidant production, photosynthesis, and growth.

3.2 ADAPTATION AND ANTIOXIDANT PRODUCTION OF PLANTS IN POLLUTED SOILS

Plants living in polluted soils have numerous opportunities to combat environmental pollution, The antioxidative defense systems in plants play an important role in reducing toxicity of heavy metals.[96] To overcome heavy metal toxicity, plants have evolved specialized defense mechanisms, such as reduced heavy metal uptake, sequestration of metals into vacuoles, binding to phytochelatins, and activation of various antioxidants.[76] Antioxidative compounds play a vital role in inhibiting the oxidation of metabolites and influence the defense system against reactive oxygen species which is a harmful by-product released during normal aerobic cellular respiration. Reactive oxygen species (ROS) are defined as specialized class of highly reactive molecules that are formed during oxygen metabolism, and in humans, they also have a role in the development of degenerative diseases. Plants produce a variety of antioxidants, including phytochemicals, which are used by the antioxidant defense system to protect them against reactive oxygen species produced during oxidative stress. Many synthetic antioxidants, such as butylated hydroxy anisole (BHA) and butylated hydroxytoluene (BHT) are used as chemical preservatives in foods to slow down their process of deterioration. There are various ways and means to mitigate the urban environmental pollution's undesirable effects caused by reactive oxygen species. Of those, the adaptation of plant species and their ability to produce antioxidant seems to be a very powerful approach to combat a range of soil pollution issues (Figure 3.1).

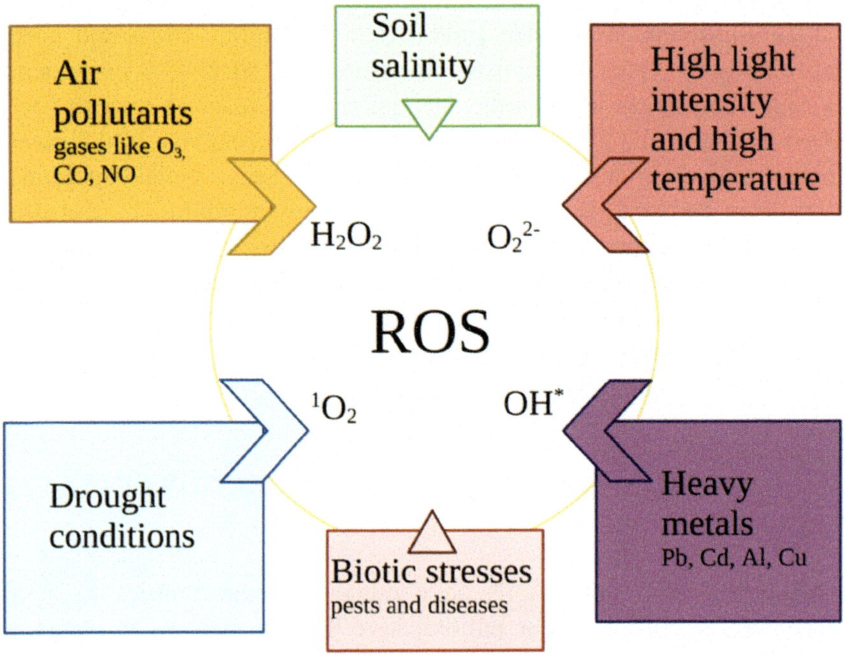

FIGURE 3.1 Various causes responsible for the generation of reactive oxygen species.

Interestingly, oxygen was formed into the early reducing atmosphere of the Earth about 2.7 billion years ago by oxygen evolving photosynthetic organisms which also released the by-product as reactive oxygen species.[38] Chloroplast and mitochondria are the powerhouse of the cell and two main sites for ROS generation within plant cell and maintain a balance between energy-linked functions and control of ROS production.[42] Reactive oxygen species are being generated constantly at basal levels under favorable conditions, but they are unable to cause any damage, as they are being scavenged by different antioxidant mechanisms.[36] The balance between ROS generation and ROS scavenging is maintained by the different types of stress factors, such as salinity, drought, extreme temperatures, heavy metals, pollution, high irradiance, pathogen infection, among others. Therefore, the survival of the plants depends on many important factors, such as change in growth conditions, severity and duration of stress conditions, and the capacity of the plants to quickly adapt to changing energy equation.[53]

Plants exposed to pollution are affected in several aspects, the severity of the damage depends on type, matrix, duration, and amount of pollutants in the environment.[16] In a recent study, it was found that soil contaminated with lead affected the growth and physiological conditions of the sunflower but when lead tolerant plant growth promoting rhizobacteria were used in the contaminated soil, it led to increase in plant growth, yield, physiology, and antioxidative activities.[72] The plant growth can also be affected by the exchangeable sodium salt present in the polluted soil which affects the availability of potassium and disturbs the ion transport. In a recent study, it was reported that consortiums of *Bacillus* strains increased the auxin production in the rhizosphere of potato plant and regulated the antioxidant enzyme production and uptake of salts in potato plant which increased the tuber yield in both normal and salt-affected soil.[85] Plants have various efficient complex enzymatic and nonenzymatic antioxidant defense systems to overcome the toxic effects of free radicals.

3.2.1 ENZYMATIC ANTIOXIDANTS

Plants have efficient enzymatic and nonenzymatic antioxidant defense to overcome the toxic effects of free radicals. The initial step is the generation of various reactive oxygen species via O^{2-}, H_2O_2 and •OH happens when the oxygen molecules accept leaked electrons from the overloaded photosystem I, and superoxide radicals or singlet oxygen are produced.[60] The superoxide radicals are converted to hydrogen peroxide molecules spontaneously or by catalytic activity superoxide dismutase; or otherwise, these radicals can reduce metal ions (such as Fe^{3+} and Cu^{2+}) present in the cell.[60] The enzymatic antioxidants include superoxide dismutase (SOD), catalase (CAT), ascorbate peroxidase (APX), monodehydroascorbate reductase (MDHAR), deydro-ascorbate reductase (DHAR), glutathione reductase (GR), and guaiacol peroxidase (GPX). Superoxide dismutase is considered as the first line of antioxidant defense as it has an ability to convert highly reactive superoxide radical to hydrogen peroxide and molecular oxygen, and play an important role in protecting the plant from the toxic effects of ROS.[93] The oxidative stress can lead to the death of the plant due to the accumulation of hydrogen peroxide, and to overcome this stress, plants are provided with hydrogen peroxide metabolizing enzymes, such as catalase, ascorbate peroxidase. Ascorbate peroxidase has higher affinity for hydrogen peroxide and reduces it to water.[83] Ascorbate peroxidase is a part of ascorbate–glutathione cycle and reduces hydrogen peroxide in the cytosol and chloroplasts. A detailed information about enzymatic antioxidants is given in Table 3.1.

TABLE 3.1 List of Enzymatic Antioxidants Present in Plants Living in Polluted Soils.

Enzymatic antioxidants	Subcellular location	Function	Reference
Superoxide dismutase (SOD)	Chloroplast, mitochondria, peroxisomes, cytoplasm, apoplasts	SOD contributed superoxide radical dismutation	7, 74
Catalase (CAT)	Peroxisomes and mitochondria	Catalase performs detoxification of hydrogen peroxide in both sensitive and tolerant varieties	28, 74
Ascorbate peroxidase (APX)	Chloroplast, cytosol, mitochondria, peroxisomes	It has high affinity for hydrogen peroxide and utilizes ascorbate for the reduction of hydrogen peroxide into water	9
Monodehydroascorbate reductase (MDHAR)	Chloroplast, stroma, cytosol	Maintains the level of ascorbic acid in plants and removes reactive oxygen species.	46, 60
Dehydroascorbate reductase (DHAR)	Seeds and roots, etiolated shoots and tissues	Maintenance of ascorbate pool and redox state	27, 31
Glutathione reductase (GR)	Chloroplast, mitochondria, cytosol	Glutathione reduction and antioxidative processes in plants	37, 88
Guaiacol peroxidase (GPX)	Mitochondria, cytoplasm, chloroplast, endoplasmic reticulum	Decomposition of indole acetic acid	28, 37

3.2.2 NONENZYMATIC ANTIOXIDANTS

To overcome oxidative stress, plants adopt enzymatic and nonenzymatic antioxidant systems, the nonenzymatic components in plant include ascorbic acid, tocopherols, glutathione, phenolics, and flavonoids, the nonenzymatic antioxidants interact with important physiochemical processes in plants and induce tolerance against abiotic stress.[41] Ascorbic acid or vitamin C plays an important role in human nutrition. Ascorbic acid is formed in mitochondria and transported to other cells through proton–electron chemical gradient and has the ability to scavenge many types of free radicals.[38] Tocopherol is a vitamin and antioxidant compound that has a role in overcoming both biotic and abiotic stress. In a recent study, it was found that alpha-tocopherol can be used on onion plant to induce salt tolerance in it.[75] Carotenoids are lipophilic antioxidants that protect the photosynthetic machinery of the plants and serve as a precursor of two plant hormones.[57] Flavonoids have been considered as reactive oxygen species scavenging enzyme in plants and have a role in scavenging oxygen and they alleviate the damages caused to the outer envelope of the chloroplast membrane.[2] A detailed description about nonenzymatic antioxidants is given in Table 3.2.

3.2.3 PROTECTIVE ROLE OF ANTIOXIDANTS

Exposure to environmental pollutants, such as ionizing and nonionizing radiation, ultraviolet radiation, elevated concentrations of ozone, nitrogen oxides, sulfur dioxide, cigarette smoke, asbestos, particulate matter, pesticides, dioxins and furans, polycyclic aromatic hydrocarbons, and many other compounds present in the environment can result in overproduction of reactive oxygen and nitrogen species[59] (Figure 3.2).

3.2.3.1 ABIOTIC STRESS CAUSED BY LEAD

According to the US Agency for Toxic Substances and Disease Registry, lead is the second most dangerous chemical.[11] Common sources of lead are lead pain, aviation fuel, metallurgical wastes, fertilizers, pesticides, lead-enriched sewage sludge, and waste waters are the main causes of lead contamination in soil.[45] Lead is a metal that is not required for plant metabolism. It has several negative consequences once it has entered the cell. These include alterations in water and mineral status, photosynthesis (chloroplast structure, pigments, enzymes, and light dependent/independent photosynthesis processes), and redox homeostasis.[6,43,90,100]. Toxicity to lead has been shown to reduce biomass output and grain yield in nearly all major field crops.[56]

TABLE 3.2 List of Nonenzymatic Antioxidants Found in Plants Living in Polluted Soils.

Nonenzymatic antioxidants	Subcellular location	Function	Reference
Ascorbic acid	Mitochondria	Regulation of physiological, biochemical processes like seed germination, fruit formation, ROS regulation	5, 38
Reduced glutathione	Cytosol, chloroplast, mitochondria, peroxisome, vacuole, apoplast	Acts as detoxifying co-substrates for enzymes like peroxidase	28
Alpha-tocopherol	Thylakoid membranes	Detoxify ROS that emerge during stress conditions and also protect embryo from ROS during germination	84
Carotenoid	Chloroplast and other nongreen plastids	Act as light harvester, scavengers, and quenchers of triplet state chlorophyll and singlet state oxygen species and also membrane stabilizers	28, 91
Flavonoid	Vacuole	It is considered as secondary antioxidant system and acts as an antioxidative agents and scavenge ROS formed in plants during stress conditions	28
Proline	Mitochondria, cytosol and chloroplast	Proline acts as an osmolyte and also activates cellular antioxidant systems and scavenge ROS	28, 47

Dias et al. investigated oxidative stress in *Pisum sativum* plants and that lead contaminated soil affects the levels of phytohormones which act as an antioxidant in plant defense.[29] As expected, they observed that the roots had higher lead-induced oxidative stress compared with the leaves, and despite the activation of glutathione reductase, ascorbate peroxidase, and superoxide dismutase, root cells during the experiment, were found to still experience oxidative stress even at the lowest lead dose. When plant cells are under reduced oxidative stress, for example, when antioxidant enzymes have been activated, a preventative antioxidant such as salicylic acid can be produced.

This is important in situations where plants already have hydrogen peroxide build up, lipid/protein peroxidation, and membrane breakdown. As a result, salicylic acid is believed to be involved in the activation of additional antioxidant defense systems when cells are already damaged by oxidants.[29]

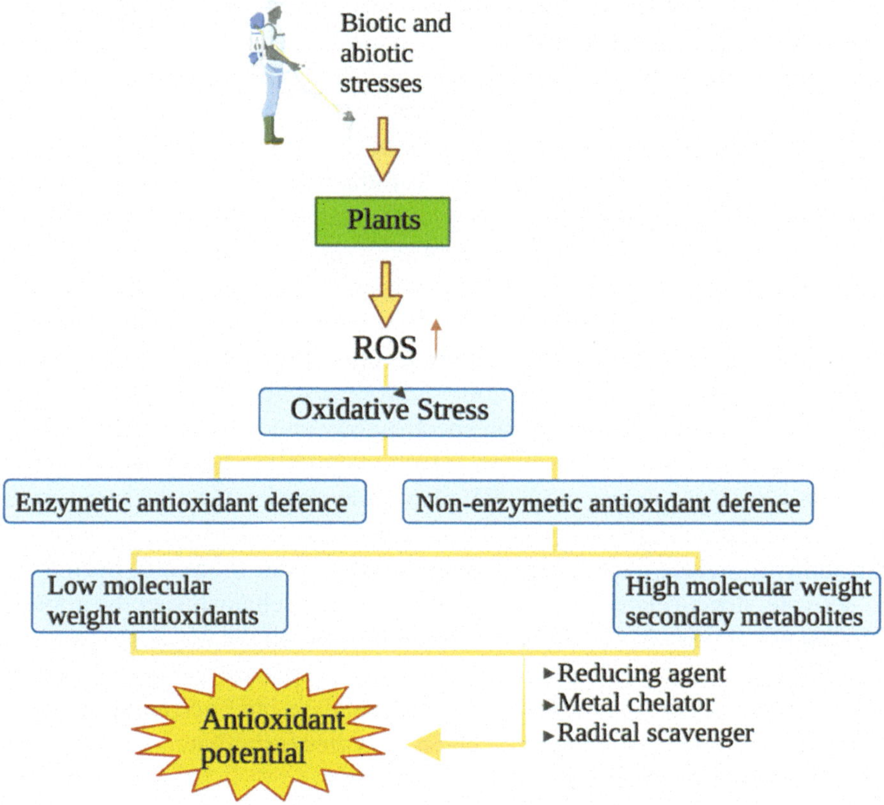

FIGURE 3.2 The antioxidant activity of plant living in polluted soils.

3.2.3.2 ABIOTIC STRESS CAUSED BY CADMIUM

Often found in rechargeable batteries, cadmium is another heavy metal causing irreversible damage in plants. Recently, it was found that the hydrogen peroxide concentration of cadmium-exposed *A. thaliana* seedlings was significantly elevated in both roots and leaves, but only in the leaves of the 5 M copper-exposed plants. Moreover, mustard (*Brassica juncea*) plants subjected to high levels of cadmium produced more hydrogen peroxide and lipid peroxidation, as well as electrolyte leakage and increased antioxidant enzyme activities, including superoxide dismutase, ascorbate peroxidase, and glutathione.[4] Hydrogen peroxide production is a frequent plant response to metal exposure and is believed to cause an increase the growth of plants.[26] The selenium supplementation reduced the buildup of hydrogen peroxide and lipid peroxidation, enhanced the activity of antioxidant enzymes, and controlled cadmium accumulation in roots and shoots. With increasing cadmium concentrations, ascorbic acid and flavonoids dropped; nevertheless, tocopherol and total phenols are increased.

On the other hand, exposure to cadmium resulted in an increase in starch production in *P. amarus* plants,[64] while Indian mustard (*Brassica juncea*) treated to Cd^{2+} in a concentration-dependent manner, reduced the plant growth and photosystem II activity with a corresponding rise in proline levels.[78] The *Monochoria hastata* plants treated with cadmium showed various visual toxicity signs, including withering, chlorosis, and leaf fall, which were more severe at 15 mg L^{-1} of cadmium.[13] Cadmium concentrations were considerably greater in root tissues than in shoot tissues, as determined by inductively coupled plasma—optical emission spectroscopy and were found to be in the following order: root > stem > leaf. Depending on the heavy metal species and concentration, certain medicinal plants may acclimate to the heavy metal stress and gradually resume normal development. Some heavy metals have the ability to reduce the toxicity of other heavy metals. It has also been found to activate stress-related reactions and defense mechanisms in the roots of *Medicago truncatula* seedlings cultivated.[63] In all genotypes examined, cadmium stress (100 M) raised reactive oxygen species levels and elevated antioxidative (total superoxide dismutase, catalase, and peroxiredoxin) and ascorbate–glutathione-related metabolic enzymes (ascorbate peroxidase and monodehydroascorbate reductase).

3.2.4 EFFECTS OF HEAVY METALS ON THE SECONDARY METABOLISM OF MEDICINAL PLANTS

Heavy metals have the ability to cause medicinal plants to produce secondary metabolites, although the effects vary depending on the plant species as well as the kind and quantity of heavy metals. The extensive study is required to determine the best heavy metal species and concentration for a certain medicinal plant. Medicinal plants that can withstand severe abiotic stress may retain their medicinal properties. An example was the discovery of the therapeutically useful chemicals phyllanthin and hypophyllanthin being produced in *Phyllanthus amarus* plants when the latter was exposed to cadmium.[64]

Plants develop a variety of nonenzymatic molecules that are engaged in heavy metal detoxification and ROS-scavenging pathways when they are exposed to heavy metals.[54] Nonenzymatic antioxidants, such as phenolics, ascorbic acid, and nonproteinthiol were found to have a significant role in cadmium detoxification. Nonenzymatic antioxidants, such as proline, tocopherol, flavonoid, glutathione, total ascorbate, and docosahexaenoic acid were also engaged in cadmium detoxification up to specific doses, that is, 150–200 M (except proline). The most common low-molecular weight antioxidants, such as ascorbic acid and docosahexaenoic acid play a critical role in the defense against metal-induced oxidative stress produced by increased levels of reactive oxygen species.[54] Hence, it is important for plants to adapt and develop new mechanisms against various abiotic, oxidative and heavy metal stress.

3.3 PHOTOSYNTHESIS ADJUSTMENTS DURING AND AFTER EXPOSURE TO POLLUTED SOILS

According to The Status of the World's Soil Resources Report (SWSR) soil pollution is a threat to global soils and the ecosystems. The imbalance in the soil compositions results in impaired nutrient uptake in the plant roots, which in turn slows down the process of photosynthesis.[34] The effects of soil pollution on plants include reduced photosynthetic rate, transpiration rate, stomatal conductance, synthesis of pigment contents, and yield of photosystem II photochemistry, amongst others (Figue 3.3). An example of the growing threat is the petroleum industry. Due to the fast development of petroleum industry during the past few years, crude oil spilling has adversely affected the environment of regions which produce petroleum. In plants, nutrient and water uptake is hindered due to petroleum contaminants in soil.[80] It causes the accumulation of reactive oxygen species due to biomembrane injury. Inhibition of photosynthesis and transpiration also occurs.[62] This leads to plant death and degradation. Han et al. monitored the photosynthetic process of *Amorpha fruticosa* seedlings in petroleum-contaminated soils at different concentrations and found that the photosynthetic rates, transpiration rate, and stomatal conductance of the seedlings decreased significantly due to petroleum-contaminated soil. Furthermore, there was a decrease in the intercellular carbon dioxide concentration and an increase in leaf relative water content of the seedlings in the contaminated soil. The Chlorophyll (A,

B, and total chlorophyll) content of the seedlings was reduced during the sampling period in contaminated soil.[40]

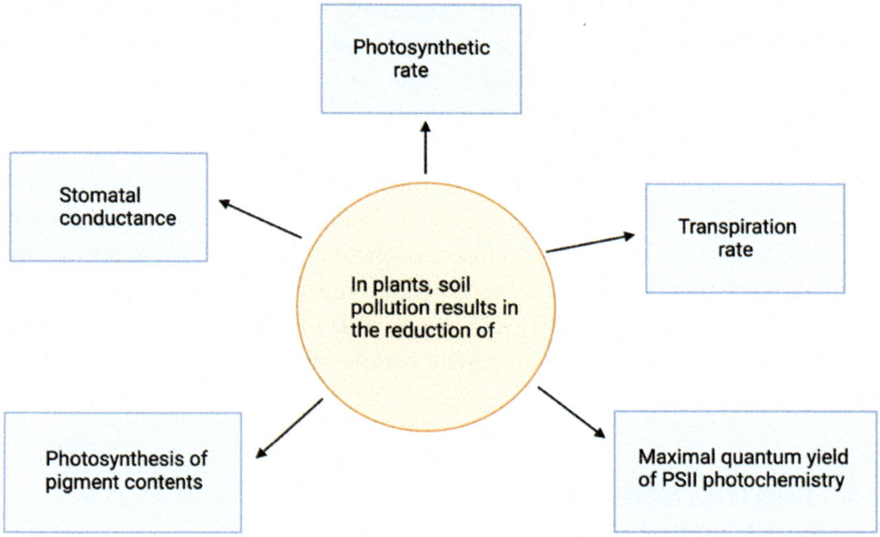

FIGURE 3.3 Photosynthesis in plants living in polluted soils.

As discussed in Section 3.2, heavy metals, such as iron (Fe), copper (Cu), cobalt (Co), and zinc (Zn) are important for plant metabolism. However, these are required in trace amounts. Excessive amount of heavy metals in plant environment can hamper the metabolic activities of plants.[39] It damages chloroplasts and disturbs the process of photosynthesis. Also, inhibition of photosynthesis occurs as metal ions interfere with the action of photosynthetic enzymes and chloroplast membranes. Heavy metal accumulation in the leaves of higher plants causes reduction in photosynthesis. This affects the functions of stomata and decreases the rate of transpiration and photosynthesis.[3]

The molecular and physiological mechanisms which make plants sensitive to heavy metals are as follows:

i) Binding of cell wall constituents and extracellular exudates for the uptake and accumulation of metals.
ii) Complexation of heavy metal ions in the cell of substances like amino acids and metallothioneins.
iii) Efflux of heavy metals from cytoplasm to extra nuclear compartments.

iv) Induction of antioxidative enzymes and accumulation of osmoproteins.
v) Modification of plant metabolism for adequate functioning of metabolic pathways and repair of damaged cell structures.[21]

In a recent study, properties of green vegetable plants (*Brassica chinensis, Brassica alboglabra* and *Chrysanthemum coronarium*), such as photosynthetic pigments, chlorophyll fluorescence was examined. These vegetables were grown under copper treatments of varied concentrations. It was found that copper contaminated soil damaged the photosynthetic pigments, negatively affected the growth of all three vegetables and hampered the gas exchange. However, photosystem II functions of all three vegetables were unaffected. Therefore, indicating that copper contamination of soil adversely affects photosynthesis, predominantly through stomatal factors.[48]

When growth inhibition and photosynthesis were studied in copper-treated *Cucumis sativus L*, photosynthesis declined in matured leaves and the surface area of young leaved had reduced. In mature leaves, net carbon dioxide assimilation was reduced at closely constant intercellular carbon dioxide mole fraction. This implies that inhibition of photosynthesis was not due to stomatal closure. No significant change was observed in maximal photochemical yield of photosystem II in dark-acclimated leaves. This suggests that light phase of photosynthesis was affected rather than dark phase. In stressed plants, the ratio of relative growth rate and leaf area was found to be lower than the control plants, while no significant difference was observed in net assimilation rate among for copper-stressed plants.[92] Another study analyzed the biochemical and physiological behavior of rice (*Oryza sativa, var. Jyoti*) treated with copper(II) oxide nanoparticles. Photosynthetic rate, stomatal conductance, transpiration rate, photosynthetic pigment contents, and maximal quantum yield of photosystem II photochemistry was reduced. Loss of photosynthetic activity was observed due to the toxicity of copper accumulation in shoots and roots.[25]

Lastly, Sheoran et al., examined the photosynthetic rate activities of main enzymes involved in photosynthetic carbon reduction cycle using the leaves of pigeonpea (*Cajanus cajan L., cv.* UPAS-120) in the presence of Cd^{2+} and Ni^{2+}. It was observed that higher concentration of Cd^{2+} and Ni^{2+} (10 mM) lowered the rate of photosynthesis. The enzyme activities were also significantly reduced. This study suggested that the decrease in photosynthesis occurs due to reduction in stomatal conductance and chlorophyll content and not because of decreased enzyme activities.[79]

3.4 GROWTH CHARACTERISTICS OF PLANTS GROWN IN POLLUTED SOILS

Studying the growth characteristics of plants can serve as a biomarker of a particular kind of pollutant or pollutants. Growth characteristics of these plants budding in the polluted soil are vital for trend analysis of its biomagnification in the food chain, planning adaptation strategies, and remedial measures.[8,33] Literatures support evidences regarding the genetic effect of pollutants on plants and resistance of plants in response,[18,30,35,68] and micro-evolutionary effects of chemical stressors.[24,49] However, the effect on food chain is not much discussed.[19,73,32,95] Although, different pollutants have different effects on the growth pattern depending upon the levels of tolerance, mean tolerance index, more detailed studies are required.[12,94] Thus, the section presents the effects of the pollutants on plants and how it affects the food chain.

Plants of the uniform genotype, or ramets rarely develop in an identical manner.[89] Growth pattern varies in both clonal and non-clonal plant. In clonal plants, the shoot and root meristems are determined by their response to the resource stimuli or local surroundings. In non-clonal plants, the intact ramets translocate to optimally quest for a heterogenous habitat. But phenotypic plasticity allows avoidance strategy toward heterogenous distribution of pollutant. The avoidance strategy may be genetically estimated like variation in the stomatal conductance that are responsible for atmospheric gases.[15,61] In response to the pollution, the natural evolutionary development of tolerant plant genotype has widely been studied but is limited to mostly mining an industrial discharge mainly.[50]

Heavy metal pollution in the soil is responsible for inducing new proteins and suppressing proteins liable for biochemical adjustment to environmental stress resulting in metallothionein-like proteins and metal-binding proteins in both metal tolerant and non-tolerant plants.[67,71,87] Such alteration at gene level may/may not be reversible and can result in mutation in plant and even in higher living being in the food chain. However, the intensity, extent, and duration of pollution are the key drivers that determine whether the plant population would survive in the short term or persist and evolve in the long term.

Nevertheless, adaptive response to pollution by plants can be altered by stressors like temperature or salinity.[33,52] Root phenotyping is also important to understand the root absorption performance and yield on soils with low fertility and aluminum toxicity.[65] Plants grown on particular polluted site like plants cultivated on soil polluted with crude oil shows inhibition of growth for 3% in shoot parts with significant decrease in chlorophyll a production and 1.5% in root parts.[82] Some varieties are resistant than other to the same pollutant (e.g., Valdai variety of Rye was tolerant to moderately polluted soil with crude oil than Krona variety). The absorption of oil particle by plants can alter the physiological features like changes in parenchyma tissue, plasma membrane, reduction in the intercellular space in cortex, alter mitosis in root. Polluted soil creates root stress with low water, and organic matter availability.[10,14] Some studies show changes in proline, nonprotein thiols, ascorbic acid, riboflavin, anthocyanins, phenolic compounds, and flavonoids in plants growing in oil polluted sites.[23,69,70] Significant changes observed from oil pollution is the formation of reactive oxygen species from oxidative stress to the roots terminating cell membrane complexes, intercellular reaction, and growth inhibition. Moreover, plant's antioxidative defense system plays a decisive role in preventing uncontrolled oxidation, life of these oxidative species, adaptation to environmental stress and phytoremediation potential. Not many literatures are available on the effect of pollution on antioxidative system.[44,51,59,98] Alteration in oxidative species and antioxidative enzymes in plants are vital and can be useful in understanding the pollutant manifestation at higher end of the food chain.

Hence, the effect of the polluted soil can range from physiological, morphological to genotypic alteration depending upon the resistance ability of the particular plant variety, exposure time, concentration of the polluted and presence of other secondary stressor like salinity and temperature.

3.5 CONCLUSION

Environment pollution is one of the biggest problems that affect all the organisms present on the earth. This chapter has described some survival aspects that plants have developed to adapt to the increase level of pollution. The growth and development of the plants are also affected due to contaminated soil. To overcome the effect of reactive oxygen species, plants adapt various defense systems. The antioxidants are produced by the plants to overcome the oxidative stress and the balance between reactive oxygen species and its scavenging is maintained by the stress factors of the plants. The plants also adapt various photosynthetic changes, such as photosynthetic rates, stomatal conductance, transpiration rates on its exposure with the pollutants in the soil. Although, heavy metals are required by the plants for their growth but are needed in microquantities and increase their concentration led to the adverse effects on the overall growth, development, and photosynthetic activity of the plants.

In conclusion, the anthropogenic activities such as urbanization, industrialization, mining, intensive farming, use of pesticides, dumping of municipal wastes, such as plastic, oils, heavy metals have evidently caused significant losses. While plants have managed to overcome and develop mechanisms for survival, from a human perspective, a large amount of work remains to be done in order to reduce the release (including unintentional) and dumping of pollutants as well as in the removal of pollutants from current natural resources and ecosystem without endangering biodiversity. Phytoremediation might be a solution in order to restore the quality of the soil, with economical and eco-friendly possibilities as simple as using weed species to re-absorb metal ions contaminated land.[81]

KEYWORDS

- **adaptation**
- **environmental pollution**
- **abiotic stress**
- **SDG 12**
- **plant survival**

REFERENCES

1. Acevedo-Whitehouse, K.; Duffus, A. L. J. Effects of Environmental Change on Wildlife Health. *Philos. Trans. R. Soc. Lond. B Biol. Sci.* **2009,** *364* (1534), 3429–3438.
2. Agati, G.; Azzarello, E.; Pollastri, S.; Tattini, M. Flavonoids as Antioxidants in Plants: Location and Functional Significance. *Plant Sci.* **2012,** *196*, 67–76.
3. Aggarwal, A.; Sharma, I.; Tripathi, B. N.; Munjal, A. K.; Baunthiyal, M.; Sharma, V. *Metal Toxicity and Photosynthesis in Photosynthesis: Overviews on Recent Progress and Future Perspectives*; IK International Publishing House (Pvt.) Limited: New Delhi, 2012.
4. Ahmad, P.; Abd Allah, E. F.; Hashem, A.; Sarwat, M.; Gucel, S. Exogenous Application of Selenium Mitigates Cadmium Toxicity in Brassica Juncea L. (Czern & Cross) by Up-Regulating Antioxidative System and Secondary Metabolites. *J. Plant Growth Regul.* **2016,** *35* (4), 936–950.
5. Akram, N. A.; Shafiq, F.; Ashraf, M. Ascorbic Acid-A Potential Oxidant Scavenger and Its Role in Plant Development and Abiotic Stress Tolerance. *Front. Plant Sci.* **2017,** *8*, 613.

6. Alamri, S. A.; Siddiqui, M. H.; Al-Khaishany, M. Y. Y.; Nasir Khan, M.; Ali, H. M.; Alaraidh, I. A.; Alsahli, A. A.; Al-Rabiah, H.; Mateen, M. Kanuascorbic Acid Improves the Tolerance of Wheat Plants to Lead Toxicity. *J. Plant Interac.* **2018,** *13* (1): 409–419.
7. Alscher, R. G.; Erturk, N.; Heath, L. S. Role of Superoxide Dismutases (SODs) in Controlling Oxidative Stress in Plants. *J. Exp. Bot.* **2002,** *53*, 1331–1341.
8. Anderson, C. J.; Kille, P.; Lawlor, A. J.; Spurgeon, D. J. Life-History Effects of Arsenic Toxicity in Clades of the Earthworm Lumbricus Rubellus. *Environ. Pollut.* **2013,** *172*, 200–207.
9. Anjum, N. A.; Sharma, P.; Gill, S. S.; Hasanuzzaman, M.; Khan, E. A.; Kachhap, K.; Mohamed, A. A.; et al. Catalase and Ascorbate Peroxidase—Representative H2O2-Detoxifying Heme Enzymes in Plants. *Env. Sci. Pollut. Res.* **2016,** *23* (19), 19002–19029.
10. Athar, H. R.; Ambreen, S.; Javed, M.; Hina, M.; Rasul, S.; Zafar, Z. U.; Manzoor, H.; et al. Influence of Sub-Lethal Crude Oil Concentration on Growth, Water Relations and Photosynthetic Capacity of Maize (Zea Mays L.) Plants. *Environ. Sci. Pollut. Res.* **2016,** *23* (18), 18320–18331.
11. ATSDR. Investigating Environmetanl Hazards to Advance Community Health, 2017.
12. Baker, A. J. M.; Grant, C. J.; Martin, M. H.; Shaw, S. C.; Whitebrook, J. Induction and Loss of Cadmium Tolerance in Holcus *Lanatus L.* and Other Grasses. *New Phytol.* **1986,** *102* (4): 575–587.
13. Baruah, S.; Bora, M. S.; Sharma, P.; Deb, P.; Sarma, K. P. Understanding of the Distribution, Translocation, Bioaccumulation, and Ultrastructural Changes of Monochoria Hastata Plant Exposed to Cadmium. *Water Air Soil Pollut.* **2017,** *228* (1). 17.
14. Bellout, Y.; Khelif, L.; Guivarch, A.; Haouche, L.; Djebbar, R.; Carol, P.; Abrous Belbachir, O. Impact of Edaphic Hydrocarbon Pollution on the Morphology and Physiology of Pea Roots (Pisum Sativum L.). *Appl. Ecol. Environ. Res.* **2016,** 14 (2), 511–525.
15. Bennett, J. H.; Hill, A. C.; Gates, D. M. A Model for Gaseous Pollutant Sorption by Leaves. *J. Air Pollut. Cont. Assoc.* **1973,** *23* (11). 957–962.
16. Bernard, F.; Brulle, F.; Dumez, S.; Lemiere, S.; Platel, A.; Nesslany, F.; Cuny, D.; Deram, A.; Vandenbulcke, F. Antioxidant Responses of Annelids, Brassicaceae and Fabaceae to Pollutants: A Review. *Ecotoxicol. Environ. Saf.* **2015,** *114*, 273–303.
17. Bickham, J. W.; Sandhu, S.; Hebert, P. D. N.; Chikhi, L.; Athwal, R. Effects of Chemical Contaminants on Genetic Diversity in Natural Populations: Implications for Biomonitoring and Ecotoxicology. *Mut. Res.* **2000,** *463* (1), 33–51.
18. Bijlsma, R.; Loeschcke, V. Genetic Erosion Impedes Adaptive Responses to Stressful Environments. *Evol. Appl.* **2012,** *5* (2), 117–129.
19. Borghi, M.; Tognetti, R.; Monteforti, G.; Sebastiani, L. Responses of Populus × Euramericana (P. Deltoides × P. Nigra) Clone Adda to Increasing Copper Concentrations. *Environ. Exp. Bot.* **2007,** *61* (1), 66–73.
20. Caverzan, A.; Casassola, A.; Brammer, S. P. Antioxidant Responses of Wheat Plants under Stress. *Genet. Mol. Biol.* **2016,** *39* (1), 1–6.
21. Chandra, R.; Kang, H. Mixed Heavy Metal Stress on Photosynthesis, Transpiration Rate, and Chlorophyll Content in Poplar Hybrids. *Forest Sci. Technol.* **2016,** *12* (2), 55–61.
22. Chen, J.; Shafi, M.; Li, S.; Wang, Y.; Wu, J.; Ye, Z.; Peng, D.; Yan, W.; Liu, D. Copper Induced Oxidative Stresses, Antioxidant Responses and Phytoremediation Potential of Moso Bamboo (Phyllostachys Pubescens). *Sci. Rep.* **2015,** *5*, 13554.
23. Chupakhina, G. N.; Maslennikov, P. V. Plant Adaptation to Oil Stress. *Russ. J. Ecol.* **2004,** *35* (5), 290–295.

24. De Coninck, D. I.; Janssen, C. R. De Schamphelaere, K. A. An Approach to Assess the Regulatory Relevance of Microevolutionary Effects in Ecological Risk Assessment of Chemicals: A Case Study with Cadmium. *Environ. Toxicol. Chem.* **2014,** *33* (2), 453–457.
25. Da Costa, M. V. J.; Sharma, P. K. Effect of Copper Oxide Nanoparticles on Growth, Morphology, Photosynthesis, and Antioxidant Response in Oryza Sativa. *Photosynthetica* **2016,** *54* (1), 110–119.
26. Cuypers, A.; Karen, S.; Jos, R.; Kelly, O.; Els, K.; Tony, R.; Nele, H.; et al. The Cellular Redox State as a Modulator in Cadmium and Copper Responses in Arabidopsis Thaliana Seedlings. *J. Plant Physiol.* **2011,** *168* (4), 309–316.
27. Dar, M. I.; Naikoo, M. I.; Khan, F. A.; Rehman, F.; Green, I. D.; Naushin, F.; Ansari, A. A. An Introduction to Reactive Oxygen Species Metabolism Under Changing Climate in Plants. In *Reactive Oxygen Species and Antioxidant Systems in Plants: Role and Regulation under Abiotic Stress*; Springer: Singapore, 2017.
28. Das, K.; Roychoudhury, A. Reactive Oxygen Species (ROS) and Response of Antioxidants as ROS-Scavengers during Environmental Stress in Plants. *Front. Environ. Sci.* **2014,** *2* (53).
29. Dias, M. C.; Mariz-Ponte, N.; Santos, C. Lead Induces Oxidative Stress in Pisum Sativum Plants and Changes the Levels of Phytohormones with Antioxidant Role. *Plant Physiol. Biochem.* **2019,** *137*, 121–129.
30. Di Battista, J. D. Patterns of Genetic Variation in Anthropogenically Impacted Populations. *Conserv. Genet.* **2008,** *9* (1), 141–156.
31. Ding, H.; Wang, B.; Han, Y.; Li, S. The Pivotal Function of Dehydroascorbate Reductase in Glutathione Homeostasis in Plants. *J. Exp. Bot.* **2020,** *71* (12), 3405–3416.
32. Disante, K. B.; Fuentes, D.; Cortina, J. Sensitivity to Zinc of Mediterranean Woody Species Important for Restoration. *Sci. Total Environ.* **2010,** *408* (10), 2216–2225.
33. Dutilleul, M.; Bonzom, J. M.; Lecomte, C.; Goussen, B.; Daian, F.; Galas, S.; Réale, D. Rapid Evolutionary Responses of Life History Traits to Different Experimentally-Induced Pollutions in Caenorhabditis Elegans. *BMC Evol. Biol.* **2014,** *14* (1), 252.
34. FAO. Soil Pollution : Hidden Reality. 2018. http://www.fao.org/3/i9183en/i9183en.pdf.
35. Forbes, V. E. *Genetics and Ecotoxicology. Current Topics in Ecotoxicology and Environmental Chemistry*, 1999.
36. Foyer, C. H.; Noctor, G. Redox Homeostasis and Antioxidant Signaling: A Metabolic Interface between Stress Perception and Physiological Responses. *Plant Cell.* **2005,** *17* (7), 1866–1875.
37. García-Caparrós, P.; De Filippis, L.; Gul, A.; Hasanuzzaman, M.; Ozturk, M.; Altay, V.; Lao, M. T. Oxidative Stress and Antioxidant Metabolism under Adverse Environmental Conditions: A Review. *Bot. Rev.* **2021,** *87*, 421–466.
38. Haida, Z.; Hakiman, M. A Comprehensive Review on the Determination of Enzymatic Assay and Nonenzymatic Antioxidant Activities. *Food Sci. Nutr.* **2019,** *7* (5), 1555–1563.
39. Hall, J. L. Cellular Mechanisms for Heavy Metal Detoxification and Tolerance. *J. Exp. Bot.* **2002,** *53* (366): 1–11.
40. Han, G.; Cui, B. X.; Zhang, X. X.; Li, K. R. The Effects of Petroleum-Contaminated Soil on Photosynthesis of Amorpha Fruticosa Seedlings. *Int. J. Environ. Sci. Technol.* **2016,** *13* (10), 2383–2392.
41. Hasanuzzaman, M.; Fujita, M.; Oku, H.; Tofazzal Islam, M. *Plant Tolerance to Environmental Stress: Role of Phytoprotectants*, Vol. 4, 2019.

42. Kasote, D. M.; Katyare, S. S.; Hegde, M. V.; Bae, H. Significance of Antioxidant Potential of Plants and Its Relevance to Therapeutic Applications. *Int. J. Biol. Sci.* **2015,** *11* (8), 982–991.
43. Kaur, G. Pb-Induced Toxicity in Plants: Effect on Growth, Development, and Biochemical Attributes. *J. Glob. Biosci.* **2014,** *3* (6), 881–889.
44. Kohli, S. K.; Khanna, K.; Bhardwaj, R.; Allah, E. F. A.; Ahmad, P.; Corpas, F. J. Assessment of Subcellular ROS and NO Metabolism in Higher Plants: Multifunctional Signaling Molecules. *Antioxidants* **2019,** *8* (12), 641.
45. Kumar, A.; Prasad, M. N. V.; Sytar, O. Lead Toxicity, Defense Strategies and Associated Indicative Biomarkers in Talinum Triangular Grown Hydroponically. *Chemosphere* **2012,** *89* (9): 1056–1065.
46. Li, J.; Li, H.; Yang, N.; Jiang, S.; Ma, C.; Li, H. Overexpression of a Monodehydroascorbate Reductase Gene from Sugar Beet M14 Increased Salt Stress Tolerance. *Sugar Tech.* **2021,** *23* (1): 45–56.
47. Liang, X.; Zhang, L.; Natarajan, S. K.; Becker, D. F. Proline Mechanisms of Stress Survival. *Antioxid. Redox Signal.* **2013,** *19* (9), 998–1011.
48. Lin, M.-Z.; Jin, M,-F. Soil Cu Contamination Destroys the Photosynthetic Systems and Hampers the Growth of Green Vegetables. *Photosynthetica* **2018,** *56* (4), 1336–1345.
49. M Oziolor, E.; De Schamphelaere, K.; Matson, C. W. 2016. Evolutionary Toxicology: Meta-Analysis of Evolutionary Events in Response to Chemical Stressors. *Ecotoxicology* **2016,** *25* (10), 1858–1866.
50. Macnair, M. R. Heavy Metal Tolerance in Plants: A Model Evolutionary System. *Trends Ecol. Evol.* **1987,** *2* (12), 354–359.
51. Maslennikov, P. V.; Chupakhina, G. N.; Skrypnik, L. N.; Feduraev, P. V.; Melnik, A. S. Assessment of the Antioxidant Potential of Plants in Urban Ecosystems under Conditions of Anthropogenic Pollution of Soils. *Russ. J. Ecol.* **2018,** *49* (5), 384–394.
52. Messiaen, M.; Janssen, C. R.; Thas, O.; De Schamphelaere, K. A. C. The Potential for Adaptation in a Natural Daphnia Magna Population: Broad and Narrow-Sense Heritability of Net Reproductive Rate under Cd Stress at Two Temperatures. *Ecotoxicology* **2012,** *21* (7), 1899–1910.
53. Miller, G.; Suzuki, N.; Ciftci-Yilmaz, S.; Mittler, R. Reactive Oxygen Species Homeostasis and Signalling during Drought and Salinity Stresses. *Plant Cell Environ.* **2010,** *33* (4), 453–467.
54. Mishra, B.; Sangwan, R. S.; Mishra, S.; Jadaun, J. S.; Sabir, F.; Sangwan, N. S. Effect of Cadmium Stress on Inductive Enzymatic and Nonenzymatic Responses of ROS and Sugar Metabolism in Multiple Shoot Cultures of Ashwagandha (Withania Somnifera Dunal). *Protoplasma* **2014,** *251* (5), 1031–1045.
55. Mishra, J.; Singh, R.; Arora, N. K. Alleviation of Heavy Metal Stress in Plants and Remediation of Soil by Rhizosphere Microorganisms. *Front. Microbiol.* **2017,** *8*, 1706.
56. Nagajyoti, P. C.; Lee, K. D.; Sreekanth, T. V. M. Heavy Metals, Occurrence and Toxicity for Plants: A Review. *Environ. Chem. Lett.* **2010,** *8*, 199–216.
57. Nisar, N.; Li.; Lu, S.; Khin, N. C.; Pogson, B. J. Carotenoid Metabolism in Plants. *Mol. Plant.* **2015,** *8* (1), 68–82.
58. Pereira, A. Plant Abiotic Stress Challenges from the Changing Environment. *Front. Plant Sci.* **2016,** *7*, 1123.
59. Poljšak, B.; Fink, R. The Protective Role of Antioxidants in the Defence against ROS/RNS-Mediated Environmental Pollution. *Oxid. Med. Cell. Longev.* **2014,** *2014*, 671539.

60. Puthur, J. T. Antioxidants and Cellular Antioxidation Mechanism in Plants. *South Indian J. Biol. Sci.* **2016,** *2*, 14.
61. Pye, J. M. Impact of Ozone on the Growth and Yield of Trees: A Review. *J. Environ. Qual.* **1988,** *17* (3), 347–360.
62. Rahbar, F. G.; Kiarostami, K.; Shirdam, R. Effects of Petroleum Hydrocarbons on Growth, Photosynthetic Pigments and Carbohydrate Levels of Sunflower. *J. Food Agric. Environ.* **2012,** *10* (1 part 2), 773–776.
63. Rahoui, S.; Martinez, Y.; Sakouhi, L.; Ben, C.; Rickauer, M.; El Ferjani, E.; Gentzbittel, L.; Chaoui, A. Cadmium-Induced Changes in Antioxidative Systems and Differentiation in Roots of Contrasted Medicago Truncatula Lines. *Protoplasma* **2017,** *254* (1), 473–489.
64. Rai, V.; Khatoon, S.; Bisht, S. S.; Mehrotra, S. Effect of Cadmium on Growth, Ultramorphology of Leaf and Secondary Metabolites of Phyllanthus Amarus Schum. and Thonn. *Chemosphere* **2005,** *61* (11), 1644–1650.
65. Rao, I. M.; Miles, J. W.; Beebe, S. E.; Horst, W. J. Root Adaptations to Soils with Low Fertility and Aluminium Toxicity. *Ann. Bot.* **2016,** *118* (4), 593–605.
66. Ravindra, K.; Sokhi, R.; Grieken, R. V. Atmospheric Polycyclic Aromatic Hydrocarbons: Source Attribution, Emission Factors and Regulation. *Atmosp. Env.* **2008,** *42* (13), 2895–2921.
67. Robinson, N. J.; Jackson, P. J. 'Metallothionein-like' Metal Complexes in Angiosperms; Their Structure and Function. *Physiologia Plantarum* 1986, *67* (3), 499–506.
68. Roelofs, D.; Morgan, J.; Stürzenbaum, S. The Significance of Genome-Wide Transcriptional Regulation in the Evolution of Stress Tolerance. *Evol. Ecol.* **2010,** *24* (3), 527–539.
69. Rusin, M.; Gospodarek, J.; Barczyk, G.; Nadgórska-Socha, A. Antioxidant Responses of Triticum Aestivum Plants to Petroleum-Derived Substances. *Ecotoxicology* **2018,** *27* (10), 1353–1367.
70. Rusin, M.; Gospodarek, J.; Nadgórska-Socha, A. Soil Pollution by Petroleum-Derived Substances and Its Bioremediation: The Effect on Aphis Fabae Scop. Infestation and Antioxidant Response in Vicia Faba L. *Agronomy* **2020,** *10* (1), 147.
71. Sachs, M. M.; Ho, T. H. D. Alteration of Gene Expression During Environmental Stress in Plants. *Annu. Rev. Plant Physiol.* **1986,** *37* (1), 363–376.
72. Saleem, M.; Asghar, H. N.; Zahir, Z. A.; Shahid, M. Impact of Lead Tolerant Plant Growth Promoting Rhizobacteria on Growth, Physiology, Antioxidant Activities, Yield and Lead Content in Sunflower in Lead Contaminated Soil. *Chemosphere* **2018,** *195*, 606–614.
73. Salemaa, M.; Monni, S. Copper Resistance of the Evergreen Dwarf Shrub Arctostaphylos Uva-Ursi: An Experimental Exposure. *Environ. Pollut.* **2003,** *126* (3), 435–443.
74. Sarker, U.; Oba, S. Catalase, Superoxide Dismutase and Ascorbate-Glutathione Cycle Enzymes Confer Drought Tolerance of Amaranthus Tricolor. *Sci. Rep.* **2018,** *8* (1), 16496.
75. Semida, W. M.; El-Mageed, T. A. A.; Howladar, S. M.; Rady, M. M. Foliar-Applied α-Tocopherol Enhances Salt-Tolerance in Onion Plants by Improving Antioxidant Defence System. *Aust. J. Crop Sci.* **2016,** *10* (7), 1030–1039.
76. Shahid, M.; Khalid, S.; Abbas, G.; Shahid, N.; Nadeem, M.; Sabir, M.; Aslam, M.; Dumat, C. Heavy Metal Stress and Crop Productivity. In *Crop Production and Global Environmental Issues*; Hakeem, K. R., Ed.; Springer International Publishing: Cham, 2015a; pp. 1–25.
77. Shahid, M.; Khalid, S.; Abbas, G.; Shahid, N.; Nadeem, M.; Sabir, M.; Aslam, M.; Dumat, C. Heavy Metals in Soils. In *Environmental Pollution,* Vol. 22; Alloway, B. J., Ed.; Dordrecht: Springer Netherlands, 2015b.

78. Sharmila, P.; Kusuma Kumari, P.; Singh, K.; Prasad, N. V. S. R. K.; Pardha-Saradhi, P. Cadmium Toxicity-Induced Proline Accumulation is Coupled to Iron Depletion. *Protoplasma* **2017,** *254* (2): 763–770.
79. Sheoran, I. S.; Singal, H. R.; Singh, R. Effect of Cadmium and Nickel on Photosynthesis and the Enzymes of the Photosynthetic Carbon Reduction Cycle in Pigeonpea (Cajanus Cajan L.). *Photosynth. Res.* **1990,** *23* (3), 345–351.
80. Shukry, W. M.; Al-Hawas, G. H. S.; Al-Moaikal, R. M. S.; El-Bendary, M. A. Effect of Petroleum Crude Oil on Mineral Nutrient Elements, Soil Properties and Bacterial Biomass of the Rhizosphere of Jojoba. *Br. J. Environ. Clim. Change* **2013,** *3* (1), 103–118.
81. Singh, A.; Prasad, S. M.; Singh, S.; Singh, M. Phytoremediation Potential of Weed Plants' Oxidative Biomarker and Antioxidant Responses. *Chem. Ecol.* **2016,** *32* (7), 684–706.
82. Skrypnik, L.; Maslennikov, P.; Novikova, A.; Kozhikin, M. Effect of Crude Oil on Growth, Oxidative Stress and Response of Antioxidative System of Two Rye (Secale Cereale l.) Varieties. *Plants* **2021,** *10* (1), 157.
83. Sofo, A.; Scopa, A.; Nuzzaci, M.; Vitti, A. Ascorbate Peroxidase and Catalase Activities and Their Genetic Regulation in Plants Subjected to Drought and Salinity Stresses. *Int. J. Mol. Sci.* **2015,** *16* (6), 13561–13578.
84. Čamagajevac, Š. I.; Pfeiffer, T. Ž.; Maronić, D. Š. Abiotic Stress Response in Plants: The Relevance of Tocopherols. In *Antioxidants and Antioxidant Enzymes in Higher Plants*; Gupta, D. K., Palma, J. M., Corpas, F. G., Eds.; Springer International Publishing: Cham
85. Tahir, M.; Ahmad, I.; Shahid, M.; Shah, G. M.; Farooq, A. B. U.; Akram, M.; Tabassum, S. A.; et al. Regulation of Antioxidant Production, Ion Uptake and Productivity in Potato (Solanum Tuberosum L.) Plant Inoculated with Growth Promoting Salt Tolerant Bacillus Strains. *Ecotoxicol. Environ. Saf.* **2019,** *178*, 33–42.
86. Tanhan, P.; Kruatrachue, M.; Pokethitiyook, P.; Chaiyarat, R. Uptake and Accumulation of Cadmium, Lead and Zinc by Siam Weed [Chromolaena Odorata (L.) King & Robinson]. *Chemosphere* **2007,** *68* (2), 323–329.
87. Tomsett, A. B.; Thurman, D. A. Molecular Biology of Metal Tolerances of Plants. *Plant Cell Environ.* **1988,** *11* (5), 383–394.
88. Trchounian, A.; Petrosyan, M.; Sahakyan, N. Plant Cell Redox Homeostasis and Reactive Oxygen Species. In *Redox State as a Central Regulator of Plant-Cell Stress Responses*; Gupta, D., Palma, J., Corpas, F., Eds.; Springer International Publishing: Cham, 2016; pp. 25–50.
89. Trenbath, B.; Harper, J. L. Population Biology of Plants. *J. Appl. Ecol.* **1978,** *15* (2), 642–644.
90. Tripathi, D. K.; Singh, V. P.; Prasad, S. M.; Dubey, N. K.; Chauhan, D. K.; Rai, A. K. LIB Spectroscopic and Biochemical Analysis to Characterize Lead Toxicity Alleviative Nature of Silicon in Wheat (Triticum Aestivum L.) Seedlings. *J. Photochem. Photobiol. B* **2016,** *154*, 89–98.
91. Uarrota, V. G.; Stefen, D. L. V.; Leolato, L. S.; Gindri, D. M.; Nerling, D. Revisiting Carotenoids and Their Role in Plant Stress Responses: From Biosynthesis to Plant Signaling Mechanisms During Stress. In *Antioxidants and Antioxidant Enzymes in Higher Plants*. Springer International Publishing: Cham, 2018.
92. Vinit-Dunand, F.; Epron, D.; Alaoui-Sossé, B.; Badot, P.-M. Effects of Copper on Growth and on Photosynthesis of Mature and Expanding Leaves in Cucumber Plants. *Plant Sci.* **2002,** *163* (1), 53–58.

93. Wang, W.; Xia, M. X.;-Chen, J.; Yuan, R.; Deng, F. N.; Shen, F. F. Gene Expression Characteristics and Regulation Mechanisms of Superoxide Dismutase and Its Physiological Roles in Plants under Stress. *Biochemistry (Moscow)* **2016,** *81* (5), 465–480.
94. Wu, L.; Bradshaw, A. D.; Thurman, D. A. The Potential for Evolution of Heavy Metal Tolerance in Plants. *Heredity* **1975,** *34* (2), 165–187.
95. Yan, Y. P.; He, J. Y.; Zhu, C.; Cheng, C.; Pan, X. B.; Sun, Z. Y. Accumulation of Copper in Brown Rice and Effect of Copper on Rice Growth and Grain Yield in Different Rice Cultivars. *Chemosphere* **2006,** *65* (10), 1690–1696.
96. Yan, Z.; Tam, N. F. Y. Effects of Lead Stress on Anti-Oxidative Enzymes and Stress-Related Hormones in Seedlings of Excoecaria Agallocha Linn. *Plant Soil* **2013,** *367* (1–2), 327–338.
97. Young, S. *Toxic Metals in Soil-Plant Systems*; Ross, S. M., Ed.; John Wiley & Sons: Chichester, (1994). £55.00 (Hardback). ISBN 0 471 94279 0. *The Journal of Agricultural Science* 124 (1), 1995; pp. Xiv + 469.
98. Zaid, A.; Wani, S. H. Reactive Oxygen Species Generation, Scavenging and Signaling in Plant Defense Responses. In *Bioactive Molecules in Plant Defense: Signaling in Growth and Stress*; Jogaiah, S., Abdelrahman, M., Eds.; Springer: Cham, 2019.
99. Zhao, S.; Liu, Q.; Qi, Y.; Duo, L. Responses of Root Growth and Protective Enzymes to Copper Stress in Turfgrass. *Acta Biologica Cracoviensia Series Botanica* **2010,** *52* (2), 7–11.
100. Zhou, J.; Zhang, Z.; Zhang, Y.; Wei, Y.; Jiang, Z. Effects of Lead Stress on the Growth, Physiology, and Cellular Structure of Privet Seedlings. *PLoS One* **2018,** *13* (3), e0191139.

CHAPTER 4

Impact of Pesticide Use in Agriculture

YUSUF OPEYEMI OYEBAMIJI[1], ISMAIL ABIOLA ADEBAYO[2,3*], MOHD NAZRI ISMAIL[4], NORAZIYAH ABD AZIZ SHAMSUDDIN[1], NOOR ZAFIRAH ISMAIL[5], and HASNI ARSAD[5]

[1]*Department of Biological Sciences and Biotechnology, Faculty of Science and Technology, Universiti Kebangsaan Malaysia, 43600 Bangi, Selangor, Malaysia*

[2]*Microbiology and Immunology Department, Faculty of Biomedical Sciences, Kampala International University, Western Campus, P. O. Box 71, Ishaka-Bushenyi, Uganda*

[3]*School of Medicine and Pharmacy, College of Medicine and Health Sciences, University of Rwanda, Rwand*

[4]*Analytical Biochemistry Research Centre, Universiti Sains Malaysia (USM), 11800, Pulau Pinang, Malaysia*

[5]*Advanced Medical and Dental Institute, Universiti Sains Malaysia, Bertam, 13200 Kepala Batas, Penang, Malaysia*

[*]*Corresponding author. E-mail: ibnahmad507@gmail.com*

ABSTRACT

Pesticide is a mixture of chemical compounds used against predators such as weeds, insect, and pest, which undermine agricultural production before and after harvest. In the modern-day agriculture, the role of pesticides is considered indispensable because of the enormous benefits that come with their application, which include increase in crop productivity and elimination

Environmental Pollution Impact on Plants: Survival Strategies under Challenging Conditions. Tariq Aftab, PhD & Khalid Rehman Hakeem, PhD (Eds.)

of pests. However, the continued use of pesticides has harmed the ecosystem as a whole. Indiscriminate pesticide use does not distinguish between targeted and nontargeted organisms, but it also kills or harms all life forms exposed to it, and it also has the potential to destroy or cause loss of biodiversity. Several useful microorganisms in the terrestrial ecosystem have been killed by pesticides. Many aquatic creatures are under threat and facing extinction due to the accumulation and migration of active toxic ingredients of pesticides into the water bodies. As a result of the detrimental roles of pesticides within the ecosystem, it is considered a major threat to global stability and a sustainable environment. This book chapter focuses on the usage of pesticides in agriculture, the significant impacts of pesticides used in enhancing the yield of crops, and the negative consequences of excessive use of pesticides on the environment and human health. Also, several approaches adopted to control pesticides usage are discussed as well.

4.1 INTRODUCTION

To sustain crop production during cultivation, almost all agricultural commodities depend on the usage of pesticides for efficient production. Usage of pesticides or other chemical substances in the modern farming system has been regarded as an essential aspect to attain maximum production success in the farming industry.[4] The word pesticide encompasses or constitutes a diverse range of chemicals, substances, or compounds used to destroy, kill, repel, and attack pests. Examples of pesticides include nematicides, rodenticides, insecticides, herbicides, molluscicides, fungicides, and other plant growth regulators. Herbicides are the most widely used pesticides, which constitute about 80% of all the pesticides.[6] Pesticides are commonly used chemical substances against a diverse range of pest which attacks crops to undermine agricultural productivity. However, pesticide use is limited to farmland and pesticide is also used in houses in fine form, poison, or spray to combat the presence of insects and other life forms such as mosquitoes, cockroaches, ticks, and bugs. As a result, pesticides are widely detected in agricultural commodities.[7] The use of pesticides has gained popularity in different sectors such as horticulture, forestry, and agriculture. In agriculture, pesticide usage has helped enhance agricultural yield by minimizing yield loss caused by plant pathogens that invade crops on the farmland. Every year, approximately 3 billion kg of pesticides are used worldwide in agricultural sectors, especially for crop production, which accounted for approximately \$40 billion.[8]

The agricultural sector has been considered the largest contributor to environmental issues such as water pollution due to pesticide overuse.[15,17] Since the advent of the green revolution, there has been a tremendous increase (multiple fold) in pesticides, which is detrimental to the green nature of the environment and human health.[20] For instance, about two-thirds of India's working class depends on agriculture as a means of sustenance and is therefore frequently exposed to pesticides. Every year, the World Health Organization reports about three million cases of severe acute poisoning globally; about 220,000 deaths are associated with pesticides exposure, with 1% of the cases being found in developed nations.[6] Pesticides contain active constituents that, despite being advantageous for agriculture purposes, could have disastrous effects on the ecosystem and habitat where they are applied.[9,12] The effect includes pollution of water, soil, food, and air.[12,13,14,] The indiscriminate use of pesticides has caused severe damage on living things such as organisms, animals, and humans beings as well as the environment.[20] Owing to the occurrence of pesticides in the environment, residues are found in groundwater,[17,21] sediments, and water ways or resources.[3,17,22] This indicates that the migration of pesticides into the environment is possible, and the residue can be detected several thousand miles away from where they are used. Usage of pesticides has raised significant concern about its impact on biodiversity and our fragile ecosystem.[6] A considerable rise in the amount of pesticide along the food cycle is known as biomagnification. This occurs when there is an accumulation of pesticides in lower species such as marine organisms, plant, and other life forms, causing the rise in the number of pesticides along the food cycle until the toxic concentrations are ingested by the higher animals.[23] In Europe, pesticides used for agricultural activities and the ones used in sectors such as industries and municipal are considered the major contributor to poor water quality in certain European nations.[24] Owing to these reasons, the European Union released directives for a diverse array of chemical compounds, including pesticides suggesting the maximum amount of residue targeted at protecting consumable waters and groundwater.[24]

4.2 PESTS AND PESTICIDES

In a broad sense, pests can describe as any animals or plants include insects, rodents, pathogens (bacteria, fungi, and viruses), and weeds, which endanger agricultural commodities and our well-being.[7] The pests that are found on farmland include weeds, mites, insects, and pathogens, while pests that are found in the residential building include rodents, mosquitoes, fleas, cockroaches, and bedbugs. Studies have shown that certain pests are perennial and can cause severe injury and destruction of valuable crops when it is not controlled. In contrast, some pests cause destruction or injury on plant species only on rare occasions when the ecosystem is disrupted.[25]

Pesticides can be defined as a mixture of chemical compounds that can be utilized to destroy, reduce, repel, or control any form of pests or population of pests.[4,5] It can also be defined as a poisonous chemical formulation of compounds or substance or biologically active ingredients deliberately discharged in a particular environment to deter, destroy, control, avert, or mitigate pests. Pesticides can be either natural (extract from a plant) or artificial (synthetic).

4.3 HISTORY OF PESTICIDES

The beginning of farming can be dated back to approximately more than 9000 years ago in a region known as the Fertile Crescent of Mesopotamia (Syria, Iraq, Turkey, and Jordan) where fertile seeds were assembled by a group of hunters.[49] As the population increases, different cash crops were cultivated in these regions. In China, crops such as rice and millet were domesticated and also, in the Sahel region of Africa, rice and sorghum were cultivated for more than 5000 years ago.[49] Domestication of others crops such as corn has been done in the Americas. As agriculture becomes a means of livelihood, especially for poor farmers who depend on farming for their daily needs, there is a need to reduce loss of crops. Pests and diseases caused 10–90% yield loss with an average of 35 to 40%, for all potential food and fiber crops. Thus, there is a need to overcome the problems encountered by plants due to pest and diseases invasion.[49]

The first pesticides used in history can be traced back to 4500 years ago in ancient Sumer which was the use of elemental sulfur to kill pests, insects, and mites.[6] while compounds such as ashes and salt are used to mitigate weed and also arsenic are used to control pest.[7] In the 1700s, an insecticide derived from tobacco leaves known as nicotine sulfate was used to control insects.[6] In the beginning of the 19th century, lead arsenate and several other stomach pesticides were considered an option and used to mitigate pest invasion of cropland.[32] In 1940, the discovery of the effects of synthetic pesticides such as BHC, dieldrin, aldrin, endrin, captan, chlordane, DDT, and 2,4-D was made.[49] All these pesticides are cheap and effective, making them widely acceptable by the farmers. The most widely used among these pesticides was DDT because of it being less toxic to living organisms and its broad spectrum.

However, in the later years, problems arise due to the heavy use of these pesticides, consequently becoming harmful to living organisms that are not intended and other target insects or pest becoming tolerant to DDT.[33] In 1962, the book titled the *Silent Spring* authored by Rachel Carson highlighted the hazardous impacts of pesticides use, which undermine the public confidence in its application.[33] Due to the harmful effects of DDT, it was banned in 1972 in the United States[34] and in 1975 in South Africa.[39] From 1970s till 1990s, pesticides such as triketone, isoxazole and triazolopyrimidine herbicides, diacylhydrazine, spinosyn, chloronicotinyl, fiprole and organophosphate insecticides, azolone, and strobilurin fungicides were produced, released, and they replaced the synthetic pesticides such as DDT in the market.[34,35,38]

Researchers such as plant breeders have intensified efforts to produce crops that will be resistant to pests attack through genetic engineering in the present era. These methods will help reduce the use of pesticides and help to solve damage to the ecosystem caused by the heavy use of pesticides.[34,35]

4.4 CLASSIFICATION OF PESTICIDES

Pesticides can be classified into five or more groups based on:

I. Toxin concentration
II. Mode of entry
III. Mode of action
IV. Target organism
V. Nature of the chemical

4.4.1 TOXIN CONCENTRATION

- **Extremely dangerous** (LD50 <5 mg/kg) Example: Aldicarb, Bromadiolone, Calcium cyanide, Captafol
- **Highly dangerous** (LD50 5–200 mg/kg) Example: Abamectin, Acrolein, Carbofuran, Coumaphos, Methiocarb
- **Moderately dangerous** (LD50 50–2000 mg/kg) Example: Acephate, Acetamiprid, Alachlor, Allethrin, Trichlorfon
- **Slightly dangerous** (LD50 >2000 mg/kg) Example: Acetochlor, Alloxydim, Ancymidol, Benalaxy, Sethoxydim

- **Relatively harmless** (LD50 >5000 mg/kg) Example: Aclonifen, Acrinathrin, Aminopyralid, Amisulbrom, Anthraquinone

4.4.2 ENTRY MODE

4.4.2.1 STOMACH/GASTRIC TOXINS

This kind of pesticides will kill the pests by penetrating the pests' body through the oral part and digestive tract and subsequently invading the stomach of the affected organisms, thus causing stomach poisoning and death of the pests. It is usually used to kill vector such as bacteria and applied to the water where filter-feeding mosquito or black fly larvae will ingest the chemical. An example of this type of pesticide is insecticides that will kill the vector by destroying the larvae's midgut (or stomach). Examples of such pesticides are malathion and thurigiensis.

4.4.2.2 CONTACT POISON/EXPOSURE

This kind of pesticide penetrates the pests' body via the skin layer upon contact and kills the pests. Examples of this kind of pesticides are fenvalerate, paraquat, and diquat dibromide.

4.4.2.3 FUMIGANTS

This kind of pesticides is commonly known to be in an aqueous form and kill the pests by penetrating the pests' body through the respiratory channel or track, thus becoming toxic to the body system. Example of this type is aluminum phosphide.

4.4.2.4 SYSTEMIC POISONS

This kind of pesticide is ingested by the targeted organisms and stays within the hosts' body fluids. Example of this type is metasystox. This kind of pesticides can also move through the different parts of the plant even to the parts that the chemicals are not applied, such as the stems and roots.[40] If the pesticide is sprayed to the root region, it will migrate to the whole part of the plant; in contrast, if it is sprayed on the leaves, it does not move to the whole part of the plant.[41] Certain pesticides such as glyphosate and 2,4-dichlorophenoxyacetic acid are applied on animals to kill pests like fleas, lice,[40] and warble grubs.

4.4.2.5 REPELLENTS

This kind of pesticides does not necessarily kill the pests but repel the pests from the host organism. Example of this type of pesticide is ethyl butylacetylaminoprpinate

4.4.3 MODE OF ACTION

4.4.3.1 PHYSICAL POISON

This is a kind of pesticides that exert physical effects such as desiccation to destroy the insect or pest. Examples include diatomaceous earth and activated clay.

4.4.3.2 PROTOPLASMIC POISON

This is a kind of pesticides that are responsible for precipitation of protein to destroy the insect. Example of such is arsenic compound.

4.4.3.3 RESPIRATORY POISON

This kind of pesticides inhibit the respiratory enzymes to exert action. Example of such is hydrogen cyanide.

4.4.3.4 NERVE POISON

This is a kind of pesticides that caused the inhibition of the nervous system or impulse. Examples of this kind of pesticides are malathion, organophosphates, and carbamates.

4.4.3.5 CHITIN INHIBITION

This is a class of pesticides that are involved in inhibiting chitin synthetase in pests by damaging the chitin synthesis. Example of this type is diflubenzuron.

4.4.4 TARGET ORGANISM

- **Termiticides**—Fight against termites. Example: Fipronil
- **Virucides**—Fight against virus. Example: Scytovirin
- **Herbicides**—Act against unwanted plants called weeds. Example: Atrazine
- **Algaecides**—Act against algae growth. Example: Copper sulfate
- **Rodenticides**—Control or kills rodents. Example: Warfarin
- **Insecticides**—Act against insects and other arthropods. Example: Aldicarb
- **Acaricides**—Act against mites that feed animals or plants. Example: Bifenazate
- **Larvicides**—Control, kill, or prevent larvae growth. Example: Methoprene
- **Nematicides**—Act against nematodes. Example: Aldicarb
- **Avicides**—Kill birds. Example: Avitrol.
- **Silvicides**—Act against woody vegetation. Example: Tebuthiuron
- **Bactericides**—Kill or control bacteria. Example: Copper complexes
- **Fungicides**—Act against different forms of fungi. Example: Azoxystrobin

4.4.5 NATURE OF THE CHEMICAL

4.4.5.1 INORGANIC PESTICIDES

(a) Arsenicals: Lead arsenate. (b) Fluorine compounds: Sodium fluoride. (c) Other inorganic compounds: sulfur, zinc phosphide.

4.4.5.2 ORGANIC PESTICIDES

(a) Oils and soaps: kerosene, diesel, natural oil, crude oil. (b) Pesticide of animal origin: Nereistoxin, Lumbrinerisheteropoda: It poisons the impulse. (c) Pesticides of plant origin which include Pyrethrum, Nicotine, Rotenone, and Neem product.

4.4.5.3 SYNTHETIC PESTICIDES

These include Organophosphorus, Carbamates, Organochlorines, and Synthetic pyrethroid.

4.4.5.4 ORGANOPHOSPHORUS

These pesticides consist of organic and phosphate compounds.[44] Since its characterization in 1937, organophosphorus was used as a pesticide and as a weapon during the war (war gases). It contains carbon that comes from phosphoric acid.[35] They are strong pesticides that inhibit the action of cholinesterase. Apart from that, they reduce insulin production and interfere in the normal metabolic pathways of carbohydrates, fats, and protein. They have genotoxic effects and adverse effects on mitochondrial function. They also disrupt both the endocrine and nervous system.[42] The effects of these pesticides on insects are similar to their effects in humans. They are biodegradable, they have minimal contribution to environmental degradation and slow pest resistance: examples include malathion, coumaphos, Monocrotophos, DDVP, and diazinon.

4.4.5.5 CARBAMATES

This class of pesticides also work exactly like organophosphorus in terms of their toxicity action mechanism, except that it can be reversed and less severe.[44] The common ones include carbaryl (Sevin) and propoxur (Baygon) used in homes, while others include aldicarb and methomyl mainly used for farming purposes.

4.4.5.6 ORGANOCHLORINES

They are majorly used as insecticides. They have been used extensively for many decades. However, several of these pesticides are no longer available in the market because of its consequence on human health, such as disruption of the endocrine system, effects on embryo development, and lipid metabolism.[42] They also have significant impact on environmental pollution, such as environmental degradation. Examples of this type of pesticides are DDT, endosulfan, chlordane, aldrin, dieldrin, and heptachlor.

4.4.5.7 SYNTHETIC PYRETHROID

These pesticides are mainly insecticides. They are regarded as newly developed natural compounds in a synthetic form derived from a plant called *Chrysanthemum cinerariaefolium*. These pesticides were developed in the 1980s to imitate the action of natural pyrethrum. Pyrethroids are known for their rapid action against insect and pest, quick biodegradation, and minimal toxic effect on mammalians.[33] They have also been altered and improved to be eco-friendly. Although some of these pyrethroids are hazardous to the nervous system, examples of synthetic pyrethroids are cypermethrin and decamethrin.

4.5 USAGE OF PESTICIDES IN AGRICULTURE

Pesticides are commonly used in three sectors that include agriculture, public health, and domestic use. After the green revolution, the transition process of modern agriculture evolution and its development are greatly enhanced by pesticides. The utilization of chemical substances such as pesticides has played an essential role in safeguarding agricultural commodities against pest attack, subsequently enhancing food safety and security[45] (Sharma & Ritu Singhvi, 2017). Worldwide, pesticide use increased in multiple folds from about two million tons per year to more than four million tons per year in 2018. Asia accounted for half of the world pesticides use with 51%, followed by Americas with 31%, Europe with 11%, and then Africa, Oceania and Australia, and New Zealand with 1.97%, 1.66%, and 1.63 %, respectively (Figure 4.1) (FAOSTAT, 2021). This signifies that regions that comprise developed nations such as the USA and many European countries consume a considerable amount of pesticides. Country-wise, China was reported to have used the highest amount of pesticides of 14 kg/ha, while India was considered the country that use the least amount of pesticides (0.57 kg/ha). Other countries such as USA, Japan, Korea, UK, France, Italy, Taiwan, Germany, and the Netherlands also used considerable amount of pesticides (Figure 4.2).[51] More than 250 kinds of pesticides are in existence that are used for agricultural and other purposes globally. This includes 20 types of nematicides, 50 kinds of herbicides, 100 kinds of insecticides, and 30 other pesticides.[5]

In 2007, the amount of pesticides used worldwide was estimated to be 2 million kg, with herbicides accounting for 950.7 million kg, insecticides accounting for 404.6 million kg, fungicides accounting for 262.17 million kg, and other pesticides such as fumigants, plant growth regulators, and nematicides accounting for 773.37 million kg.[46,47] Similarly, in 2006, herbicides

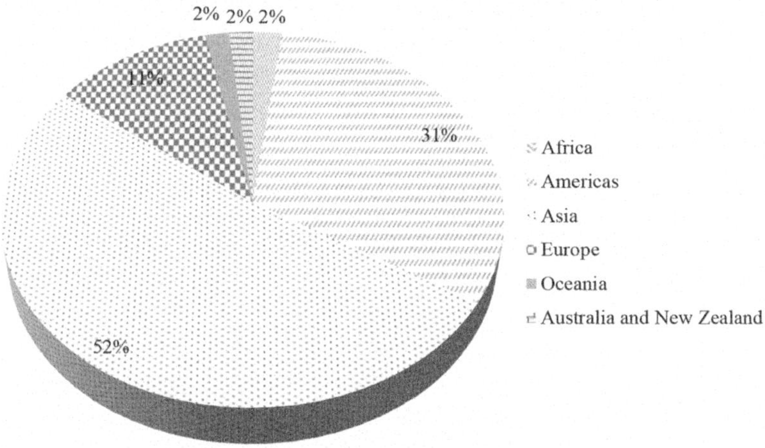

FIGURE 4.1 Worldwide consumption of pesticides by region.

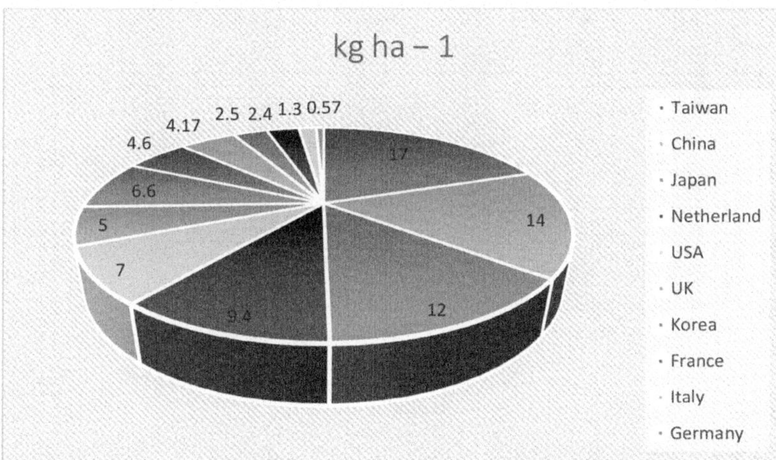

FIGURE 4.2 Worldwide consumption of pesticides by country.

were also the most widely used pesticides worldwide (Table 4.1).[16] The use of different forms of pesticides such as insecticides, herbicides, and other forms such as nematicides and fungicides varies significantly from one region to another (Figure 4.3) (FAOSTAT, 2021). For instance, in Africa, insecticides and herbicides are common due to the widespread occurrence of insects and predators such as weeds on the farmland. Similarly, in the Americas, Europe, Oceania and Australia and New Zealand, herbicides and insecticides are widespread compared to Asia, where very few insecticides and herbicides are used. However, this region (Asia) prefers the use of other forms of pesticides such as fungicides. Based on each country, the use of various forms of pesticide varies considerably based on the types of plant pathogens encountered.

TABLE 4.1 Amount of Pesticide Used Around the World in 2006.

Pesticides type	World market (million blbs)	World market (%)
Herbicides	2018	39
Insecticides	955	18
Fungicides	519	10
Other	1705	33
Total	5197	100

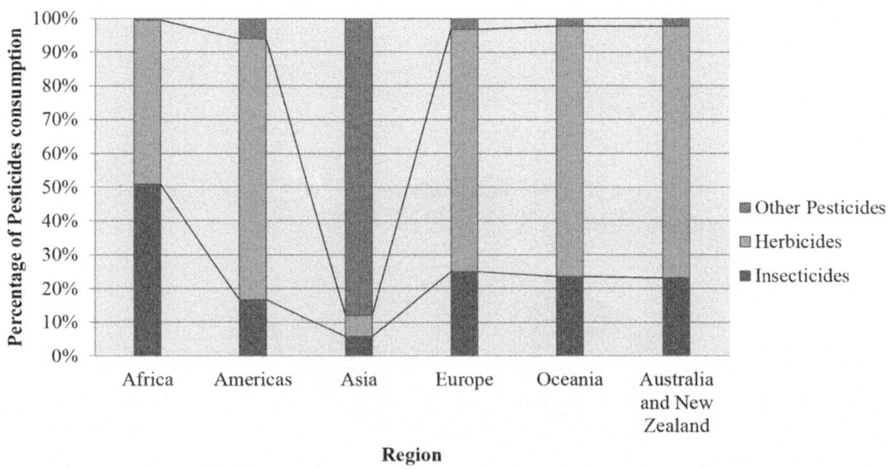

FIGURE. 4.3 Consumption pattern of pesticides in worldwide.

For instance, in Nepal, fungicides is the most popular pesticides in Nepal (45%),[46] while in other parts of Asia, such as India, insecticides accounted for approximately 76% of pesticides used, which is far beyond the total amount of insecticides used globally.[6] However, the use of fungicides is significantly less in India as compared to Nepal. Pesticides consumption has been intensified, particularly for exportable food commodities such as rice, vegetables, wheat, cotton, orange, soybeans, sorghum, corn, grapes, apples, sugarcane, and tomatoes (Figure 4.4).[10] However, in India, cotton accounted for the highest share of pesticides consumption (37%), followed

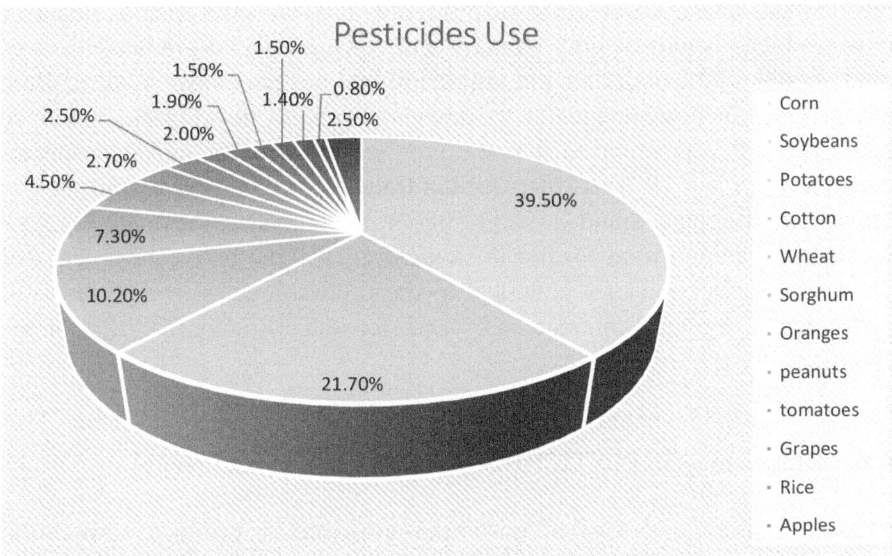

FIGURE. 4.4 Pesticides use in crops plantation.
Note: "Other Crops" include lettuce, pears, sweet corn, barley, peaches, grapefruit, pecans, and lemons.

by paddy (20%), vegetables (9%), wheat (4%), and other species (7%).[6] The most popular chemical used includes DDT, HCH (only gamma-HCH is allowed), and malathion that accounted for more than half of the pesticides consumption worldwide.[48] DDT is still utilized in several nations such as India. The majority of DDT are used to kill vectors such as tsetse fly and mosquito-vectors transmitting malaria in many Africa countries with limited and shortage availability of inexpensive chemicals to counter the effect of these vectors in the region.[50] More so, several nations, especially China and India, use HCH, an organochlorine insecticide, in both health and agriculture sectors. This type of insecticides that is HCH is regarded as one of the pesticides which is highly detrimental and causes deletion of the ecosystem and human health.[1,48] In India, significant amount of pesticides residues are found in harvested food commodities due to excessive usage of chemical substances and a high rate of illiteracy among the farmers regarding the use of pesticides. However, in most developed nations such as Japan, USA, Taiwan, Japan, and several European countries, farmers are well educated and aware of the application and usage of pesticides, subsequently resulting in minimal crop loss.[51] However, in India this is not the case. Farmers in India use pesticides excessively to increase crop yield.[51,53] More so, in Bangladesh, the frequent and excessive use of pesticides has caused a serious negative impact on the health of both workers and consumers and gradually loss of biodiversity.[5,54,55] Continued and indiscriminate pesticides use in agriculture has resulted in diverse species of pests and insects developing resistance against it.[45] It was reported that the number of pesticide-resistant species between 1965 and 1977 has increased in three folds from 182 to 364 such as brown planthoppers found in paddy fields in Malaysia.[5] Pesticide residues present in feed and food for livestock and humans can be a consequence of the direct application of a chemical to the food source by the presence of pollutants in the environment or by transfer and bio-magnification of the chemical along a food chain.[45]

4.6 BENEFITS OF PESTICIDES USAGE

Yearly, thousands of tons of pesticides are used to produce agricultural commodities to control pest attack. Without the use of pesticides, plants tend to loosen their fertility due to exhaustion and crop productivity will decline due to the activity of pest.[57] Crops yield loss reported in certain provinces worldwide indicates that more than 40% of produce are affected due to pest invasion of agricultural commodities.[58,60] Other pests such as insects, microbes, and weeds caused a decline in crops to output up to 15%, 13%, and 13%, respectively, while pest invasion of crops at the postharvest stage such as storage, processing caused another 10% loss globally.[61] It was reported in Punjab (a province in India) that the invasion of cotton by an insect called *P. solenopsi* causes dramatic yield loss in four districts within the province.[61,69] Apart from cotton, *P. solenopsi* has also been reported to infect several economic important crops such as grape, tomatoes, apple, banana, and okra, consequently undermining plant production[70] (Jindal & Dhawan, 2010**).** This also causes hitches in price of commodities consequently affecting farmers in the global markets.[58]

Due to these reasons, there is need for plant protection products such as pesticides to increase production, availability, and supply of food around the world. Pesticides also help to enhance crop yield with specific applications under conditions of good agricultural practice (GAP) and pesticides help safeguard plant from vector-borne diseases that include dengue fever, malaria, and schistosomiasis.[48,71,72] The benefits of pesticides use are direct or primary benefits and secondary or indirect benefits. The primary consequences of pesticides usage are the immediate effect of pesticides use within the environment such as protection of crop, animals and to protect human activities and structure against harmful organisms within the environment. Secondary or indirect impacts of pesticides use are the long-terms benefits that are not witnessed instantly but seen as after a period of time such as improved nutrition, reduction in maintenance expenses, enhancement of farmer's quality of life, national benefits (improved economy and international trade via export of food commodities). However, the long-terms consequences (benefits) of pesticides are complex to establish, but they can be the compelling reasons for pesticides usage.[48,73,74,79,]

Globally, increase in the plant production has been attributed to many factors such as development of high yielding cultivars, use of modern machine, and fertilizer application (e.g. pesticides). For more than six decades, pesticides have been introduced to the agricultural market while yield of agricultural crops has increased tremendously in multiple folds. These have been achieved through the use of pesticides to mainly control or mitigate the invasion of pesticides on cropland which brings about yield loss. The use of pesticides by farmers has helped increase food production and availability throughout the year at an affordable price. In most developed and under-developed nations, pesticides application in the modern farming system has enhanced the production of food and helped reduce agricultural loss experienced on farmland.[79] For example, countries such as the United Kingdom witnessed increase in wheat production up to 110 kg/ha every year which was attributed to increase in pesticide usage.[80] Similarly, the USA saw increase in corn production during the Green Revolution due to the use of pesticides among other factors.[81] More so, numerous nations including Russia have documented massive increase in the overall crop yield.[79] The use of chemical compounds such as pesticides has been considered an essential part of the process to improve production by attacking external agents such as diseases, insects, weed, and pest which cause decline in plant yield, consequently reducing the quantity of harvestable produce.[74] Webster et al.[82] reported that considerable rise in crop productivity was witnessed due to the application of pesticides and also stated that economic losses without the availability of pesticides will be considerably high.[7] Estimates from previous reports indicated that around 10–20% increase in the yield of bread grains was observed due to the use of herbicides.[7]

More so, pesticides use helps to reduce crop loss to pathogen such as weeds, insect, diseases, human vector-borne which are considered one of the main factors which causes decrease in the crop production. According to Zadoks & Schein,[85] various loss levels may be differentiated, e.g. direct and indirect losses, or primary and secondary losses, indicating that pests not only endanger crop yield and reduce the farmer's net income, but may also affect the supply of food and feed as well as the economies of rural areas and even countries.[84] Several crop commodities such as rice, groundnuts, and maize are susceptible to mycotoxins infestation and insecticides are required to prevent toxins transfer to the plant species by the insect. Mycotoxins such as aflatoxin are considered carcinogen in nature which causes liver cancer, kidney cancer, and other forms as well as causes abnormal growth in children and suppresses the immune system of the human body. The uses of pesticides help to mitigate insect-mediated aflatoxin contamination.[7] Predators such as weed were reported to be one of the major factors causing decline in tomato yield to about 47%.[86] The study conducted by Behera and Singh[86] reported that use of herbicides in different combinations helps to tackle the weed challenges faced during tomato production by reducing the total dry weight of weeds by 93% consequently increasing production. Severe weeds infestation especially during early germination or nursery stage causes decline in crop yield up to 40%. This simply pinpoints the benefits of herbicides in both the labor and economic aspect.[74] In addition, tomato Ebola outbreak that was witnessed in Nigeria has helped show the significance of pesticides in modern-day agriculture.[25]

Furthermore, the application of pesticides is considered one of the most effective ways to control the vectors activity and impact within the environment. The use of insecticides kills vectors, curbs diseases, and helps save millions of lives. An example is the annual outbreak of malaria which causes the death of approximately 5000 people worldwide. However, the use of insecticides has greatly helped to curb such outbreak.[7,34,36] Also, Bhatia in 2004 documented that vector killing diseases such as malaria was regarded as one of the main causes of death in many developing nations.[74] This has reinforced the crucial need of insecticides use.

Pesticides help farmers to produce safe, high-quality food at low cost. It also assists farmers to harvest a wide variety of nutritious food throughout the year which are deemed fit for human health and consumption. Crops such as vegetables and fruits, which are rich in nutrients, are produced in abundance and at a relatively cheap price. Also, agricultural products from livestock such as milk, which are rich in nutrient and essentials for infant growth, are produced in abundance due to lower costs to produce food and animal feed.[11] This indicates that the use of pesticides has a great impact on the quantity and quality of crop produced. For instance, studies conducted in USA showed that the production of fruit and vegetables will decrease by more than 90% without the use of pesticides. Also, contamination of foods is reduced through the use of pesticides.[11] Finally, proper use and management of pesticides can assist in improving both the farmers and families' income through increased production.[34]

4.7 IMPACT OF PESTICIDES ON HUMAN HEALTH

The usage of pesticides has caused serious hazards in regards to the environment degradation and human health. They do not only harm targeted organism but are also harmful to nontarget species such as humans. People such as the applicators and farmers who have direct contact with the pesticides are mostly affected, followed by the farmers' immediate family and followed by other people who ingest agricultural commodities with high amount of pesticides residues.[25] Most susceptible individuals to pesticide exposure are the children, partially owing to biological factors.[25,37] Several studies have established that the use of pesticides in little amount or excessive use may lead to several illnesses such as abnormal neural development,[46,87–90] cancer,[91] and noncancer complications.[92] The use of pesticides also causes altered growth, genotoxicity, fetal death, acute and chronic neurotoxicity, birth defects, habitat destruction,[46] and occupational skin diseases.[93,94] According to reports, pesticides poisoning affects approximately 1 million to 5 million people worldwide on a yearly basis causing about 20,000 deaths of workers in the agricultural sectors.[25] It is sadden that despite the limited use of pesticides by developing nations, they accounted for 99% of the fatalities.[25]

Between 1970s and 1980s, the indiscriminate pesticides use in several provinces in Indonesia caused serious health problems such as acute and chronic human pesticide poisoning as well as several environmental issues.[5] Also, the first-ever pesticides-related accident occurred in India in a province known as Kerala in 1953 where 108 individuals gave up the ghost.[95] Another finding reported that approximately 10 000 people die due to pesticide poisoning each year out of 375 000 exposed individuals.[5] Despite the consequences of pesticides application, it is still widely used in several developing nations.[46] The use of pesticides is frequently applied indiscriminately, resulting in unintended exposure to other species including the developing and young species that are mostly vulnerable to the harmful effects of pesticides.[23] In addition, specific individuals such as pregnant women and developing population are also considered more vulnerable to the effects of pesticides as compared to others.[97] Even little amount of exposure at the developmental and other critical stages may be detrimental to organisms health-wise.[23]

4.8 HUMAN EXPOSURE TO PESTICIDES

Human exposure to pesticides can be through either direct or indirect means. Direct exposure usually happens through agricultural, household use, or occupational such as agricultural workers working in the cropland, workers in the chemical or fertilizer industries,[,59,62–68104] store keepers of chemical substances, and immediate families who use pesticides to kill malaria parasites. Indirect exposure usually occurs through consumption of food and water that are preserved by the use of chemical substances. Moreover, workers required to mix the chemical content for pesticides formulation are constantly exposed to high level of risk associated with the chemical substances.[104] More so, the inability and negligence on the part of agricultural workers in regards to the instructions regarding the use of pesticide, premature harvest of crop, and not following hygiene norms such as the use of protective equipment, constant hand wash before eating and after the application of pesticides make them vulnerable to pesticide exposure.[103,104] In the topical areas, exposure to pesticides increases because of the rise in the surrounding temperature and humidity causing particles to remain in the air, and mixed with water molecules and then dispersed by the wind to the urban areas.[104] Furthermore, the usage of pesticides near major routes such as roads and on golf courses may result in the general public being exposed to it voluntarily and unintentionally.[102] Nonoccupational pesticide leftover in air, food, and drinking water exposes people to low doses over time and is chronic.[59] During pesticides-handling stages, many factors influence human exposure. The degree of exposure may also be influenced by various forms of formulations of pesticide. Pesticides in the form of a solution are vulnerable to the skin contact either directly such as splashing or something spillage and indirectly such as when the pesticides are on clothing material. Other forms of pesticides such as solidified form can produce dust which can result in both the eye and the face being exposed to contamination as well as respiratory track contamination. Various kinds of pesticide packaging such as opening of pesticides bags can also have the risks of exposure. Furthermore, container size including bottle, cans, and other types may affect the potential for splashing and spillage.[59] There are several ways in which humans can be exposed to pesticides such as through the air, soil, water, fauna, flora.[101,102] and tropic chains to consumption of contaminated food.[100,103,104] The transportation of pesticides within the entire human body occurs via the blood transmission and can be eliminated via several routes such as the skin, exhaled air, and urine.[59,102] Pesticides can enter through the human body via four different routes that include oral, eye, dermal, and respiratory pathways (Table 4.2). The routes of exposure (dermal, oral, or respiratory (inhalation)) determine the nature and the severity of harmful effect of the pesticides within the human body. As one might expect, the risk of pesticide contamination rises with the increase in the concentration.[99,102] A survey conducted in Bangladesh at three different locations of vegetable farms revealed that during pesticides spraying, 11% of workers experienced eye problem, 37% of workers had sensation problem, 13% felt dizziness, 18% had skin problems such as itching, and 28% encountered problems in breathing.[98]

TABLE 4.2 Route of Pesticide Exposure.

Routes of exposure	Short description	Case study
Dermal	• This is the most common type of pesticides exposure route. • It can occur due to spill, spray drift, or sprinkle when stacking, mixing, or arranging as well as cleaning of pesticides without wearing protective equipment[96,97]	In the United States, skin diseases only accounted for 34,400 cases in 2012 surpassing occupational respiratory illnesses.[101] Furthermore, a case of two agricultural workers admitted to the hospital in severe pain due to extensive chemical burns on the skin, due to Ducatalon[35]
Oral	• This is the ingestion of pesticide intentional or accidentally via the mouth by splashing or • Negligence in washing the hand both before and after consuming food substances.	A case of a 22-year-old man who intended to commit suicide by drinking pesticides specifically paraquat (50 mL). This resulted in development of symptoms such as dysphagia, hemoptysis, sore throat, retrosternal pain, and blistering and ulceration of the mouth and tongue[35]
Eye	• Splashing of pesticides into the eye by accident • Working on granular pesticides form in a windy environment • Using contaminated hand to rub the eye[23]	According to study conducted by Friends of the Earth from 1988 to 1992, more than 400 workers and farmers who work in sheep dip were reported to have been affected by pesticides exposure.[2]
Respiratory pathway	• Long-term pesticides exposure • Vapors of fumigants, dust, or mist • Utilization of improper or outdated pesticide application kits[23]	The death of about 5000 people in a region called Bhopal due to exposure to methyl isocyanate from a pesticide factory which was unintentional.[45]

Pesticide effects are highly variable. The effect may be: (i) acute, showing symptoms within few days after exposure to the chemical substance or (ii) chronic, showing symptoms after many months or years of exposure. Generally, the use of herbicides is known to be more dangerous in regard to chronic poison while the uses of insecticides are more toxic acutely. Acute and chronic effects of pesticide exposure on human health are discussed below.

4.8.1 ACUTE EFFECTS OF EXPOSURE TO PESTICIDES

Acute effects of pesticides are symptoms witnessed or experienced within a short period of time after exposure to chemical substance (pesticide) either through dermal, oral, eye, or respiratory pathways. The symptoms are stinging of the eyes and skin, appearance of the rash and blisters on the skin, headache, throat and nose irritation, itching of the skin, diarrhea, nausea vomiting, abdominal pain, blurred, and dizziness. Globally, approximately 3 million cases of acute poisoning are witnessed annually due to pesticides exposure, 2 million of the total cases are attempted suicide cases while the remaining are unintentional or workplace-related cases.[41,43] Moreover, in 1980s, 220,000 deaths were reported globally out of 3 million cases of acute poisoning recorded.[46,93] However, this is considered outdated due to the increase use of pesticides in the present decade. Currently, there is no accurate and comprehensive data of acute poisoning around the world, although there are scattered reports of cases in different provinces around the world affirming the existence of acute poisonings both in non-work-related places and workplace with mild to severe effects that make it a global threat to human life.[52]

4.8.2 CHRONIC EFFECTS OF EXPOSURE TO PESTICIDES

Chronic effects of pesticides are signs and symptoms developed on a long term (months, years) after prolonged exposure to pesticides. These types of pesticide effect are more detrimental as compared to acute effects that were proven and confirmed via laboratory analysis. This is mostly caused by exposure to pesticides which are sublethal in a repeated and continuous basis over a long duration of time.[41] Symptoms of chronic effects include reduced coordination, cancer and depression, reproductive problems such as poor infant growth, birth defects, miscarriages and memory disorders, shortened attention span.[45] The proof for pesticides carcinogenic effects has been linked to experimental and epidemiological research conducted.

Prostate cancer, lung cancer, and hematopoietic system have been linked to epidemiological studies, while exposure to pesticides through occupation and environment has been associated with childhood cancer.[52] A survey conducted in four districts in Punjab (a province in India) revealed that six to thirteen different kinds of pesticides (Malathion, Aldrin, HCH, DDT, Endosulfan, Chlorpyrifos, Monocrotophos) were found in nearly all blood samples tested. Long-term pesticide exposure has resulted in rise in cancer cases in several districts in Punjab.[20] The use of pesticides has also been linked to and correlated to multiple chronic illnesses such as multiple myelomas, sarcomas, testicles cancer, ovaries cancer, intestines cancer, breasts cancer, pancreas cancer, and brain tumors.[19,25,37] However, when it comes to chromic effects of pesticides exposure, it is difficult to accurately assess pesticides exposure at a particular time when research is conducted, due to prolonged period of time for the symptoms to develop. Just of recent, it was discovered that exposure to pesticides could also cause certain health problems such as endocrine disruption, neurodevelopmental toxicity, and immunotoxicity.[18,52]

4.9 IMPACT ON THE ENVIRONMENT

Pesticides-related environmental issues are regarded as a major threat in both advanced nations, underdeveloped, and developing nations. Pesticides are extremely dangerous and bioaccumulative causing damage to ecosystem components such as water, soil, air, and environment.[48] Apart from being useful in controlling weeds and insects, they are also lethal to a wide range of species such as beneficial insects, fish, birds, and nontarget plants.[74] They are frequently biomagnified via the food cycle, affecting the well-being of organisms at higher trophic levels.[48,,78] Pesticides pollution level in different environmental media has been reported in various parts of the world which has been summarized in Tables 4.3 and 4.4. Certain pesticides such as dichlorodiphenyltrichloroethane (DDT), aldrin, and heptachlor comprise persistent organic pollutants that are not degradable and thus persist in the ecosystem for a long time.[45,51,102] Pesticides release into the environment and their active constituents that are developed to effectively mitigate the activities of weeds and pests lay the groundwork for the possibility of pesticides having widespread effects on nontarget organisms[48] and becoming a threat to the entire ecosystem. The migration or dispersion of pesticides from the applied location into other unintended locations can occur via several transfer pathways such as volatilization, run-off, leaching, spray drift, and adsorption (Figure 4.5).[34,83] Several factors influence the impacts of pesticides on the environment. These include the amount of pesticides used (which is partially dependent on the machinery precision), the kind of pesticides used, and the location of the land sprayed as well as the climatic conditions.[2] It has been established that both climatic conditions and soil factors are the two predominant factors that determine the rate of pesticide effect on the environment.[59]

TABLE 4.3 The Status of Pesticide Pollution Reported in Air.

Pesticides	Mean	Study area	Location	References
Methamidophos	6.37 microg m($^{-3}$)	Near Potato field	Prince Edward Island, Canada	105
α-endosulfan	307 pg m^{-3}	Taihu Lake Region	China	106
p,p′-DDD	36 pg m^{-3}	Taihu Lake Region	China	106
heptachlor (HEPT)	53 pg m^{-3}	Taihu Lake Region	China	106
p,p′-DDT	124 pg m^{-3}	Taihu Lake Region,	China	106
hexachlorobenzene (HCB)	47 pg m^{-3}	Taihu Lake Region,	China	106
Chlorothalonil	284 ng/m^3	Potato Field	Prince Edward Island, Canada	107
Diazinon	–	Urban site	Mississippi, USA	112
methyl parathion	–	Agricultural site	Mississippi, USA	112

TABLE 4.4 The Status of Pesticide Pollution Reported in Water.

Pesticides	Pesticides level	Location	References
acetamiprid	1.1 mg/L	Netherlands	76
Metolachlor	56 microg/L	Portugal	75
thiamethoxam	0.4 mg/L	Netherlands	76
Atrazine	30 microg/L	Portugal	75
Alachlor	13 microg/L	Portugal	75
(2,4-dichlorophenoxy)acetic acid	75 ng/L	Canada	56
Dichlorprop	11 ng/L	Canada	56
Mecoprop	3 ng/L	Canada	56
Diazinon	77.6–101.6 microg/L	Iran	29
Malathion	55.7–75.9 microg/L	Iran	29
α-HCH	1.28 ng/L	China	77
β- HCH	1.83 ng/L	China	77
Heptachlor epoxide	2.82 ng/L	China	77

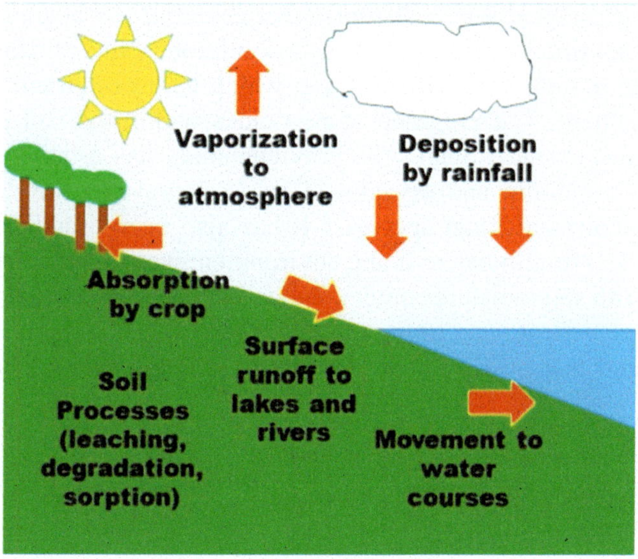

FIGURE 4.5 Pesticide behavior in the natural environment.

Source: Reprinted from Ref. [34]. https://creativecommons.org/licenses/by/4.0/

4.9.1 *IMPACTS ON TERRESTRIAL BIODIVERSITY*

Continuous and excessive pesticides use causes increase in the accumulation of chemical toxic substances within the soil profile. Many active chemical ingredients used for the development and formulations of pesticides persist in the soil for a long time. Previous studies have reported the occurrence of pesticide residues within the soil profile in different concentrations including acetochlor (16.6 g/kg), acifluorfen (10.2 g/kg), 2,4-Dichlorophenol (27 g/kg), Metolachlor (15.4 g/kg), and Norflurazon (13.9 g/kg).[102] The continuous existence of pesticides residues within the soil profile may allow the pesticides to undergo several processes such as migration, degradation, and adsorption which have a detrimental effect on the soil microflora and soil properties.[41] This also hinders and kills several beneficial soil microbes that help in nutrient recycle in the ecosystem. The application of pesticides in the terrestrial environment affects the development of plants species which helps in the nitrogen fixing process. The use of toxic pesticides such as methyl parathion, pentachlorophenol, and DDT when degraded in the terrestrial environment interferes with the activities of legume rhizobium which reduces the process of nitrogen fixation consequently affecting crop yield.[48] Apart from interfering with the nitrogen fixation processes, the persistence of pesticides also disrupts other essential biochemical processes such as ammonification by inhibiting soil microbes and nitrification.[8,41,108] Many plants possess mycorrhizal fungi which help in nutrient uptake and root development. The presence of pesticides within the soil profile may be detrimental to these fungi. The previous study has established that the presence of trifluralin and oryzalin in the soil prevents the development of some mycorrhizal fungi species.[74] It has also been established that the ability of 2,4-D to mineralize in the environment was found to be positively correlated with soil moisture content, indicating that Mediterranean soils that are arable are delicate and prone to pesticides residues accumulation.[5,109,110] This indicated that pesticides play a role in mineralization of soil organic matter.

4.9.2 IMPACTS ON AIR

Pesticides pollution of the air can occur and trigger either via post-application volatilization or pesticide drift (Figure 4.5).[102,111] Furthermore, the wide spread of pesticides in residential homes, offices is caused by air currents generated through cooling, ventilation systems, and heating. In the tropospheric ozone level, pesticides make up of 6% of the constituents.[102] According to a study by Coupea et al.,[112] a substantial concentration of pesticide accumulation (methyl parathion and diazinon) was detected in the atmosphere in agricultural and urban site in the United States. This indicates the detrimental consequences of high amount pesticide on agricultural production and the ecosystem. In other studies, it has also affirmed the high concentration of pesticides present in air. This includes studies conducted by White et al.,[107] and Garron et al.,[105] near potato field in Prince Edward Island, Canada revealing that the concentrations of Methamidophos and Chlorothalonil in the air were 6.37 microg m(−3) and 284 ng/m^3 respectively. This is considered a major threat to biodiversity.

4.9.3 IMPACTS ON AQUATIC BIODIVERSITY

When pesticides contaminate water bodies such as ground water and surface water, they cause harm to the living creatures such as flora and fauna in the aquatic environment and are also detrimental to human health, especially when the water is intended for consumption purposes. Pesticides used for agricultural production are exposed to fauna and flora in the aquatic ecosystem either directly via leaching, surface run-off, drift, drainage[24,26–2879] or indirectly via food cycle.[79] The ability of pesticides to have detrimental effects on aquatic organisms depends on four factors that include time of exposure, amount of pesticides applied, environmental persistence, and toxicity.[31] The presence of pesticides (such as Heptachlor epoxide, α-HCH, Dichlorprop, Atrazine) in water bodies has been reported by many scientists in different countries around the world. For instance, 1.28 ng/L, 1.83 ng/L, and 2.82 ng/L of α-HCH, β-HCH, and Heptachlor epoxide were detected in surface water in China.[77] Similarly, another study conducted in Iran reported that 77.6–101.6 microg/L and 55.7–75.9 microg/L of diazinon and malathion, respectively, were found in surface water.[29] Moreover, in Canada, a study conducted detected the presence of dichlorprop and mecoprop at 11 ng/L and 3 ng/L respectively in surface drinking water.[56] Furthermore, according to the US Geological Survey that reported that all stream in USA and over 85% of wells are contaminated by pesticides.[98] In another study, 19 of the 20 river basins had 2,4-D and Trifluralin in their water samples.[74] Continuous increase in the concentration of pesticides in aquatic ecosystem especially organochlorine and organophosphorus often leads to water pollution.[29,30,57] Water bodies contaminated with pesticides pose a serious threat to life forms in the aquatic habitat, because it has the potential to harm the flora presence in the habitat, reduce dissolved oxygen level in water, and bring about changes in both the behavioral and physiological character of the fish population,[7] loss of weight, reduced reproduction rate, decreased ability to withstand intense temperature and inability to escape predators.[31] When pesticides contaminate water bodies such as groundwater, a long time is required to clean up the contamination.[45]

4.10 CONCLUSION

The use of pesticides has a long history in agricultural processes and practices in many countries around the world. Most of the pesticides are classified based on toxin concentration, mode of entry, mode of action, target organism, and chemical nature. Enormous benefits have been accomplished in both agricultural production and public health via a wide range of pesticides. Moreover, the benefits achieved through the use of pesticides provide compelling and persuasive evidence for its continuous application to increase agricultural productivity, eradicate pest-related diseases, and control the activities of weeds and other parasites in the agricultural field. In regards to public health, pesticides are used to eradicate pests such as ticks, mites, mice, rats, and mosquitoes found in residential homes, shops, offices, and others. During the application of pesticides on targeted organisms, certain factors such as the behavior of pesticides (run-off, migration, degradation) in both the aquatic and terrestrial ecosystem must be considered. Despite the fact that pesticides are designed to act on specific organisms and pose minimal risk to both the terrestrial and aquatic ecosystems and humans, several reports have raised concerns about the potential risk associated with pesticide exposure on workers who are directly exposed through their occupation and the general public who are exposed through pesticide residues in food commodities and drinking water. Pesticides have polluted the environment due to their indiscriminate use and behavior in the environment.

Although the impact of pesticides usage cannot be fully eradicated in the environment, the impact of pesticide can be controlled and mitigated by following various preventive measures that include using protection and regulation policy in pesticide application, alternative cropping methods, or using spraying equipment and maintaining hygiene in the work environment. Similarly, pesticide formulations that are more effective, safe, and eco-friendly could minimize the consequence of pesticide usage. Reduction in pesticide risks can also be achieved by following the stipulated precaution and instruction regarding the use of different chemicals recommended by the producer. Public awareness regarding the use of pesticides and the potential risks associated should be intensified. Furthermore, the integrated pest management (IPM) program has proven more effective in reducing the hazard caused by pesticides usage since its adoption as compared with standard chemical or control treatment.[5,113,114]

In the future, the focus should be on determining more sustainable and eco-friendly approaches in controlling agriculture pests. This includes the utilization of chemical substances with natural bio-control agents such as natural predator animals, insects, viruses, beneficial bacteria, and nematodes. This will help in sustaining the environment.[7] Development of pest-resistant plant varieties through plant breeding approach and genetic engineering technique will help minimize the use of pesticides in agricultural production, subsequently enhancing ecosystem sustainability. Furthermore, both government and nongovernmental agencies should devote more efforts to strengthening research and development (R&D) in key areas such as sustainability, eco-friendliness, and breeding.[102]

KEYWORDS

- **pesticides**
- **agricultural production**
- **chemical compounds**
- **environment**
- **toxicity**

REFERENCES

1. Srivastava, A.; Shivanandappa, T. Haxachlorocylohexane Differentially Alters the Antioxidant Status of the Brain Regions in Rat. *Toxicology* **2005,** *214*, 123–130.
2. Skinner, J. A.; Lewis, K. A.; Bardon, K. S.; Tucker, P.; Catt, J. A.; Chambers, B. J. An Overview of the Environmental Impact of Agriculture in the U.K. *J. Environ. Manage.* **1997,** *50*, 111–128.
3. Allinson, G.; Zhang, P.; Bui, A.; Allinson, M.; Rose, G.; Marshall, S.; Pettigrove, V. Pesticide and Trace Metal Occurrence and Aquatic Benchmark Exceedances in Surface Waters and Sediments of Urban Wetlands and Retention Ponds in Melbourne, Australia. *Environ. Sci. Pollut. Res. Int.* **2015,** *22*, 10214–10226.
4. Margni, M.; Rossier, D.; Crettaz, P.; Jolliet, O. Life Cycle Impact Assessment of Pesticides on Human Health and Ecosystems. *Agric. Ecosyst. Environ.* 2001, 93, 379–392.
5. Mohd Fuad, M. J.; Junaidi, A. B.; Habibah, A.; Hamzah, J.; Toriman, M. E.; Lyndon, N.; Selvadurai, S.; Azima, A. M. *Adv. Nat. Appli. Sci.* **2012,** *6* (1), 65–70.
6. Jawale, C. A.; Rajput, K. H.; Ugale, B. J. Assessing the Impact of Pesticides: An Overview. *Int. J. Life Sci.* **2017,** *5* (3), 474–479.
7. Mahmood, I.; Imadi, S. R.; Shazadi, K.; Gul, A.; Hakeem, K. R. *Effects of Pesticides on Environment*; Springer International Publishing: Switzerland, 2016; pp. 254–266.
8. Hussain, S.; Siddique, T.; Saleem, M.; Arshad, M.; Khalid, A. Impact of Pesticides on Soil Microbial Diversity, Enzymes, and Biochemical Reactions. *Adv. Agronom.* **2009,** *102*, 160–190.
9. Shang, Y.; Hasan, M. D. K.; Ahammed, G. J.; Li, M.; Yin, H.; Zhou, J. Applications of Nanotechnology in Plant Growth and Crop Protection: A Review. *Molecules* **2019,** *24* (2558), 2–23.
10. Fernandez-Cornejo, J.; Nehring, R.; Osteen, C.; Wechsler, S.; Martin, A.; Vialou, A. *Pesticide Use in U.S. Agriculture: 21 Selected Crops, 1960–2008*; United States Department of Agriculture. Economic Information Bulletin Number, 2014, pp. 1–86.
11. Pesticides Facts. Importance and Benefits of Pesticides. https://pesticidesfacts.org/topics/necessity-of-pesticides/ (accessed Apr 4, 20201).
12. Pazikowska-Sapota, G.; Galer-Tatarowicz, K.; Dembska, G.; Wojtkiewicz, M.; Duljas, E.; Pietrzak, S.; Dzierzbicka-Glowacka, L. A. The Impact of Pesticides Used at the Agricultural Land of the Puck Commune on the Environment of the Puck Bay. *Peer J.* **2020,** *8*, e8789.

13. Grygiel, K.; Sadowski, J.; Snopczyński, T.; Wysocki, A. Herbicide Residues in Agricultural Products and the Soil. *J. Ecol. Health Environ.* **2012,** *16* (4), 159–163.
14. Bento, C. P. M.; Goossens, D.; Rezaei, M.; Riksen, M.; Mol, H. G. J.; Ritsema, C. J.; Geissen V. Glyphosate and AMPA Distribution in Wind-Eroded Sediment Derived from Loess Soil. *Environ. Pollut.* **2017,** *220*, 1079–1089.
15. Foley, J. A.; DeFries, R.; Asner, G. P.; Barford, C.; Bonan, G.; Carpenter, S. R.; Chapin, F. S.; Coe, M. T.; Daily, G. C.; Gibbs, H. K.; Helkowski, J. H.; Holloway, T.; Howard, E. A.; Kucharik, C. J.; Monfreda, C.; Patz, J. A.; Prentice, I. C.; Ramankutty, N.; Snyder, P. K. Global Consequences of Land Use. *Science* **2005,** *309*, 570–574.
16. Shi, Z. Pesticide Pollution in China. 2017, 1–28
17. Rosic, N.; Bradbury, J.; Lee, M.; Baltrotsky, K.; Grace, S. The Impact of Pesticides on Local Waterways: A Scoping Review and Method for Identifying Pesticides in Local Usage. *Environ. Sci. Pol.* **2020,** *106*, 12–21.
18. Colborn, T. Neurodevelopment and Endocrine Disruption. *Environ. Health Perspect.* **2004,** *112*, 944949
19. Alavanja, M. C., Hoppin, J. A.; Kamel, F. Health Effects of Chronic Pesticide Exposure—Cancer and Neurotoxicity. *Annu. Rev. Public Health.* **2004,** *25*, 155–197.
20. Bhardwaj, T.; Sharma, J. P. Impact of Pesticides Application in Agricultural Industry: An Indian Scenario. *Int. J. Agric. Food Sci. Technol.* **2013,** *4* (8), 817–822.
21. Dabrowski, J. M.; Shadung, J. M.; Wepener, V. Prioritizing Agricultural Pesticides Used in South Africa Based on their Environmental Mobility and Potential Human Health Effects. *Environ. Int.* **2014,** *62*, 31–40.
22. Landis, D. A.; Gardiner, M. M.; van der Werf, W.; Swinton, S. M. Increasing Corn for Biofuel Production Reduces Biocontrol Services in Agricultural Landscapes. *Proc. Natl. Acad. Sci. U S A.* **2008,** *105*, 20552–20557.
23. Singh, N. S.; Sharma, R.; Parween, T.; Patanjali, P. K. *Pesticide Contamination and Human Health Risk Factor*. Springer International Publishing AG, 2018, pp. 49–68.
24. Hildebrandt, A.; Guillamo'n, M.; Lacorte, S.; Tauler, R., Barcelo, D. Impact of Pesticides Used in Agriculture and Vineyards to Surface and Groundwater Quality (North Spain). *Water Rese.* **2008,** *42*, 3315–3326.
25. Joshua, O. Pesticides Use and Health in Nigeria. *Ife J. Sci.* **2016,** *18* (4), 981–991.
26. Kolpin, D. W.; Barbash, J. E.; Gilliom, R. J. Occurrence of Pesticides in Shallow Groundwater of the United States: Initial Results from the National Water-Quality Assessment Program. *Environ. Sci. Technol.* **1998,** *32* (5), 558–566.
27. Guzzella, L.; Pozzoni, F.; Giuliano, G.; Herbicide Contamination of Surficial Groundwater in Northern Italy. *Environ. Pollut.* **2006,** *142* (2), 344–353.
28. Papastergiou, A.; Papadopoulou-Mourkidou, E.; Occurrence and Spatial and Temporal Distribution of Pesticide Residues in Groundwater of Major Corn-Growing Areas of Greece (1996–1997). *Environ. Sci. Technol.* **2001,** *35* (1), 63–69.
29. Fadaei, A.; Dehghani, M. H.; Nasseri, S.; Mahvi, A. H.; Rastkari, N.; Shayeghi, M. Organophosphorous Pesticides in Surface Water of Iran. *Bull Environ Contam Toxicol.* **2012,** *88*, 867–869.
30. Pedersen, J. A.; Yeager, M. A.; Suffe, I. H. Organophosphorus Insecticides in Agricultural and Residential Run-Off: Field Observations and Implications for Total Maximum Daily Load Development. *Environ. Sci. Technol.* **2006,** *40*, 2120–2127.
31. Helfrich, L. A.; Weigmann, D. L.; Hipkins, P.; Stinson, E. R. Pesticides and Aquatic Animals: A Guide to Reducing Impacts on Aquatic Systems. In: Virginia Polytechnic

Institute and State University. https://pubs.ext.vt.edu/420/420-013/420-013.html (accessed Apr 2, 2015, 2009).
32. Weddle, P. W.; Welterb, S. C.; Thomson, D. History of IPM in California pears—50 Years of Pesticide use and the Transition to Biologically Intensive IPM. *Pest Manag Sci.* **2009,** 1–6.
33. Özkara, A.; Akyıl, D.; Konuk, M. Pesticides, Environmental Pollution, and Health. *Intech.* 2016, 1–27.
34. Tudi, M.; Daniel Ruan, H.; Wang, L.; Lyu, J.; Sadler, R.; Connell, D.; Chu, C.; Phung, D. T. Agriculture Development, Pesticide Application and Its Impact on the Environment. *Int. J. Environ. Res. Public Health.* **2021,** *18*, 1112.
35. Bernardes, M. F. F.; Pazin, M.; Pereira, L. C.; Dorta, D. J. Impact of Pesticides on Environmental and Human Health. In *Toxicology Studies—Cells, Drugs and Environment*; IntechOpen: London, UK, 2015; pp. 195–233.
36. Ross, G. Risks and benefits of DDT. *Lancet* **2005,** 366 (9499), 1771–1772.
37. Zahm, S. H.; Ward, M. H. Pesticides and Childhood Cancer. *Environ. Health Perspect.* **1998,** *106* (3): 893–908.
38. Zhang, K.; Zhang, B. Z.; Li, S. M.; Zeng, E. Y. Regional Dynamics of Persistent Organic Pollutants (POPs) in the Pearl River Delta, China: Implications and perspectives. *Environ. Pollut.* **2011,** *159*, 2301–2309.
39. Barnhoorn, I. E. J.; Bornman, M. S.; Jansen van Rensburg, C.; Bouwman, H. DDT Residues in Water, Sediment, Domestic and Indigenous Biota from a Currently DDT-Sprayed Area. *Chemosphere* **2009,** *77*, 1236–1241.
40. Hassaan, M. A.; El Nemr, A. Pesticides Pollution: Classifications, Human Health Impact, Extraction and Treatment Techniques. *Egypt. J. Aquat. Res.* **2020,** *46*, 207–220.
41. Yadav, I. C.; Dev, N. L. Pesticides Classification and its Impact on Human and Environment. *Environ. Sci. Engg.* **2017,** *6*, 139–158.
42. Nicolopoulou-Stamati, P.; Maipas, S.; Kotampasi, C.; Stamatis, P.; Hens L. Chemical Pesticides and Human Health: The Urgent Need for a New Concept in Agriculture. *Front. Public Health.* **2016,** *4*, 148.
43. Singh, B.; Mandal, K. Environmental Impact of Pesticides Belonging to Newer Chemistry. In *Integrated Pest Management*; Dhawan, A. K.; Singh, B.; BrarBhullar, M., Arora, R., Eds; Scientific Publishers: Jodhpur, India, 2013; pp. 152–190.
44. Barr, D. B.; Needham, L. L. Analytical Methods for Biological Monitoring of Exposure to Pesticides: A Review. *J. Chromatogr. B* **2002,** *778*, 5–29.
45. Kumar, S.; Sharma, A. K.; Rawat, S. S.; Jain, D. K.; Ghosh, S. Use of Pesticides in Agriculture and Livestock Animals and its Impact ON Environment of India. Asian *J. Environ. Sci.* **2013,** *8* (1), 51–57.
46. Sushma, D.; Dipesh, R.; Lekhendra, T.; Ram, S. S. A Review on Status of Pesticides Use in Nepal. *Res. J. Agric. Forest. Sci.* **2015,** *3* (3), 26–29.
47. Padmavathi, M. Eco Toxins and Their Impacts on Human Health, *Res. J. Agric. Forest. Sci.* **2013,** *1* (4), 10–17.
48. Sharma, A. K.; Sharma, D.; Chopra, A. K. An Overview of Pesticides in the Development of Agriculture Crops. *J. Appl. Nat. Sci.* **2020,** *12* (2), 101–109.
49. Unsworth, J. History of Pesticide Use. IUPAC-International Union of Pure and Applied Chemistry, Mai. 2010. http://agrochemicals.iupac.org/index.php?option=com_sobi2&sobi2Task=sobi2Details&catid=3&sobi2Id=31 (accessed Feb. 15, 2021).

50. Sánchez-Bayo, F. Impacts of Agricultural Pesticides on Terrestrial Ecosystems. In *Ecological Impacts of Toxic Chemicals*, 2011; pp. 63–87.
51. Yadav, I. C.; Devi, N. L.; Syed, J. H.; Cheng, Z.; Li, J.; Zhang, G.; Jones, K. C. *Current Status of Persistent Organic Pesticides Residues in Air, Water, and Soil, and their Possible Effect on Neighbouring Countries: A Comprehensive Review of India*; Elsevier, 2015; pp. 1–32.
52. Blair, A.; Ritz, B.; Wesseling, C.; Freeman, L. B. Pesticides and Human Health. *Occup. Environ. Med.* **2014,** *0* (0), 1–4.
53. Sarkar, S. K.; Bhattacharya, B. D.; Bhattacharya, A.; Chatterjee, M.; Alam, A.; Satpathy, K. K.; Jonathan M. P. Occurrence, Distribution and Possible Sources of Organochlorine Pesticide Residues in Tropical Coastal Environment of India: An Overview, *Environ. Int.* **2008,** *34*, 1062–1071.
54. Parveen, S.; Nakagoshi, N. An Analysis of Pesticide Use for Rice Management in Bangladesh. *J. Int. Dev. Cooperat.* **2001,** *8* (1): 107–126.
55. Mourato, S.; Huxtley, T. H. Evaluation Health and Environmental Impacts of Pesticide Use: Implications for the Design of Ecolabels and Pesticide Taxes. *Environ. Sci. Technol.* **2000,** *34* (8), 1456–1461.
56. Donald, D. B.; Cessna, A. J.; Sverko, E.; Glozier, N. E. Pesticides in Surface Drinking-Water Supplies of the Northern Great Plains. *Environ. Health Perspect.* **2007,** *115* (8), 1183–1191.
57. Davydov, R.; Sokolov, M.; Hogland, W.; Glinushkin, A.; Markaryan, A. The Application of Pesticides and Mineral Fertilizers in Agriculture. *MATEC Web Conf.* **2018,** *245*, 11003.
58. Damalas, C. A. Understanding Benefits and Risks of Pesticide Use. *Sci. Res. Essay.* **2009,** *4* (10), 945–949.
59. Damalas, C. A.; Eleftherohorinos, I. G. Pesticide Exposure, Safety Issues, and Risk Assessment Indicators. *Int. J. Environ. Res. Public Health*, **2011,** *8*, 1402–1419.
60. Oerke, E. C.; Dehne, H. W.; Schonbeck, F.; Weber, A. *Crop Production and Protection: Estimated Losses in Major Food and Cash Crops*; Elsevier, 1994.
61. Dhaliwal, G. S.; Jindal, V.; Dhawan, A. K. Insect Pest Problems and Crop Losses: Changing Trends. *Indian J. Ecol.* **2010,** *37* (1), 1–7.
62. Van der Werf, H. M. G. Assessing the Impact of Pesticides on the Environment. *Agr. Ecosyst. Environ.* **1996,** *60*, 81–96.
63. Atreya, K. Health Costs from Short-Term Exposure to Pesticides in Nepal. *Soc. Sci. Med.* **2008,** *67*, 511–519.
64. Tariq, M. I.; Afzal, S.; Hussain, I.; Sultana, N. Pesticides Exposure in Pakistan: A Review. *Environ. Int.* **2007,** *33*, 1107–1122.
65. Soares, W. L.; Porto, M. F. D. Estimating the Social Cost of Pesticide Use: An Assessment from Acute Poisoning in Brazil. *Ecol. Econ.* **2009,** *68*, 2721–2728.
66. Maroni, M.; Fanetti, A. C.; Metruccio, F. Risk Assessment and Management of Occupational Exposure to Pesticides in Agriculture. *Med. Lav.* **2006,** *97*, 430–437.
67. Pimentel, D. Environmental and Economic Costs of the Application of Pesticides Primarily in the United States. *Environ. Dev. Sustain.* **2005,** *7*, 229–252.
68. Wilson, C.; Tisdell, C. Why Farmers Continue to Use Pesticides Despite Environmental, Health and Sustainability Costs. *Ecol. Econ.* **2001,** *39*, 449–462.
69. Dhawan, A. K.; Saini, S. First Record of Phenacoccus Solenopsis Tinsley (Homoptera: Pseudococcidae) on Cotton in Punjab. *J. Insect Sci.* **2009,** *22* (3):309–310.

70. Mohindru, B.; Jindal, V.; Dhawan, A. K. Record of Parasitoid on Mealy Bug Phenacoccus Solenopsis in Tomato. *Indian J. Ecol.* **2009,** *36* (1), 101–102.
71. Stone, B. Developments in Agricultural Technology. *China Quart.* **1988,** *110*, 767–822.
72. Kroes, R.; Muller, D.; Lambe, J.; Lowik, M. R. H.; van, Kleiner, J.; Massey, R.; Mayer, S.; Urieta, I.; Verger, P.; Visconti, A. Assessment of Intake from the Diet. *Food Chem. Toxicol.* **2002,** *40*, 327–385.
73. Cooper, J.; Dobson, H. The Benefits of Pesticides to Mankind and the Environment. *Crop Protect.* **2007,** *26*, 1337–1348.
74. Aktar, M. W.; Sengupta D.; Chowdhury, A. Impact of Pesticides Use in Agriculture: Their Benefits and Hazards. *Interdisc Toxicol.* **2009,** *2* (1): 1–12.
75. Cerejeiraa, M. J.; Vianab, P.; Batistaa, S.; Pereiraa, T.; Silvaa, E.; Valerio, M. J.; Silvaa, A.; Ferreirab, M.; Silva-Fernandesa, A. M. Pesticides in Portuguese Surface and Ground Waters. *Water Res.* **2003,** *37*, 1055–1063.
76. Sjerps, R. M. A.; Kooij, P. J. F.; Loon, A. V.; Van Wezel, A. P. Occurrence of Pesticides in Dutch Drinking Water Sources. *Chemosphere* **2019,** *235*, 510e518.
77. Zhou, R.; Zhu, L.; Yang, K.; Chen, Y. Distribution of Organochlorine Pesticides in Surface Water and Sediments from Qiantang River, East China. *J. Haz. Mater.* **2006,** *A137*, 68–75.
78. Gomes, J.; Lloyd, O. L.; Revitt, D. M. The Influence of Personal Protection, Environmental Hygiene and Exposure to Pesticides on the Health of Immigrant Farm Workers in a Desert Country. *Int. Arch. Occup. Environ. Health* **1999,** *72*, 40–45.
79. Maksymiv, I. Pesticides: Benefits and Hazards. *J. Vasyl Stefanyk Precarpathian Natl. Univ.* **2015,** *2* (1), 70–76.
80. Austin R. B. Yield of Wheat in the United Kingdom: Recent Advances and Prospects. *Crop Sci.* **1999,** *39* (6), 1604–1610.
81. Kucharik, C. J.; Ramankutty, N. Trends and Variability in US Corn Yields Over the 20th Century. *Earth Interac.* **2005,** *9* (2005), 1–29.
82. Webster, J. P. G.; Bowles, R. G.; Williams, N. T. Estimating the Economic Benefits of Alternative Pesticide Usage Scenarios: Wheat Production in the United Kingdom. *Crop Prot.* **1999,** *18*, 83.
83. Robinson, D. E.; Mansingh, A.; Dasgupta, T. P. Fate and Transport of Ethoprophos in the Jamaican Environment. *Sci. Tot. Environ.* **1999,** *238*, 373–378.
84. Oerke, E. C. Century Review Crop Losses to Pests. *J. Agric. Sci.* **2006,** *144*, 31–43.
85. Zadoks, J. C.; Schenin, R. D. *Epidemiology and Plant Disease Management*; Oxford University Press: Oxford, 1979.
86. Behera, B.; Singh, S. G. Studies on Weed Management in Monsoon Season Crop of Tomato. *Indian J. Weed. Sci.* **1999,** *31* (1–2), 67.
87. Garry, V. F.; Harkins, M. E.; Erickson L. L.; Long Simpson, L. K.; Holland, S. E.; Burroughs, B. L. Birth Defects, Season of Conception, and Sex of Children Born to Pesticide Applicators Living in the Red River Valley of Minnesota, USA. *Environ. Health. Perspect.* **2002,** *110*, 441–449.
88. Garry, V. F.; Harkins, M. E.; Lyubimov, A.; Erickson, L. L.; Long, L. Reproductive Outcomes in the Women of the Red River Valley of the North. I. The Spouses of Pesticide Applicators: Pregnancy Loss, Age at Menarche, and Exposure to Pesticides. *J. Toxicol. Environ. Health. A* **2002,** *65* (11), 769–786.
89. Loffredo, C. A.; Silbergeld, E. K.; Ferencz, C.; Zhang J. Association of Transposition of the Great Arteries in Infants with Maternal Exposures to Herbicides and Rodenticides. *Am. J. Epidemiol.* **2001,** *153* (6), 529–536.

90. Shaw, G. M.; Wasserman, C. R.; O'Malley, C. D.; Nelson, V.; Jackson R. J. Maternal Pesticide Exposure from Multiple Sources and Selected Congenital Anomalies. *Epidemiology* **1999,** *10* (1), 60–66.
91. Basil, K. L.; Vakil, C.; Sanborn, M.; Cole, D. C.; Kaur, J. S.; Kerr, K. J. Cancer Health Effects of Pesticides. *Can. Fam. Physician* **2007,** *53* (10), 1704–1711.
92. Sanborn, M.; Kerr, K. J.; Sanin L. H.; Cole D. C.; Basil, K. L.; Vakil, C. Non-Cancer Health Effects of Pesticides. *Can. Fam. Physician.* **2007,** *53* (10), 1712–1720.
93. Spiewak, R. Pesticides as a Cause of Occupational Skin Diseases in Farmers. *Ann. Agri. Environ. Med.* **2001,** *8* (1), 1–5.
94. Grewal, A. S.; Singla, A.; Kamboj, P.; Dua, J. S. Pesticide Residues in Food Grains, Vegetables and Fruits: A Hazard to Human Health. *J. Med. Chem. Toxicol.* **2017,** *2* (1). 1–7.
95. Parameswari, E.; Davamani, V.; Ilakiya, T.; Arulmani, S.; Prithiv Raj, V. *Imp. Pest. Environ.* **2020,** *2* (5), 136–138.
96. Salvatore, A. L.; Bradman, A.; Castorina, R.; Camacho, J.; López, J.; Barr, D. B.; Eskenazi, B. Occupational Behaviors and Farmworkers' Pesticide Exposure: Findings from a Study in Monterey County, California. *Am. J. Indus. Med.* **2008,** *51* (10): 782–794.
97. Kumar, V.; Kumar, P. Pesticides in Agriculture and Environment: Impacts on Human Health. I *Contaminants in Agriculture and Environment: Health Risks and Remediation*; 2019, 77–91.
98. Sharma, D. R.; Thapa, R. B.; Manandhar, H. K.; Shrestha, S. M.; Pradhan, S. B. Use of Pesticides in Nepal and Impacts on Human Health and Environment. *J. Agric. Environ.* **2012,** *13*, 1–8.
99. Meenakshi; Sharon, P.; Bhawana, M.; Anita, S.; Gothecha, V. K. A Short Review on How Pesticides Affect Human Health. *Int. J. Ayurvedic Herbal Medic.* **2012,** *5*, 935–946.
100. Bolognesi, C.; Parrini, M.; Bonassi, S.; Lanello, G.; Salanito, A. Cytogenetic Analysis of a Human Population Occupationally Exposed to Pesticides. *Mutat Res.* **1993,** *285*, 239–249.
101. Anderson, S. E.; Meade, B. J. Potential Health Effects Associated with Dermal Exposure to Occupational Chemicals. *Environ. Health Insights.* **2014,** *8*, 51–62.
102. Kim. K.; Kabir, E.; Jahan, S. A. Exposure to Pesticides and the Associated Human Health Effects. *Sci. Tot. Environ.* **2017,** *575*, 525–535.
103. Falck G.; Hirvonen, A.; Scarpato, R.; Saarikoski, S.; Migliore, L.; Norppa, H. Micronuclei in Blood Lymphocytes and Genetic Polymorphism GSTM1, GSTT1 and NAT2 in Pesticide Exposed Greenhouse Workers. *Mutat. Res.* **1999,** *441,* 225–237.
104. Martínez-Valenzuela, C.; Gómez-Arroyo, S.; Villalobos-Pietrini, R.; Waliszewski, S.; Calderón-Segura, M. E.; Félix-Gastélum, R.; Álvarez-Torres, A. Genotoxic Biomonitoring of Agricultural Workers Exposed to Pesticides in the North of Sinaloa State, Mexico. *Environ. Int.* **2009,** *35*, 1155–1159.
105. Garron, C. A.; Davis, K. C.; Ernst, W. R. Near-Field Air Concentrations of Pesticides in Potato Agriculture in Prince Edward Island. *Pest Manag. Sci.* **2009,** *65* (6), 688–696.
106. Qiu, X.; Zhu, T.; Li, J.; Pan, H.; Li, Q.; Miao, G.; Gong, J. Organochlorine Pesticides in the Air Around the Taihu Lake, China. *Environ. Sci. Technol.* **2004,** *38* (5), 1368–1374.
107. White, L. M.; Ernst, W. R.; Julien, G.; Garron, C.; Leger, M. Ambient air Concentrations of Pesticides Used in Potato Cultivation in Prince Edward Island, Canada. *Pest Manag. Sci.* **2006,** *62* (2), 126–136.

108. MunozLeoz, B.; RuizRomera, E.; Antiguedad, I.; Garbisu, C. Tebuconazole Application Decreases Soil Microbial Biomass and Activity. *Soil Biol. Biochem.* **2011,** *43,* 2176–2183.
109. Singh, N.; Singh, S. B. Translocation and Degradation of Pyrazosulfuron-Ethyl in Rice Soil. *Pest Manag. Sci.* **2011,** *67* (11), 1451–1456.
110. Bouseba, B.; Zertal, A.; Beguet, J.; Rouard, N.; Devers, M.; Martin, C.; Martin-Laurent, F. Evidence for 2.4-D Mineralization in Mediterranean Soils: Impact of Moisture Content and Temperature. *Pest Manag. Sci.* **2009,** *65,* 1021–1029.
111. Rull, R. P.; Ritz, B. Historical Pesticide Exposure in California Using Pesticide Use Reports and Land-Use Surveys: An Assessment of Misclassification Error and Bias. *Environ. Health Perspect.* **2003,** *111* (13), 1582–1589.
112. Coupea, R. H.; Manning, M. A.; Foreman, W. T.; Goolsby, D. A.; Majewskid, M. S. Occurrence of Pesticides in Rain and Air in Urban and Agricultural Areas of Mississippi, April-September 1995. *Sci. Tot. Environ.* **2000,** *248,* 227–240.
113. Reddy, G. V. P.; Comparative Effect of Integrated Pest Management and Farmers' Standard Pest Control Practice for Managing Insect Pests on Cabbage (Brassica Spp). *Pest Manag. Sci.* **2011,** *67* (8), 980–985.
114. Kronmann, P.; Pradel, W.; Cole, D.; Taipe, A.; Forbes, G. A. Use of the Environmental Impact to Estimate Health and Environmental Impacts of Pesticide Usage in Peruvian and Ecuadorian Potato production. *J. Environ. Protect.* **2011,** *2* (5), 581–591.

CHAPTER 5

Effect of Environmental Pollution on the Generation of Reactive Oxygen and Nitrogen Species in Plant Tissues

SUMMIA REHMAN[1*], SUBZAR AHMAD NANDA[1], SHAZIA QURAT UL AIEN[2], ISHFAQ UL REHMAN[1], and TAJAMUL ISLAM[1]

[1]*Department of Botany, University of Kashmir Srinagar 190006, J&K, India*

[2]*Department of Zoology, University of Kashmir Srinagar 190006, J&K, India*

[*]*Corresponding author. E-mail: sumaiyarehman348@gmail.com*

ABSTRACT

Pollution is linked with various deleterious consequences on plant growth and development. Environmental pollutants, like heavy metals, pesticides nanomaterials, boost the development of reactive oxygen and nitrogen species (RONS) within plant tissues. Under natural conditions, plants collaborate carefully with the surrounding atmosphere, soil, and water, but unfortunately RONS extremely alter their biochemistry. Investigators have completely studied the importance of RONS in plants and oxidative variations to cellular constituents. For instance, the reactive nitrogen species, i.e., nitric oxide (NO) has been shown to reduce the oxidative stress marker accumulation in wheat roots up to 6 hours. Other RONS such as hydrogen peroxide, superoxide anion, hydroxyl radicals, and nitrogen (II) oxide have been found to participate in numerous metabolic pathways. Engineered nanoparticles exerting toxicity at greater concentrations could mimic the effects of respective heavy metals on reactive oxygen species. This chapter provides an overview of the effect of environmental pollution on the generation of reactive oxygen and nitrogen species (RONS), their metabolism and regulation in plant tissues.

Environmental Pollution Impact on Plants: Survival Strategies under Challenging Conditions. Tariq Aftab, PhD & Khalid Rehman Hakeem, PhD (Eds.)

5.1 INTRODUCTION

The reactive oxygen and nitrogen species (RONS) is a combined term for two types of radicals namely ROS (reactive oxygen species) and the RNS (reactive nitrogen species) that are made in response to biotic and ecological strains in plants. Numerous environmental pollutants affect their production and influence their proper functioning and utility. Pollutants like lethal metal oxides, heavy metals, and nano-particles cause deep distressing conditions. Research shows that they cause distress in plants by encouraging the production of reactive oxygen and nitrogen species (RONS) inside their tissues.[1] The ROS includes oxygen radicals, such as hydroxyl (OH^-) and superoxide (O_2^-) radicals, and other nonradicals like singlet oxygen ($^1O_2$), hydrogen peroxide (H_2O_2). Likewise, RNS comprises nitrogen radicals like nitric dioxide ($NO_2\cdot$) and nitric oxide (NO•), and nonradicals like dinitrogen tetroxide (N_2O_4) and nitrous acid (HNO_2). Over the past three decades and more so, studies show a greater link between environmental pollutants and plant tolerance. More specifically, it has been plant tolerance against that toxicity which has been studied intensely.[2] From such studies it has become obvious that both ROS and NO at low concentrations help in protecting cellular homeostasis. This tendency in turn makes them key participants in the normal cellular signal transduction.[3] Furthermore, soil impurities cause oxidative stress in organisms which is a specific kind of plant reactions to diverse environmental elements. Besides, soil impurities lead to a shortage of some mineral salts like phosphate.[5] Likewise in soil, mixing of pesticides creates RONS intensively.[4] And lead pollutant causes severe impacts. Plants uptake lead, mostly by their root systems[6] and it induces the heavy-metal-related oxidative stress in them.[6]

Apart from heavy metals, there is another category of environmental pollutants comprised of nanomaterials. Nanomaterials is a group of environmental pollutions possessing potential to induce oxidative stress in plants. It has been found that on daily basis plants face severe kind of nanomaterials. The prime nanomaterials include Al_2O_3, CeO_2, SiO_2, TiO_2, ZnO, carbon nanotubes, fullerenes, and quantum dots.[7] On account of their synthetic character these nanomaterials are called engineered nanomaterials (ENMs) and they cause severe damage to the plants. ENM like Al_2O_3 (400–4000 mg/L) causes root elongation.[8] Inhibition of root hydraulic conductivity is the outcome of Fe_3O_4.[9] Decrease in biomass is caused by TiO_2,[10] CuO results in chlorotic symptoms during seedling exposure, inhibition of root elongation, and reduced biomass.[11] SiO_2 has been found to cause increase in grain weight and shoot biomass.[12] Cheng[13] observed root elongation caused by Gd_2O_3.

More interestingly, plants, considered primary producers on the surface of earth, are developing their own strategy to overcome, to a greater extent, the increasing concentrations of environmental pollutants. They seem to be resilient to growing quantities of heavy metals. It has been observed that they display a unique strategy of hyper accumulating these heavy metals. They usually try to contain these heavy metals within their vacuoles. Most often, if not always, they attach the heavy metals to cell walls or else absorb the heavy metal ions into the cell apoplast.[14] Nonetheless, the heavy metal pollutants are rendering plants less productive and lean in growth with a greater susceptibility to various diseases. The increasing use of heavy metals in nanoparticles has caused its own environmental issues. Their increasing use with an eye on increasing crop production and number of other industrial applications has become a trend in the last few decades. All eventually are going to damage the environment in multiple ways without making things much visible in a shorter span of time.[15] Their gradual impact is causing a heavy metal pollution spectrum[16] leading to the constant exposure of flora to these intruding nanoforms widely distributed in soil, water, and aerosol.

Heavy metals upon their accumulation in the soil disturb soil biology as well as its broader ecology. Studies show a great multitude of effects of pollution caused by heavy metals. The heavy metal pollution, for instance, has been found a prime factor in inhibiting the plant growth. Besides, they have an alarming effect on crop production and produce. In worst situations they even cause death of plants.[17] Studies also show the phototoxic type of effects of heavy metals. Generally ascribed to pro-oxidative effects of heavy metal ions, it is believed that phototoxic effects disturb various biochemical and physiological processes.[17]

Nickel, for instance, is considered one of the important heavy metals. It is taken up easily by plants on account of its being as an essential trace metal. Its excessive quantity causes unexpected results. Its consumption beyond the need causes within plant cells an upsurge of ROS. The upsurge like situation affects their survival through extensive physicochemical as well as genomic damages.[18] The overwhelming impact of nickel consumption shows that plant systems suffer sudden fluxes of NiO-NP in water, soil, or aerosol.[19]

5.2 ENVIRONMENTAL POLLUTION AND GENERATION OF RONS

The process of formation of reactive oxygen and nitrogen species (RONS) is attributed to soil contamination and environmental pollution. Soil contamination caused by heavy metals is significantly creating RONS. In recent decades the unending use of nanomaterials has enhanced further this process of soil contamination and environmental pollution that adds tremendously in the production and proliferation of RONS. Likewise, pesticides possess potential to produce unimaginable stock of RONS. Additionally, the nitrogen oxides yield enough RONS to deteriorate further the soil quality.[1] It has been reported that the endogenous sources of RONS include xanthine oxidase, auto-oxidant reactions, inflammations, redox cycles, oxidative reactions in phagocytic cells, and cytochrome P_{450} reactions. Environmental toxins, chemicals, and smog are considered some of the other important exogenous sources of RONS.[20]

The plant mainly absorbs heavy metals through the root system, but absorption through the lamina is also possible. It has been shown that different kinds of RONS are generated by diverse heavy metals in various plants for instance the induction of H_2O_2 in *Piston sativum, Ipomoea batatas, Vicia faba, Lapinus luteus, Cucumis sativus, Withania somnifera, Lycopersicon lycoopersicum,* and *Alocasia macrorrhiza* by Zn, Pb, Mn, Cu, and Cd,[21,22,24.26,27,28] similarly the induction of superoxide ($O_2^{\cdot-}$) in *Mytilus galloprovincialis, Lapinus luteus, Cucumis sativus, Withania somnifera,* and *Alocasia macrorrhiza* by Zn, Pb, Mn, Cu, and Cd[21,23,25,26,27] and induction of hydroxyl (.OH) radicals in *Alocasia macrorrhiza* by Cd.[29]

Rizwan[30] argues that plants under the impact of heavy metals induce oxidative challenges. These oxidative challenges adopt pathways specific to a particular metal and conditions. Such effects culminate in an unbalanced pattern of production and neutralization of ROS such as hydroxyl ion, H_2O_2, and superoxides.[31] Studies show that H_2O_2 is an uncharged nonradical, having both oxidizing and reducing properties and is selectively reactive. All these properties make it important in the mitigation of energy-efficient stress.[32] It is equally worthwhile mentioning that in exposed tissues heavy metals like nickel, due to being a nonredox active metal, indirectly lead to an increase in intracellular ROS levels. For Nickel, this becomes possible by inhibiting specific enzymes or else by diminishing cellular Glutathione (GSH) pool. Sometimes, nickel accumulation affects NADPH oxidase and turns over the cellular antioxidant profiles. All eventually develops a ROS furor.[33] Similarly, the production process of various nitric oxide (NO) mechanisms witnesses an accumulation due to augmented ROS.[3]

5.3 RONS PRODUCTION AND REGULATION

RONS production is also caused by both abiotic and biotic stresses in addition to various environmental pollutants discussed above. In most of the plants, both abiotic and biotic stresses culminate in the formation of RONS.[1] It has been reported that some plant organelles like mitochondria and chloroplasts manufacture RONS.[34] A few studies show that the main RONS formation largely takes place in apoplast provoked by NADPH oxidase.[35] Hydroxyl radical, one of the RONS species, is considerably highest in terms of reactivity toxicity.[36] RONS production is a complex process shaped by a set of factors. It is believed that plants in the presence of special enzymes, like ascorbate in the chloroplast stroma (enzyme concentration 2–3 mM), or secondary metabolites, such as the hypecins in *Hypericum hirsutum*, reduce RONS levels.[37] The physiological concentration of the superoxide anion radical is considered to be ca. 10−11 M,[38] while the value for hydrogen peroxide is ca. 10−7 M,[39] for hydroxyl radical ca. 10−10 M[40] and for singlet oxygen ca. 10−9 M.[41]

RONS production causes severe implications. It has been examined that in the presence of RONS heavy metals occurring in the environment become toxic and harmful for organisms.[42] More importantly, the strongly polluted soil increases the toxic effect unprecedentedly. Through the process of increasing toxicity, human life is also getting affected. Harmful effects caused by heavy metals like cadmium and lead have been attested in a number of studies.[43] Such effects are increasing as the plants amid the extreme metallic environments get adopted only to become hyperaccumulators.[44]

For plants, pollution caused by heavy metals is something making them a composite of great variability surviving in challenging conditions. Heavy metals cause carbohydrate concentrations in plants by the tangible levels of RONS-induced oxidative stress. Lehner[45] ascribes pollen abortion in *Triticum aestivum* to the reaction of RONS with carbohydrates. Moreover, the visible decrease in the pollen viability of *Oryza sativa* is believed to be an outcome of the reaction between RONS and carbohydrates.[46] Similarly, RONS affect the carbohydrate transporter genes as heavy metals reduce the oxidative stress in plant cells.[47] What is more alarming is the process in which RONS affect the carbohydrates structures as on account of heavy metals, plant cells undergo oxidative stress like conditions.[48]

Pourrut[49] argues that the heavy metals have also some indirect effects. It has been observed that heavy metals have more toxic effects on macromolecules in plant cells. ROS on account of being initiators of the toxicity causes more damage. They even affect process like protein oxidation as they tend to modify the qualitative composition of cell proteins. It has been reported that heavy metals more directly bind to proteins and eventually cause a denature effect.[50] In another study, Beltagi reported exposure to lead in *Vicia faba* affects protein composition of root cells.[51] Heavy metal pollution causes structural modifications in enzymes. Enzymes such as serine hydroxymethyl transferase, isocitrate lyase, cysteine proteinase, and arginine decarboxylase undergo structural transformation induced by heavy metal pollution.[52] Heavy metals take a toll of soil quantity and its capability to endure plant diversity, their development, beneficial activity, diversity, and growth of soil-microorganisms as well as other species.[53] Additionally, ROS induces DNA damages, base deletions, cross-links, base modifications, strand breaks, and pyrimidine dimers.[54] Specific oxidation also, at the mRNA level, has been reported.[55] Moreover, heavy metals induced ROS causes lipid peroxidation as well as inhibits the protein functions.

Growing parts of plants on being exposed to the lead generate ROS that affects their growth and development severely.[56] It has been found that due to Lead ions ROS generation process accelerates in *Oryza sativa* and exposure to Pb_2^+ increases lipid peroxides of ca. 177%[57] which in turn causes peroxidation of lipids, decreases saturated fatty acids and multilayered oxidative damages.[58] In a study crops like rice seedlings and soybean have reported induced peroxidase activity which is classified as a development of stress enzyme in plants.[59] Release of peroxidase found in cell wall is essential for Lead-induced peroxidase activity,[60] interestingly the increased superoxide dismutase (SOD) synthesis is also a lead-induced activity.[61] Lead causes oxidative stress conditions to accelerate activities of glutathione reductase and ascorbate peroxidase in plant cells.[62]

5.4 ROLE OF RONS IN PLANT METABOLISM

RONS play a vital role in the metabolic rate of plants. For example, they play a role in disease resistance,[63] regulate apoptosis,[64] control plant growth,[65] induce mitogen-activated protein kinase cascades,[66] and also improve seed inactivity.[67] Numerous enzymes, like respiratory burst oxidase homologues (Rbohs), can yield ROS.[68] RONS when formed in a cell are considered a signaling molecule.[69] Hydrogen peroxide, a byproduct of Rbohs, has a vital role in cell signaling. Recent findings about the mechanisms of reactions between several metabolic products, RONS, and biological molecules have revealed that hydrogen peroxide has a role in cell signaling and acts as an indicator of stress in plants.[69] H_2O_2 plays a role in defense reactions, regulation of cell expansion, and closure of stomata.

In plants, one of the ways of defense reactions involves intense production of RONS, which is the result of enhanced enzymatic action by amine oxidases, cell-wall-bound peroxidase, and Rbohs.[70] Regardless of the common faith that hydrogen peroxide is the foremost molecule in cell signaling, it must be noted that free radicals have a role in thiol-based redox signaling.[71]

A free radical bioactive molecule, Nitric Oxide (NO), plays an important part in plant signaling systems.[72] NO has a role to improve heavy metal stress, as described in cases of arsenic, cadmium, copper aluminum exposures.[73] Comparable opinions were described in cases of exposures to silver and zinc oxide nanoparticles.[74]

In recent times, the relations between ·NO and ·NO-derived molecules with unsaturated fatty acids in plants have been defined.[75] This is the landmark in the field because so far in animals only these interactions between ·NO and unsaturated fatty acids have been studied. The findings from such studies revealed that nitro-fatty acids are novel key mediators of ·NO metabolism in plants.[75] Plants adjust to environmental variations[76] and the consequence of such variations is alterations in the ·NO concentrations in plant tissues. ·NO has a role in the developmental practices of plants.[77] ·NO has a role in limiting the invasiveness of pathogens. During water scarcity, water stress encourages greater manufacture of intrinsic proteins that are generated only in the presence of ·NO. Remarkably, ·NO can react antagonistically against RONS.[78] ·NO occurs as an antioxidant in cytotoxic conditions in plant cells.

5.5 BIOCHEMICAL MECHANISMS OF PLANT PROTECTION AGAINST THE HEAVY METAL ACTIVITY

Plants are continuously exposed to heavy metals because of greater action of humans in nature; as a result plants have established the biochemical mechanisms of defense against the heavy-metal toxicity. In the epidermis heavy-metal detoxification processes occur,[79] likewise in the cuticule,[80] and in trichomes.[81] Plants grow continually to safeguard their cells against the oxidative damage caused by the heavy metals.[82] Particular plant defense mechanism against the RONS formation is determined by the type of plant species, the kind of metal involved, and also on the plant's maturity. Vacuolar sequestration is the principal defense mechanism against the heavy metals' toxicity, which allows us to control the contrary effects of the heavy metals.[49] In reaction to the adverse consequences of heavy metals the thickness of roots increases, resulting in a reduction in the infiltration of the metals.[83] Chromium and copper, also called redox-active metals, may openly distress the reactive oxygen species (ROS) framework of a cell by instinctively generating an oxidative burst, whereas non-redox metals, like mercury, nickel cadmium, can encourage toxicity both by blocking the functional groups and by exchanging cations from essential biomolecules.[84] Andrade[85] witnessed an increased cell wall thickness in *Enteromorpha flexuosa* upon exposure to copper. Along plasma membranes of Sesbania root cells the deposits of lead have been shown.[86] Krzesłowska[87] stated a reduced penetration of lead into root cells of *Funaria hygrometrica* owing to enlarged cell wall thickness.

The subjects discussed above discuss information on the mechanisms over which RONS are created, for instance, as a consequence of agricultural development, and also the processes that support their damaging influence on the oxidative stress in cells. The oxidative stress has proved to encourage the expansion of diseases of numerous etiologies and to obscure treatments,[88] that is why understanding the production processes and the action of RONS in plant and human environments is very significant.

KEYWORDS

- **environmental pollution**
- **reactive oxygen and nitrogen species**
- **engineered nanoparticles**

REFERENCES

1. Drzeżdżon, J.; Jacewicz, D.; Chmurzyński, L. The Impact of Environmental Contamination on the Generation of Reactive Oxygen and Nitrogen Species—Consequences for Plants and Humans. *Environ. Int.* **2018,** *119*, 133–151.
2. Shah, F. U. R.; Ahmad, N.; Masood, K. R.; Peralta-Videa, J. R. Heavy Metal Toxicity in Plants. In *Plant Adaptation and Phytoremediation*; Springer: Dordrecht, 2010; pp. 71–97.
3. Romero-Puertas, M. C.; Sandalio, L. M. Nitric Oxide Level is Self-Regulating and Also Regulates its ROS Partners. *Front. Plant Sci.* **2016,** *7*, 316.
4. Gülçin, I.; Elmastaş, M.; Aboul-Enein, H. Y. Determination of Antioxidant and Radical Scavenging Activity of Basil (Ocimum basilicum L. Family Lamiaceae) Assayed by Different Methodologies. *Phytother. Res.* **2007,** *21* (4), 354–361.
5. Gill, S. S.; Tuteja, N. Reactive Oxygen Species and Antioxidant Machinery in Abiotic Stress Tolerance in Crop Plants. *Plant Physiol. Biochem.* **2010,** *48* (12), 909–930.
6. Kaur, G. Pb-induced Toxicity in Plants: Effect on Growth, Development, and Biochemical Attributes. *J. Glob. Biosci.* **2014,** *3* (6), 881–889.
7. Zhu, X.; Wang, J.; Zhang, X.; Chang, Y.; Chen, Y. Trophic Transfer of TiO2 Nanoparticles from Daphnia to Zebrafish in a Simplified Freshwater Food Chain. *Chemosphere* **2010,** *79* (9), 928–933.

8. Garg, N.; Manchanda, G. ROS Generation in Plants: Boon or Bane? *Plant Biosyst.* **2009,** *143* (1), 81–96.
9. Van Breusegem, F.; Dat, J. F. Reactive Oxygen Species in Plant Cell Death. *Plant Physiol.* **2006,** *141* (2), 384–390.
10. Lopez-Delgado, H.; Dat, J. F.; Foyer, C. H.; Scott, I. M. Induction of Thermotolerance in Potato Microplants by Acetylsalicylic Acid and H_2O_2. *J. Exp. Bot.* **1998,** *49* (321), 713–720.
11. Thannickal, V. J.; Fanburg, B. L. Reactive Oxygen Species in Cell Signaling. *Am. J. Physiol. Lung. Cell. Mol. Physiol.* **2000,** *279* (6), L1005–L1028.
12. Kobayashi, M.; Ohura, I.; Kawakita, K.; Yokota, N.; Fujiwara, M.; Shimamoto, K.; Yoshioka, H. Calcium-Dependent Protein Kinases Regulate the Production of Reactive Oxygen Species by Potato NADPH Oxidase. *Plant Cell* **2007,** *19* (3), 1065–1080.
13. Cheng, G.; Cao, Z.; Xu, X.; Van Meir, E. G.; Lambeth, J. D. Homologs of gp91phox: Cloning and Tissue Expression of Nox3, Nox4, and Nox5. *Gene* **2001,** *269* (1–2), 131–140.
14. Vázquez, M. D.; Barceló, J.; Poschenrieder, C. H.; Madico, J.; Hatton, P.; Baker, A. J. M.; Cope, G. H. Localization of Zinc and Cadmium in Thlaspi Caerulescens (Brassicaceae), a Metallophyte that can Hyperaccumulate both Metals. *J. Plant Physiol.* **1992,** *140* (3), 350–355.
15. Rizwan, M.; Ali, S.; Ali, B.; Adrees, M.; Arshad, M.; Hussain, A.; Waris, A. A. Zinc and Iron Oxide Nanoparticles Improved the Plant Growth and Reduced the Oxidative Stress and Cadmium Concentration in Wheat. *Chemosphere* **2019,** *214*, 269–277.
16. Mustafa, G.; Komatsu, S. Toxicity of Heavy Metals and Metal-Containing Nanoparticles on Plants. *Biochimica et Biophysica Acta* **2016,** *1864* (8), 932–944.
17. Gratão, P. L.; Polle, A.; Lea, P. J.; Azevedo, R. A. Making the Life of Heavy Metal-Stressed Plants a Little Easier. *Funct. Plant Biol.* **2005,** *32* (6), 481–494.
18. Pinto, M.; Soares, C.; Pinto, A. S.; Fidalgo, F. Phytotoxic Effects of Bulk and Nano-Sized Ni on Lycium Barbarum L. Grown in Vitro—Oxidative Damage and Antioxidant Response. *Chemosphere* **2019,** *218*, 507–516.
19. Ovais, M.; Khalil, A. T.; Ayaz, M.; Ahmad, I. Metal Oxide Nanoparticles and Plants. In *Phytonanotechnology: Challenges and Prospects*; Elsevier: Amsterdam, **2020,** pp. 123–141.
20. Wang, Z.; Li, R.; Liu, Y.; Liu, X.; Chen, W.; Xu, S.; Wang, C. Protective Effects of Edaravone Combined Puerarin on Inhalation Lung Injury Induced by Black Gunpowder Smog. *Int. Immunopharmacol.* **2015,** *26* (1), 125–132.
21. Turchi, A.; Tamantini, I.; Camussi, A. M.; Racchi, M. L. Expression of a Metallothionein A1 Gene of Pisum Sativum in White Poplar Enhances Tolerance and Accumulation of Zinc and Copper. *Plant Sci.* **2012,** *183*, 50–56.
22. Kim, Y. H.; Lee, H. S.; Kwak, S. S. Differential Responses of Sweetpotato Peroxidases to Heavy Metals. *Chemosphere* **2010,** *81* (1), 79–85.
23. Koutsogiannaki, S.; Evangelinos, N.; Koliakos, G.; Kaloyianni, M. Cytotoxic Mechanisms of Zn2+ and Cd2+ Involve Na+/H+ Exchanger (NHE) Activation by ROS. *Aqu. Toxicol.* **2006,** *78* (4), 315–324.
24. Shahid, M.; Dumat, C.; Silvestre, J.; Pinelli, E. Effect of Fulvic Acids on Lead-Induced Oxidative Stress to Metal Sensitive Vicia Faba L. Plant. *Biol. Fert. Soils* **2012,** *48* (6), 689–697.
25. Kopyra, M.; Gwóźdź, E. A. Nitric Oxide Stimulates Seed Germination and Counteracts the Inhibitory Effect of Heavy Metals and Salinity on Root Growth of Lupinus Luteus. *Plant Physiol. Biochem.* **2003,** *41* (11–12), 1011–1017.

26. Shi, Q.; Zhu, Z. Effects of Exogenous Salicylic Acid on Manganese Toxicity, Element Contents and Antioxidative System in Cucumber. *Environ. Exp. Bot.* **2008,** *63* (1–3), 317–326.
27. Khatun, S.; Ali, M. B.; Hahn, E. J.; Paek, K. Y. Copper Toxicity in Withania Somnifera: Growth and Antioxidant Enzymes Responses of In Vitro Grown Plants. *Environ. Exp. Bot.* **2008,** *64*(3), 279–285.
28. Liu, N.; Lin, Z. F.; Lin, G. Z.; Song, L. Y.; Chen, S. W.; Mo, H.; Peng, C. L. Lead and Cadmium Induced Alterations of Cellular Functions in Leaves of Alocasia Macrorrhiza L. Schott. *Ecotoxicol. Environ. Safe.* **2010,** *73* (6), 1238–1245.
29. Wang, L.; Yang, L.; Yang, F.; Li, X.; Song, Y.; Wang, X.; Hu, X. Involvements of H2O2 and Metallothionein in NO-Mediated Tomato Tolerance to Copper Toxicity. *J. Plant Physiol.* **2010,** *167* (15), 1298–1306.
30. Rizwan, M.; Ali, S.; Qayyum, M. F.; Ok, Y. S.; Adrees, M.; Ibrahim, M.; Abbas, F. Effect of Metal and Metal Oxide Nanoparticles on Growth and Physiology of Globally Important Food Crops: A Critical Review. *J. Hazard. Mater.* **2017,** *322*, 2–16
31. Cuypers, A.; Hendrix, S.; Amaral dos Reis, R.; De Smet, S.; Deckers, J.; Gielen, H.; Keunen, E. Hydrogen Peroxide, Signaling In Disguise During Metal Phytotoxicity. *Front. Plant Sci.* **2016,** *7*, 470.
32. Das, K.; Roychoudhury, A. Reactive Oxygen Species (ROS) and Response of Antioxidants as ROS-Scavengers During Environmental Stress in Plants. *Front. Environ. Sci.* **2014,** *2*, 53.
33. Valko, M.; Jomova, K.; Rhodes, C. J.; Kuča, K.; Musílek, K. Redox-and Non-Redox-Metal-Induced Formation of Free Radicals and their Role in Human Disease. *Arch. Toxicol.* **2016,** *90* (1), 1–37.
34. Minibayeva, F.; Dmitrieva, S.; Ponomareva, A.; Ryabovol, V. Oxidative Stress-Induced Autophagy in Plants: The Role of Mitochondria. *Plant Physiol. Biochem.* **2012,** *59*, 11–19.
35. Potocký, M.; Pejchar, P.; Gutkowska, M.; Jiménez-Quesada, M. J.; Potocká, A.; de Dios Alché, J.; Žárský, V. NADPH Oxidase Activity in Pollen Tubes is Affected by Calcium Ions, Signaling Phospholipids and Rac/Rop GTPases. *J. Plant Physiol.* **2012,** *169* (16), 1654–1663.
36. Pranczk, J.; Jacewicz, D.; Wyrzykowski, D.; Chmurzynski, L. Analytical Methods for Determination of Reactive Oxygen Species. *Curr. Pharmaceut. Anal.* **2014,** *10* (4), 293–304.
37. Mazimba, O.; Nana, F.; Kuete, V.; Singh, G. S. Xanthones and Anthranoids From the Medicinal Plants of Africa. In *Medicinal Plant Research in Africa*; Elsevier, 2013; pp. 393–434.
38. D'Autréaux, B.; Toledano, M. B. ROS as Signalling Molecules: Mechanisms that Generate Specificity in ROS Homeostasis. *Nat. Rev. Mol. Cell Biol.* **2007,** *8* (10), 813–824.
39. Karuppanapandian, T.; Moon, J. C.; Kim, C.; Manoharan, K.; Kim, W. Reactive Oxygen Species in Plants: Their Generation, Signal Transduction, and Scavenging Mechanisms. *Aust. J. Crop Sci.* **2011,** *5* (6), 709–725.
40. D'Autréaux, B.; Toledano, M. B. ROS as Signalling Molecules: Mechanisms that Generate Specificity in ROS Homeostasis. *Nat. Rev. Mol. Cell Biol.* **2007,** *8* (10), 813–824.
41. Krieger-Liszkay, A. Singlet Oxygen Production in Photosynthesis. *J. Exp. Bot.* **2005,** *56* (411), 337–346.
42. Rai, R.; Agrawal, M.; Agrawal, S. B. Impact of Heavy Metals on Physiological Processes of Plants: With Special Reference to Photosynthetic System. In *Plant Responses to Xenobiotics*; Springer: Singapore, 2016; pp. 127–140.

43. Inal, A.; Gunes, A.; Zhang, F.; Cakmak, I. Peanut/Maize Intercropping Induced Changes in Rhizosphere and Nutrient Concentrations in Shoots. *Plant Physiol. Biochem.* **2007,** *45* (5), 350–356.
44. Kramer, U. Metal Hyperaccumulation in Plants. *Annu. Rev. Plant Biol.* **2010,** *61*, 517–534.
45. Lehner, A.; Mamadou, N.; Poels, P.; Come, D.; Bailly, C.; Corbineau, F. Changes in Soluble Carbohydrates, Lipid Peroxidation and Antioxidant Enzyme Activities in the Embryo During Ageing in Wheat Grains. *J. Cereal Sci.* **2008,** *47* (3), 555–565.
46. Fu, G. F.; Jian, S. O. N. G.; Xiong, J.; Li, Y. R.; Chen, H. Z.; Le, M. K.; Tao, L. X. Changes of Oxidative Stress and Soluble Sugar in Anthers Involve in Rice Pollen Abortion Under Drought Stress. *Agric. Sci. China* **2011,** *10* (7), 1016–1025.
47. Møller, I. M.; Rogowska-Wrzesinska, A.; Rao, R. S. P. Protein Carbonylation and Metal-Catalyzed Protein Oxidation in a Cellular Perspective. *J. Proteom.* 2011, *74* (11), 2228–2242.
48. Zadak, Z.; Hyspler, R.; Ticha, A.; Hronek, M.; Fikrova, P.; Rathouska, J.; Stetina, R. Antioxidants and Vitamins In Clinical Conditions. *Physiol. Res.* **2009,** *58* (1), 13–17.
49. Pourrut, B.; Shahid, M.; Dumat, C.; Winterton, P.; Pinelli, E. Lead Uptake, Toxicity, and Detoxification in Plants. *Rev. Env. Contam. Toxicol.* **2011,** *213*, 113–136.
50. Tamás, M. J.; Sharma, S. K.; Ibstedt, S.; Jacobson, T.; Christen, P. Heavy Metals and Metalloids as a Cause for Protein Misfolding and Aggregation. *Biomolecules* **2014,** *4* (1), 252–267.
51. Beltagi, M. S. Phytotoxicity of Lead (Pb) to SDS-PAGE Protein Profile in Root Nodules of Faba Bean (Vicia faba L.) Plants. *Pak. J. Biol. Sci.* **2005,** *8* (5), 687–690.
52. Kovalchuk, I.; Titov, V.; Hohn, B.; Kovalchuk, O. Transcriptome Profiling Reveals Similarities and Differences in Plant Responses to Cadmium and Lead. *Mut. Res.* **2005,** *570* (2), 149–161.
53. Jin, Z.; Li, Z.; Li, Q.; Hu, Q.; Yang, R.; Tang, H.; Li, G. Canonical Correspondence Analysis of Soil Heavy Metal Pollution, Microflora and Enzyme Activities in The Pb–Zn Mine Tailing Dam Collapse Area of Sidi Village, SW China. *Environ. Earth Sci.* **2015,** *73* (1), 267–274.
54. Whitacre, D. M. (Ed.), *Reviews of Environmental Contamination and Toxicology*. Springer, 2012.
55. El-Maarouf-Bouteau, H.; Meimoun, P.; Job, C.; Job, D.; Bailly, C.; Role of Protein and mRNA Oxidation in Seed Dormancy and Germination. *Front. Plant Sci.* **2013,** *4*, 77.
56. Srivastava, R. K.; Pandey, P.; Rajpoot, R.; Rani, A.; Dubey, R. S. Cadmium and Lead Interactive Effects on Oxidative Stress and Antioxidative Responses in Rice Seedlings. *Protoplasma* **2014,** *251* (5), 1047–1065.
57. Verma, S.; Dubey, R. S. Lead Toxicity Induces Lipid Peroxidation and Alters the Activities of Antioxidant Enzymes in Growing Rice Plants. *Plant Sci.* **2003,** *164* (4), 645–655.
58. Kaur, G.; Singh, H. P.; Batish, D. R.; Mahajan, P.; Kohli, R. K.; Rishi, V. Exogenous Nitric Oxide (NO) Interferes with Lead (Pb)-Induced Toxicity by Detoxifying Reactive Oxygen Species in Hydroponically Grown Wheat (Triticum Aestivum) Roots. *PLoS One* **2015,** *10* (9), e0138713.
59. Bela, K.; Horváth, E.; Gallé, Á.; Szabados, L.; Tari, I.; Csiszár, J. Plant Glutathione Peroxidases: Emerging Role of the Antioxidant Enzymes in Plant Development And Stress Responses. *J. Plant Physiol.* **2015,** *176*, 192–201.
60. Ralph, J.; Bunzel, M.; Marita, J. M.; Hatfield, R. D.; Lu, F.; Kim, H.; Steinhart, H. Peroxidase-Dependent Cross-Linking Reactions of p-Hydroxycinnamates in Plant Cell Walls. *Phytochem. Rev.* **2004,** *3* (1), 79–96.

61. Malar, S.; Manikandan, R.; Favas, P. J.; Sahi, S. V.; Venkatachalam, P. Effect of Lead on Phytotoxicity, Growth, Biochemical Alterations and Its Role on Genomic Template Stability in Sesbania Grandiflora: A Potential Plant for Phytoremediation. *Ecotoxicol. Environ. Saf.* **2014,** *108*, 249–257.
62. Thakur, S.; Singh, L.; Zularisam, A. W.; Sakinah, M.; Din, M. F. M. Lead Induced Oxidative Stress and Alteration in the Activities of Antioxidative Enzymes in Rice Shoots. *Biologia Plantarum* **2017,** *61* (3), 595–598.
63. Kranner, I.; Minibayeva, F. V.; Beckett, R. P.; Seal, C. E. What is Stress? Concepts, Definitions and Applications in Seed Science. *New Phytol.* **2010,** *188* (3), 655–673.
64. Blokhina, O.; Fagerstedt, K. V. Oxidative Metabolism, ROS and NO Under Oxygen Deprivation. *Plant Physiol. Bioch.* **2010,** *48* (5), 359–373.
65. Arasimowicz-Jelonek, M.; Floryszak-Wieczorek, J.; Gwóźdź, E. A. The Message of Nitric Oxide in Cadmium Challenged Plants. *Plant Sci.* **2011,** *181* (5), 612–620.
66. Jaspers, P.; Kangasjärvi, J. Reactive Oxygen Species in Abiotic Stress Signaling. *Physiologia Plantarum* **2010,** *138* (4), 405–413.
67. Lounifi, I.; Arc, E.; Molassiotis, A.; Job, D.; Rajjou, L.; Tanou, G. Interplay Between Protein Carbonylation and Nitrosylation in Plants. *Proteomics* **2013,** *13* (3–4), 568–578.
68. Mittler, R.; Vanderauwera, S.; Gollery, M.; Van Breusegem, F. Reactive Oxygen Gene Network of Plants. *Trends Plant Sci.* **2004,** *9* (10), 490–498.
69. Garg, N.; Manchanda, G. ROS Generation in Plants: Boon or Bane? *Plant Biosyst.* **2009,** *143* (1), 81–96.
70. Kärkönen, A.; Kuchitsu, K. Reactive Oxygen Species in Cell Wall Metabolism and Development in Plants. *Phytochemistry* **2015,** *112*, 22–32.
71. Sies, H.; Berndt, C.; Jones, D. P. Oxidative stress. *Annu. Rev.*, 2017.
72. Domingos, P.; Prado, A. M.; Wong, A.; Gehring, C.; Feijo, J. A. Nitric Oxide: A Multitasked Signaling Gas in Plants. *Mol. Plant* **2015,** *8* (4), 506–520.
73. Nabi, R. B. S.; Tayade, R.; Hussain, A.; Kulkarni, K. P.; Imran, Q. M.; Mun, B. G.; Yun, B. W. Nitric Oxide Regulates Plant Responses to Drought, Salinity, and Heavy Metal Stress. *Environ. Exp. Bot.* **2019,** *161*, 120–133.
74. Tripathi, D. K.; Singh, S.; Singh, S.; Srivastava, P. K.; Singh, V. P.; Singh, S.; Chauhan, D. K. Nitric Oxide Alleviates Silver Nanoparticles (AgNps)-Induced Phytotoxicity in Pisum Sativum Seedlings. *Plant Physiol. Biochem.* **2017,** *110*, 167–177.
75. Mata-Pérez, C.; Sánchez-Calvo, B.; Padilla, M. N.; Begara-Morales, J. C.; Valderrama, R.; Corpas, F. J.; Barroso, J. B. Nitro-Fatty Acids in Plant Signaling: New Key Mediators of Nitric Oxide Metabolism. *Redox Biol.* **2017,** *11*, 554–561.
76. Franks, S. J.; Weber, J. J.; Aitken, S. N. Evolutionary and Plastic Responses to Climate Change in Terrestrial Plant Populations. *Evol. Appl.* **2014,** *7* (1), 123–139.
77. Sanz, L.; Albertos, P.; Mateos, I.; Sánchez-Vicente, I.; Lechón, T.; Fernández-Marcos, M.; Lorenzo, O. Nitric Oxide (NO) and Phytohormones Crosstalk During Early Plant Development. *J. Exp. Bot.* **2015,** *66* (10), 2857–2868.
78. Lounifi, I.; Arc, E.; Molassiotis, A.; Job, D.; Rajjou, L.; Tanou, G. Interplay Between Protein Carbonylation and Nitrosylation in Plants. *Proteomics* **2013,** *13* (3–4), 568–578.
79. Freeman, J. L.; Zhang, L. H.; Marcus, M. A.; Fakra, S.; McGrath, S. P.; Pilon-Smits, E. A. Spatial Imaging, Speciation, and Quantification of Selenium in the Hyperaccumulator Plants Astragalus Bisulcatus and STANLEYA PINNATA. *Plant Physiol.* **2006,** *142* (1), 124–134.

80. Robinson, B. H.; Lombi, E.; Zhao, F. J.; McGrath, S. P. Uptake and Distribution of Nickel and Other Metals in the Hyperaccumulator Berkheya Coddii. *New Phytol.* **2003,** *158* (2), 279–285.
81. Küpper, H.; Lombi, E.; Zhao, F. J.; McGrath, S. P. Cellular Compartmentation of Cadmium and Zinc in Relation to Other Elements in the Hyperaccumulator Arabidopsis Halleri. *Planta* **2000,** *212* (1), 75–84.
82. Benekos, K.; Kissoudis, C.; Nianiou-Obeidat, I.; Labrou, N.; Madesis, P.; Kalamaki, M.; Tsaftaris, A. Overexpression of a Specific Soybean GmGSTU4 Isoenzyme Improves Diphenyl Ether and Chloroacetanilide Herbicide Tolerance of Transgenic Tobacco Plants. *J. Biotechnol.* **2010,** *150* (1), 195–201.
83. Krzesłowska, M.; Lenartowska, M.; Samardakiewicz, S.; Bilski, H.; Woźny, A. Lead Deposited in the Cell Wall of Funaria Hygrometrica Protonemata is Not Stable—A Remobilization Can Occur. *Environ. Poll.* **2010,** *158* (1), 325–338.
84. Schutzendubel, A.; Polle, A. Plant Responses to Abiotic Stresses: Heavy Metal-Induced Oxidative Stress and Protection by Mycorrhization. *J. Exp. Bot.* **2002,** *53* (372), 1351–1365.
85. Andrade, L. R.; Farina, M.; Amado Filho, G. M. Effects of Copper on Enteromorpha Flexuosa (Chlorophyta) In Vitro. *Ecotoxicol. Env. Saf.* **2004,** *58* (1), 117–125.
86. Sharma, N. C.; Sahi, S. V.; Jain, J. C. Sesbania Drummondii Cell Cultures: ICP-MS Determination of the Accumulation of Pb and Cu. *Microchem. J.* **2005,** *81* (1), 163–169.
87. Krzesłowska, M.; Lenartowska, M.; Mellerowicz, E. J.; Samardakiewicz, S.; Woźny, A. Pectinous Cell Wall Thickenings Formation—A Response of Moss Protonemata Cells to Lead. *Environ. Exp. Bot.* **2009,** *65* (1), 119–131.
88. Torres, M. A. ROS in Biotic Interactions. *Physiologia Plantarum* **2010,** *138* (4), 414–429.

CHAPTER 6

Plants and Microbe-Assisted Bioremediation of Heavy Metal Pollution in the Environment

UZMA AZEEM[1*], GURPAUL SINGH DHINGRA[2], and RICHA SHRI[3]

[1]*Government College Malerkotla, Malerkotla-148023, Punjab, India*

[2]*Department of Botany, Punjabi University, Patiala 147002, Punjab, India*

[3]*Department of Phramaceutical Sciences and Drug Research, Punjabi University, Patiala 147002, Punjab, India*

[*]*Corresponding author. E-mail: uzmaazeem2@gmail.com*

ABSTARCT

Heavy metals represent the elements with various biological functions and have industrial value too. However, in the past few decades, due to various natural phenomena like (rock weathering, erosion of soil, volcanoes, hurricanes, etc.) and anthropogenic processes, the concentrations of heavy metals increased rapidly to alarming levels. Heavy metals are toxic and persistent in nature. The pollution of the ecosystem with heavy metals is ubiquitous and disturbs the ecosystem homeostasis to the detriment of humans. The execessive heavy metals concentrations in the agricultural produce lead to their entry into the food chain and consequent biomagnification. Thereby, heavy metal ecosystem contamination is posing risk to ecosystem stability and food security, and needs immediate attention. Many physical and chemical remediation techniques are in use to remmediate heavy metal-contaminated environments. However, these are costly, less effective, not ecofriendly and

Environmental Pollution Impact on Plants: Survival Strategies under Challenging Conditions. Tariq Aftab, PhD & Khalid Rehman Hakeem, PhD (Eds.)

cause secondary pollution. To overcome these limitations, a more effective, ecofriendly, and economic remediation technique known as bioremediation is gaining much appreciation. Bioremediation involves the use of plants/ microbes alone or in combination with one another for more effective remediation.

6.1 INTRODUCTION

Heavy metals are a group of metals having high density and atomic number more than 20. Recently, heavy metal pollution is attaining heights due to natural (weathering of earth's crust, soil erosion, forest fires, volcanic eruptions, hurricanes, etc.) and anthropohenic (rapid increase in industrialization, urbanization, mining, sewage irrigation, and excessive use of pesticides and fertilizers) activities. This rise in heavy metal contents is hazardous to life on earth and stability of ecosystem.[1,2] Heavy metals exceeding optimum levels are harmful and adversely affect plant physiology such as photosynthesis, respiration, and enzymatic activities.[3–5] Heavy metal ions can enter the food chain and exhibit the phenomenon of bioaccumulation posing risk to human health even at trace levels.[6] Excessive concentrations of heavy metals disrupt the homeostasis of the living system leading to various diseases like Menkes disease, Alzheimer's disease, and cancers.[7] There are several remediation techniques to restore the natural condition of heavy metal contaminated ecosystems. The choice and implementation of available remediation techniques depend on the cost, time taken, usefulness against elevated levels of heavy metal contamination, general likelihood, and agreement to implement the method, market availability, durability, and effectiveness for multimetal contaminated sites.[8] The chemical precipitation, ion-exchange, and membrane filtration techniques work on the principle of decreasing the metal bioavailability by converting heavy metals into harmless form. The chemical precipitation, ion-exchange methods are easy, efficient, and successful in removing the heavy metals from the polluted environments. However, chemical precipitation generates sludge of low-density, causes secondary pollution and problematic to remove trace metals in bigger regions. Most of the ion-exchange methods can be employed only in acidic environment.[9–11] Membrane filtration techniques are highly efficient to remove heavy metals, but the membrane material production is generally difficult.[12] To overcome the limitations of these methods, a simple, low-cost, and ecofriendly remediation technique known as bioremediation is the most viable option. Bioremediation includes utilization of plants and microbes (algae, bacteria, and fungi) to reduce the toxicants/pollutants including the heavy metals from the polluted environments. Plants and microbes possess various heavy metal resistance mechanisms making them capable of tolerating heavy metal toxicity. Plants are consecrated with natural morphological or biochemical defense mechanisms. Thick cuticle, trichomes, cell walls, and mycorrhizal symbiosis are the biophysical barriers during heavy metal excess.[13] If the heavy metals cross these barriers and enter the cell, plants overcome the adverse effects of reactive oxygen species (ROS) with the assistance of enzymatic antioxidants and nonenzymatic antioxidants.[14–16] Phytochelatins produced from glutathione by phytochelatin synthase also defend plants from drought, salinity, herbicides, and heavy metal toxicity. Another important class of proteins with crucial role in plant defense against heavy metal stress is metallothioneins with greater affinity for an array of heavy metals. In addition to heavy metal detoxification, metallothioneins participate in maintaining the redox equilibrium, repairing plasma membrane, cell growth and proliferation, repairing damaged DNA and ROS scavenging.[17] The plant microbe bioremediation system includes the participation of mycorrhizae and especially plant growth promoting rhizospheric (PGPR) bacteria. Mycorrhizal association improves the mineral contents of plants and their tolerance toward heavy metal-induced stress.[18] The heavy metal resistance of microbes can be credited to several extracellular functional groups, extracellular and intracellular compounds, membrane transporters, and enzymes.[19] Plants/microbes can act alone or exhibit integrated remediation of heavy metal polluted environments in combination with one another.[20] The bioremediation potential varies with the living organism selected for the process, type of soil, physical properties, and level and type of heavy metal levels in the polluted area.[21]

6.2 SOURCES AND EFFECTS OF HEAVY METAL POLLUTION

Heavy metals occur in the D horizon of soil (the bedrock part with most of the mineral matter) and have high density, melting points, and boiling points. Because of greater density and toxic nature at low amounts, these are recognized as serious environmental pollutants of immediate concern.[22] The soils formed from the parent material comprising of metal excess in soil bedrocks exhibit higher heavy metal concentrations naturally.[23,24] There is an increasing demand of heavy metals in various modern-day technologies and therefore production of heavy metals through mining becomes significantly high.[25] Heavy metals present in the form of mineral ores in the earth's crust are mined following mineral processing methods.[26,27] Many other anthropogenic activities contributing toward heavy metal pollution are industrialization, urbanization, and unsafe agricultural methods like excessive application of fertilizers, pesticides, livestock manure as soil fertilizer to enhance the crop yield and use of waste water for irrigation.[28] Aquatic environments become contaminated with heavy metals from volcanoes, weathering of rocks, eutrophication, and soil leaching.[2] Dust exposes the heavy metals to atmosphere and rain brings acid drainage and particulate dust storms in heavy metal contaminated regions. In this way, heavy metal pollution contaminates the atmosphere, lithosphere, and the hydrosphere.[29] Heavy metals, compared to organic compounds, are somewhat more resistant to either biological or chemical degradation. Therefore, these exhibit high persistence for a long time after entering the environment.[30] Agricultural crops grown in heavy metal polluted soils generally accumulate heavy metals above the permissible limits.[31] The low contents of heavy metals are permissible in the ecosystem and living organisms without any side effect. However, exceeding the permissible limits, heavy metals will hamper ecology and fertility of soil, plant growth, metabolism, and physiology of living organisms. The water bodies, seas/oceans, rivers, and ground water bodies contaminated with heavy metals pose threat to aquatic life. Heavy metals at higher concentrations enter the food chain and accumulate in the living organisms including humans and pose risk to their survival and health. This phenomenon is commonly known as biomagnification.[32–36] The intake of heavy metal contaminated food seems to be the main gateway (>90%) of human exposure in contrast to physical contact or entry through respiratory tract.[37,38] Bioaccumulation of heavy metals is hazardous for humans causing detrimental health impacts such as cancer, neurotoxicity, kidney failure, respiratory disorders, cardiac disease, lungs disorders, etc. and sometimes leads to death.[39] Heavy metals in the aquatic habitat form particulates that reach the bottom sediments adversely affect the aquatic biota including microbes, the diatoms, and the macrophytes.[40] To avoid or reduce the impact of heavy metal pollution, bioremediation is gaining much attention which includes phytoremediation and microbial-assisted remediation.

6.3 PYTOREMEDIATION

Phytoremediation is a global need to restore the heavy metals contaminated ecosystems.[41] This is a green biotechnology tool for sustainable rehabilitation of environment.[42] Mostly, plants with efficacy to accumulate huge concentrations of heavy metals are preferred for phytoremediation. These plant species are called hyperaccumulators and they compete for the heavy metals to accumulate.[43–45] Radionuclides are also removed from soil, sediments, and water resources with phytoremediation.[46] Phytoremediation of air pollutants occurs in the plant rhizosphere as air pollutants may deposit in the soil during leaf fall and precipitation.[47] The plant species commonly used for phytoremediation are listed in Table 6.1. Plants need macroelements and microelements for their growth and metabolism. The common non-accumulator plant species need micronutrients at a concentration of 10 mg/L, sufficient for their growth and metabolism. The high amount of heavy metals is phytotoxic and plants grown in heavy metal contaminated soil exhibit phytotoxicity symptoms. However, many plant spp. are capable of tolerating greater heavy metal quantities and are called metallophytes. These are of three types; metal indicators, metal excluders, and metal accumulators. Metal indicator plants take up higher quantities of heavy metals and accumulate them in roots and shoots. They display phytotoxicity symptoms and act as heavy metal pollution indicators.[48] Metal excluders are the plant species which accumulate large quantities of heavy metals in their roots but prevent their transportation to the aerial parts. Therefore, these are best suited for phytostabilization.[49–52] The most common metal excluders are *Silene paradoxa* and *Bidens pilosa* that exclude As, Cd, Co, Cr, Cu, Fe, Mn, Ni, Pb, and Zn.[53,54] The plants of third kind, used widely in phytoremediation, are called as metal accumulators. The latter include the plant species that accumulate heavy metals efficiently in their foliage with no phytotoxicity.[55,56] The most preferred characteristics of plant species used in phytoremediation are adaptation to local environment, length of plant roots, efficacy to flourish in the present soil type, capability to accumulate or convert the pollutant into less harmful form, rapid growth, no or less planting and maintenance problem, and absorption of higher quantities of water.[57] Trees with long roots and high transpiration rate are grown at the contaminated sites. These trees absorb large quantities of underground water as well as the contaminants present in it.[58,59] This uptake of water prevents the surface contaminants from entering the below ground water table and lowers the contamination of water table.[60] *Populus*, *Eucalyptus*, and *Salix* species are well known for use in this kind of hyraulic control.[61–64] The introduction of non-native plant species is generally avoided. Mostly, hybrid poplars, willows, cotton wood trees, grasses (like *Rye, Sorghum*, and *Fescue*), legumes (e.g., *Clover*, *Alfalfa*, and *Cowpea*), various aquatic and wetland plants are commonly utilized in phytoremediation of polluted soil and water. These species have fast growth rate, extensive and

TABLE 6.1 The Most Common Plant Species With Phytoremediation Efficacy and the Heavy Metals Removed by Them.

Heavy metal(s)	Plant species	Medium	References
	Phytoextraction/phytoaccumulation		
As	*Corrigiola telephiifolia*, *Eleocharis acicularis*, *Horedeum vulgare*, *Pteris cretica*, and *P. vittata*	Soil and water	74–79
As and Hg	*P. vittata*	soil	80–83
Ni	*Alyssum bertolonii*, *A. murale*, *Berkheya coddii*, *Brassica juncea*, and *Isatis pinnatiloba*		84–89
Cd	*Agave americana*, *B. juncea*, *Furcraea foetida*, *Solanum photeincarpum*, *Thlaspi caerulescens*, and *Youngia japonica*		90–95
Cd, Co, and Ni	*B. napus*, *Elymus elongatus*, and *Zea mays*		96
Cd, Cr, Cs, Cu, Ni, Pb, U, and Zn	*B. juncea*		97
Cd, Pb, and Zn	*Helianthus annuus*		98, 99
Cd and Zn	*Arabidopsis halleri*		100, 101
Cd, Pb, and Zn	*Arabis paniculata*		102
Cu, Fe, Pb, and Zn	*Euphorbia cheiradenia*		103
Cu, Pb, and Zn	*Hordeum hirta*		104
Fe	*Eclipta alba* and *Alternanthera philoxeroide*	Pulp and paper waste	105
Hydrocarbons	*Jatropha curcas* and *Vetiveria zizanioides*	Petroleum hydrocarbon contaminated soil	106
Mn	*Schima superba*	Quartz	107
Pb	*Lolium italicum*, *Medicago sativa*, *Sedum alfredii*, and *Zygophyllum fabago*	Soil and water	108–110
Zn	*E. cheiradenia*, *Sedum alfredii*, and *Th. Caerulescens*	Soil	111–113
Se	*Astragalus racemosus*		114

TABLE 6.1 *(Continued)*

Heavy metal(s)	Plant species	Medium	References
V	*Nicotiana tabacum*		115
U	*Cannabis sativa*		116
	Phytodegradation		
As	*Arundo donax*, *Canna glauca*, *Colocasia esculenta*, *Cyperus papyrus*, *P. vittata*, and *Typha angustifolia*	Soil	117–119
Hg, methyl-Hg and Hg	*Azolla caroliniana*, *B. juncea*, *Liriodendron tulipifera*, and *Lupinus* sp.	Water and mine tailings	120–122
	Phytostabilization		
As	*Cynodon dactylon*	Soil	123
Ag, As, Cr, and Sb	*Solanum tuberosum*		124
Cd	*Desmostachya bipinnata*, *Dichanthium annulatum*, *N. tabacum*, *Populus cathayana*, *P. prezwaskii*, *P. yunnanensis*, *Tamarindus indica*, and *Z. mays*		125–128
Cd, Cu, and Pb	*Amaranthus spinosus*		129
Cd, Cu, and Se	*Lemna minor*		130
Cd and Ni	*Wedelia trilobata*		131
Cr	*Azara microphylla*, *Azolla pinnata*, *A. filiculoides*, and *Wolffia globosa*		132,133
Cu	*Salix jiangsuensis* and *S. babylonica*		134
Fe	*Clerodendrum indicum*		135
Fe, Cu, Zn and Hg	*M. aquaticum*, *Ludwigina palustris*, and *Mentha aquatic*		136
Fe and Zn	*J. curcas*		137
Ni	*M. aquatic*, *M. sylvestris*, *Riccinus communis*		138,139
Pb	*N. tabacum* and *V. zizanioides*		140
Pb, Cd, Cr and Cu	*Solanum nigrum* and *Spinacia oleracea*		141

TABLE 6.1 *(Continued)*

Heavy metal(s)	Plant species	Medium	References
Re	*Salvinia natans* and *Trifolium repens*		142
U	*B. juncea* and *H. annuus*		143
Zn	*Lemna gibba* and *Scirpus mucronatus*		144,145
	Phytovolatilization		
As and Cd	*Lupinus albus*	Soil	146
As, Cd, Co, Cr, Mn, Ni, and Zn	*Typha latifolia*	Indutrial sludge	147
As, sulfide gold	*Eucalyptus cladocalyx, E. melliodora, E. polybractea*, and *E. viridis*	Mine tailings	148
Cd	*Acanthus ilicifolius, Lupinus uncinatus, Quercus ilex*, and *Silphium perfoliatum*	Soil	149–152
Cd and Ni	*Picea abies* and *Populus tremula*		153
Cd, Mn, Pb, and Zn	*A. philoxeroides, Artemisia princeps, Bidens frondosa, B. pilosa, C. dactylon, Digitaria sanguinalis, Erigeron canadensis*, and *Setaria plicata*	Mine tailings	154
Cd, Pb and Zn	*Lolium perenne* and *T. repens*	Soil	155
Cd and Zn	*Sorghum bicolor*		156
Cu	*Festuca rubra* and *Phragmites australis*		157,158
Hg	*Arabidopsis thaliana, H. vulgare, Liriodendron tulipifera*, and *N. tabacum*		159,160
Pb	*Athyrium wardii* and *Thysanolaena maxima*	Soil and mine tailings	161,162
Pb and Zn	*Artemisia roxburghiana, A. tangutica, Coriaria sinica, Cynoglossum lanceolatum, Miscanthus nepalensis, Oxyria sinensis, Plantago depresa*, and *Potentilla saundesiana*	Mine tailings	163
Se	*Allium cepa, Astragalus bisulcatus, Beta vulgaris, Brassica juncea, B. oleracea* var. *botrytis, B. oleracea* var. *capitata, B. oleracea* var. *Italica, Oryza sativa, Phaseolus vulgaris, Lactuca sativa*, and *T. latifolia*	Soil	164–167

TABLE 6.1 *(Continued)*

Heavy metal(s)	Plant species	Medium	References
Zn	*Solanum nigra*	Soil	178
	Phytofiltration		
As and Sb	*Ph. Australis*	Soil	169
Cd, Co, Cr, Cu, Fe, Mn, Ni, Pb and Zn	*Pistia stratiotes*	Wetlands	170
Cd, Pb and Zn	*Eichhornia crassipes*	Water	171
Cd and Zn	*Arundo donax*		172
Cr, Pb and Ni	*P. stratiotes*		173,174
Cs and U	*P. vulgaris*		175
Pb	*A. pinnata*, *Carex pendula*, and *Plectranthus amboinicus*		176–178
U	*H. annuus* and *P. Vulgaris*		179,180

deep root system, and high transpiration rate (responsible for uptake of large amounts of water).[65] Aquatic plants are capable of accumulating organic and inorganic contaminants. For example, floating hydrophytes like *Eichhornia crassipes*, *Pistia stratiotes*, and *Lemna minor* are prominent heavy metal accumulators.[66] Ferns and sea weeds act as metal accumulators in aquatic and submerged habitats.[67,68] In heavy metal remediation of polluted water bodies, the plant species such as *Lepironia articulata*, *L. minor*, *Hydrocotyle umbellate*, *Polygonum amphibium*, *Oenathe javanica*, *Hydrilla verticillata*, *Eichhornia crassipes*, and *P. stratiotes* are widely used. Submerged hydrophytes such as *Potamogeton malaianus*, *Nymphoides peltata*, *E. crassipes*, and *H. verticillata* have been reported as effective accumulators of Co, Cr, Cu, Mn, Ni, Pb, Ti, V, and Zn from moderately contaminated lakes.[69] The efficiency of plant species toward accumulation and tolerance of high quantities of metal contaminants without phytotoxicity has been well documented in hydroponics systems and pot culture experiments.[70] A plant species may exhibit preference to uptake some metals over others. The duck weed (*L. minor*) has been observed accumulating Cd, Cu, and Se in good amounts, while Cr in a moderate quantity. *E. crassipes* has been found to remediate Ag, Cd, Cr, Cu, Pb, and Se from waste water, while As, Cr, and Hg in hydroponics. *Lepironia articulate* and *O. javanica* exhibit Pb and Hg accumulation in greater amounts respectively as compared to other heavy metals.[71] The oxalates and polyphenolic exudates of *Lemna* sp. and *Myriophyllum aquaticum* form complexes with heavy metals.[72] Therefore, macrophytes can be used for heavy metals removal from living as well as dead states in the aquatic ecosystems and the methods for living and dead states are called as bioaccumulation and biosorption respectively.[73]

6.3.1 MECHANISMS OF PHYTOREMEDIATION

Plants follow different mechanisms to bioremediate polluted sites from excess of heavy metals and organic pollutants as shown in Figure 6.1. Phytoextraction/phytoaccumulation is used to remove organic and inorganic pollutants from land, sediments, aquatic habitats, and sludges. Phytostabilization is an effective technique to remove heavy metals and chlorinated solvents occurring in soil, sediments, and sludges. Phytovolatilization works in case of chlorinated solvents and inorganic compounds in ground water, soil, sediments, and sludges. Phytofiltration operates against toxic metals and organic compounds found in surface, waters, and waste water.[181] Phytodegradation includes degradation of chlorinated solvents, and various organic and inorganic contaminants.[182]

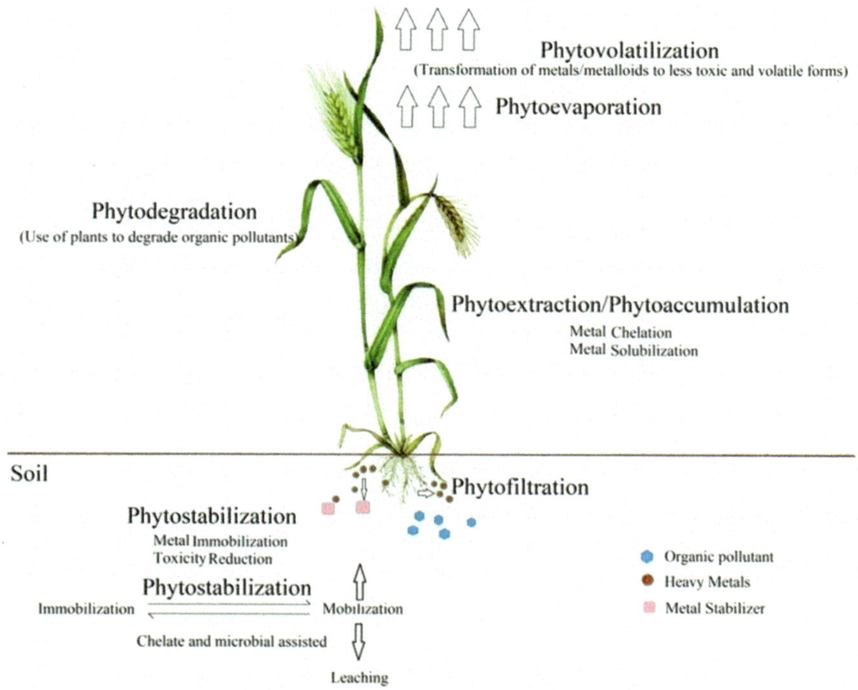

FIGURE 6.1 Phytoremediation techniques followed by plants.

6.3.1.1 PHYTOEXTRACTION/PHYTOACCUMULATION

In this technique, roots of plants absorb heavy metals and transport these metals to the aerial parts for accumulation. The effectiveness of phytoextraction depends on the plant species used, the extent of plant tolerance to elevated levels of heavy metals, efficacy of plant species to drastically absorb and move heavy metals from roots to above ground plant parts.[183] Plants with greater efficiency of accumulation i.e. hyperaccumulators take up large amounts of metal contaminats, transport them to the above ground parts, and accumulate approximately 100–1000 times greater quantities of heavy metal contaminats than the non-hyperaccumulators without any apparent phytotoxicity. Hyperaccumulator plant species have greater efficacy to take up heavy metals, higher upward transport of metal ions, higher capacity to detoxify and sequester huge quantities of heavy metals, rapid growth, and enhanced root growth. Plant taxa identified as hyperaccumulators mainly belong to *Asteraceae*, *Brassicaceae*, *Caryophylaceae*, *Cyperaceae*, *Euphorbiaceae*, *Fabaceae*, *Lamiaceae*, *Poacea*, and *Violaceae*.[184–186] Currently, >400 species of plants (0.2% of all angiosperms) have been found as heavy metals accumulators.[187] Some plant species are hyperaccumulators while some are moderate accumulators. *Brassica* spp. including *B. juncea*, *B. napus*, and *B. rapa* are moderate accumulators of Cd and Zn and are used in various systems including hydroponics.[188] *Thlaspi* sp., *Arabidopsis* sp., and *Sedum alfredii* are recognized as hyperaccumulators of which *Thlaspi* sp. can accumulate multiple heavy metals.[189] This capability of *Thlaspi* spp. can be attributed to the root structure, cell membrane carriers, and organic ligands in root exudates.[190] The halophytic plants make use of the mechanism of phytoextraction to remove salts from salt contamianted soils to make these soils capable of supporting plant growth.[191] This is called Phytodesalination. Halophytic plants exhibit better adaptions to cope with heavy metal pollution in comparison to glycophytic plants.[192] Phytoextraction/phytoaccumulation is commonly classified into two types. One is the chelate-assisted phytoextraction in which artificial chelates enhance the heavy metal ions mobility to make them amenable to plants. The second type is the continuous phytoextraction which is a natural plant efficiency to extract/accumulate and remove the toxicants.[193]

6.3.1.2 PHYTOFILTRATION

It involves rhizofiltration, blastofiltration, and caulofiltration. In rhizofiltration, plant roots, uptake, accumulate, and precipitate heavy metals present in the soil.[194] The hydroponically raised plants are grown in polluted water. These plants uptake and accumulate heavy metals. These plants are then picked and disposed off.[195] Rhizofiltration is generally used for Pb, Cd, Cu, Ni, Zn, and Cr which accumulate in roots.[196] Both land plants and aquatic plants are utilized in rhizofiltration (in situ as well as ex situ methods). However, terrestrial plants are more suited than aquatic plants as the former use natural solar driven pumps to take up heavy metals.[197] As the pollutants are not moved upward to aerial parts, non-accumulators can also be used in rhizofiltration.[198] Blastofiltration includes heavy metals removal from polluted water using young seedlings. Young seedlings are more suitable for remediation as after germination, a significant increase in surface-to-volume ratio has been observed.[199]

6.3.1.3 PHYTOSTABILIZATION

In this technique, heavy metal bioavailability and mobility in soil decreases because of their stabilization by plants.[200] Adsorption and precipitation of toxic metals occurs into less soluble forms such as carbonates and sulphides. Toxic metals bind to organic compounds and get accumulated in root tissues in the rhizosphere. Plants increase microbial populations generally heterotrophic that promote plant growth in contaminated soil and stabilize the toxic metals.[201–202] Mycorrhizae also play role in heavy metal phytostabilization similar to ericoid and ectomycorrhizal fungi.[203] Mycorrhizal hyphae have polyphosphate capable of binding with heavy metals up to saturation with >60% metals retained in apoplast cell walls.[204–208] Heavy metal immobilization inside roots occurs by combining with pectins of the cell wall and by strong electrostatic interaction with the negatively charged cell surfaces.[209] Many plant species secrete redox enzymes that reduce the valence of heavy metals and lower their toxicity such as transformation of toxic Cr(IV) to less toxic Cr(III).[210] Phytostabilization may operate weakly in greatly polluted soils. Under such conditions, cultivation of metal tolerant plant species adapted to the local environment is the most useful option. Metal excluder plants are able to survive in soils polluted with heavy metals showing no bad impacts on growth, low metal contents in aerial parts and high to very high in roots.[211–212] Plants exclude metals through different approaches involving mycorrhizae, cell wall, and plasma membrane. Mycorrhizae perform mechanisms like higher plants such as extracellular binding or confinement in the vacuole.[213–214] Different hypotheses exist for metal exclusion mechanism such as (1) metal binding to cell wall, (2) secretion of ligands with metal binding potential, and (3) establishment of redox and pH barriers at the plasma membrane.[215] The function of cell wall in metal tolerance is controversial. Some researchers give small credit to cell wall while others think that cell wall accumulated metals stay associated with protein or silicate.[216] Metal tolerant plants possess efficiency to tolerate higher concentrations of heavy metals because of metal chelation. The extracellular complexation of Al with citric acid and malic acid accounts for metal tolerance in wheat.[217] Al resistance of *Arabidopsis* is due to root exudation of organic acids.[218] Plants used in this remediation technique must have elaborative root system and less upward movement of metals.[219] This method is generally employed for the restoration of soils polluted due to high quantities of As, Cd, Cr, Cu, Pb, Zn, etc. This technique does not involve dumping of the hazardous waste. The heavy metals immobilization results in preservation of surface and groundwater resources.[196] Phytostabilization is a good remediation option for contaminated soils if phytoextraction is not possible to apply or desirable. It actually lowers the contamination level of adjacent media/area. Moreover, it is suitable for sites where technical or regulatory limitations interfere with the selection and implementation of the best suited remediation techniques.

This technique is beneficial to prevent off-site heavy metals movement from barren contaminated soil. Various methods are involved in limiting the movement of heavy metals such as reduced leaching due to upward movement of water by transpiration, decreased erosion of soil because of soil stabilization by roots of plants, and reduced run off because of aerial part of vegetation. Additionally, no contaminated secondary waste is produced in this method. This remediation technique is beneficial in the restoration of ecosystem as it enhances the fertility of soil. As heavy metals stabilization occurs in soil, regular monitoring of the site is essential to ensure optimal phytostabilizing conditions.[220]

6.3.1.4 PHYTOVOLATILIZATION

The heavy metal contaminated water is absorbed by plant roots from soil and is transported through the xylem. In the aerial parts of plants, these heavy metals are converted into volatile forms which are of low toxicity and are released in the environment such as As, Hg, and Se.[221–224] Remediation of contaminated groundwater, soil, sediments, and sludges is generally done with this technique.[225] Various abiotic factors like temperature and intensity of light show influence on the efficiency of the foliar tissues to release heavy metals into the atmosphere.[226–228] Moreover, the plant transpiration rate can also affect phytovolatilization.[229] Phytovolatilization exhibits the accumulation of contaminants in the edible plant products. Another limitation of this technique is that it incompletely removes the pollutants. It lowers the toxicity of the pollutants and these are transferred to atmosphere.[230] However, some reports indicate that volatile compounds cause little or no environmental risk as these compounds get diluted and unfurl in the atmosphere.[231] Phytovolatilization brings negligible erosion, no disposal of contaminated plant biomass and negligible disturbance of sites.[232] Phytoevaporation is a closely related approach to phytovolatilization in which the contaminants get assimilated and transpired either in the same or changed form.[233]

6.3.1.5 PHYTODEGRADATION

In this technique, plant enzymes break down the organic contaminants into harmless forms. Plants contain nitroreductases and halogenases which are able to degrade organic pollutants.[234] Phytodegradation is applicable to organic pollutants as heavy metals are non-biodegradable. Various enzymes with phytodegradation efficacy are dehalogenase (conversion of chlorinated compounds), peroxidase (transformation of phenolic components), nitrilase (conversion of cyanated aromatic compounds), nitroreductase (conversion of explosives and other nitrate compounds), and phosphatase (conversion of organophosphate pesticides).[184,235]

6.3.1.6 PHYTO-MICROBIAL REMEDIATION SYSTEM

Plants assist to remove heavy metals and organic pollutants alone or with the help of rhizospheric microbes.[236,237] Plant exudates stimulate microbial activity (nearly 10–100 times) to degrade organic contaminants in the rhizosphere. This is known as phytostimulation. These exudates secreted by plants are rich in carbohydrates, amino acids, and flavonoids that provide C and N to the soil microbes and stimulate their activities. Moreover, plants also secrete certain enzymes holding the potential to degrade organic contaminants in soil.[238] Microbes especially bacteria residing in the plant rhizosphere assist in heavy metal removal (Figure 6.2). These microbes are resistant to multiple chemical pollutants at co-contaminated sites, that is, adulterated with multiple heavy metals.[239] These rhizobacteria can be extracellular or intracellular based on their interaction with root cells of the host plant.[240,241] Mostly, these are extracellular PGPR represented by *Agrobaciterium*, *Azospirillium*, *Azotobacter*, *Chromobacterium*, *Caulobacter*, *Erwinia*, *Flavobacterium*, *Micrococcus*, *Pseudomonas*, and *Serratia*.[242] Microbial association is beneficial for plant growth in various ways such as regulation of stomatal opening and closing, osmotic regulation, and mineral uptake. They also increase host plant metal tolerance. Microbes also exhibit mechanisms of metal solubility and bioaccumulation. Rhizobacteria produce extracellular polymeric substances (EPS), plant growth promoting substances, and metal chelaters such as siderophores, organic acids, and biosurfactants.[243] Under Fe limiting conditions, siderophores of microbial origin solubilize Fe present in the soil and make it available to plants.[244] Phytosiderophores also form complexes with Fe(III) present in the rhizosphere soil through their amino and carboxyl groups. These complexes can pass through the root plasma membrane.[245–247] The phosphate-solubilizing microbes mainly the rhizospheric bacteria such as *Azotobacter*, *Bacillus*, *Beijerinckia*, *Burkholderia*, *Enterobacter*, *Erwinia*, etc. solubilize phosphorous and make it available for uptake by plants. These phosphate-solubilizing rhizospheric bacteria not only make phosphorous available to plants under stress but also enhance biological nitrogen (N2) fixation and make many trace elements available through plant growth promoting

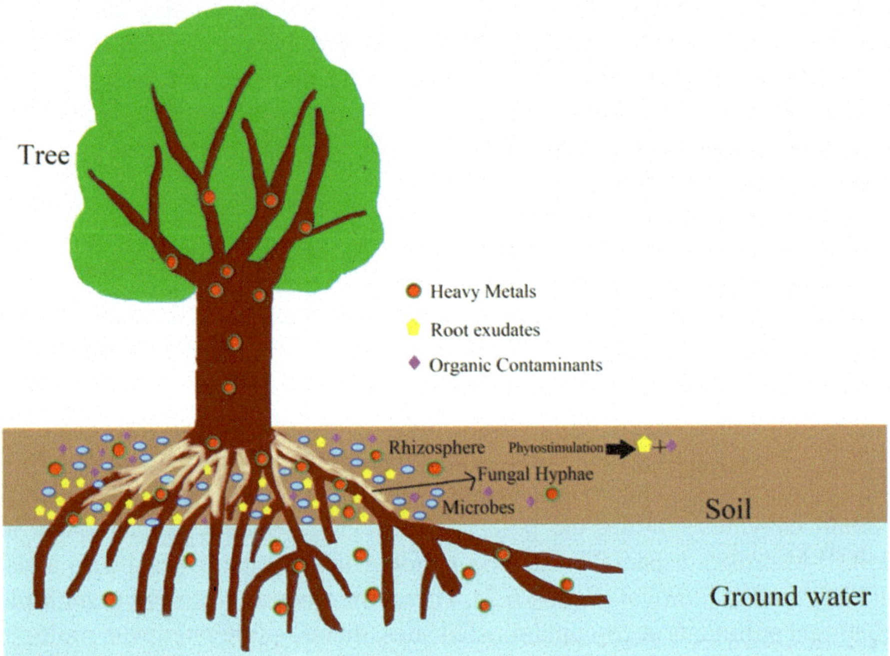

FIGURE 6.2 Phyto-microbial remediation system.

substances.[242,248–250] PGPR produce 1-aminoacyclopropane-1-carboxylate deaminase under the conditions of drought, salinity, or heavy metal stress. This enzyme causes breakdown and sequestering of 1-aminoacyclopropane-1-carboxylate reduces ethylene levels and makes plants able to cope with abiotic stress.[251–253] The enzyme, 1-aminoacyclopropane-1-carboxylate deaminase breaks 1-aminoacyclopropane-1-carboxylate to α-ketobutyrate and ammonia in plants that in turn develops an extensive root system. In this way, ethylene accumulation is reduced which could otherwise lead to plant death.[249,254] Indole-3-acetic acid (IAA) exhibits a significant role in plants and rhizobacteria interactions. IAA promotes growth of plants by enhancing growth of root system and also defends plants from abiotic stress. IAA is produced by PGPR by utilizing tryptophan from plant roots in the rhizospheric area. IAA then enters the plant cells, promotes plant growth, or reduces ethylene levels by stimulating 1-aminoacyclopropane-1-carboxylate synthase.[255] IAA promotes length of roots, increases surface area, makes soil nutrients readily available, increases root exudates supplying to the rhizospheric bacteria.[248,256] In the last decade, significant research has been done to explore the physiological and molecular mechanisms of arbuscular mycorrhizal fungi (AMF) in the amelioration of toxic effects of heavy metals in plants, acquisition of nutrients, and improvement in plant performance under heavy metal stress. Mycorrhizal fungi consist of vesicles that uptake and store huge quantities of heavy metals. A glycoprotein, glomalin, secreterd by AMF chelates with metals and hence lowers the uptake of heavy metals in plants and improves soil quality. They also play a role in phytoremediation by enhancing the surface area for absorption, antioxidative activity and, chelating heavy metals and stimulating synthesis of proteins that in turn decrease the hazards of ROS.[18]

6.3.2 FACTORS AFFECTING PHYTOREMEDIATION

Phytoremediation is an excellent technique for revegetation of the heavy metal polluted environments. This technique is aesthetically pleasing, ecofriendly, less destructive, less expensive, applicable to broad spectrum of contaminated environments and can be in situ or ex situ. Despite various advantages, phytoremediation is time consuming and depends on the amount of biomass produced, plant root depth, soil composition, level of contamination, plant age, vegetation type, and environmental conditions.[39,257] The selection of plant species for phytoremediation purpose is the most improtant thing that affects the success of heavy metal remediation. The plant species used for phytormediation are preferably the hyperaccumulators. The properties of the medium from where the heavy metals have to be removed influence the phytoremediation effiecency of the plant species. The characteristics of soil such as (pH, texture, cation exchange capacity, and nutrient content), temperature, plant root exudates affect the properties and metal availability to plants.[258–263] The metal uptake ability of plant roots is dependent on the rhizobiological activity, amount and type of root exudates, pH, moisture content, temperature, level of competing metal ions, and metal ions concentration in soil solution.[264–265] The local environment such as temperature, drought, precipitation, relative humidity, soil moisture content, and vapor pressure deficit at the crop canopy has great influence on plant growth, transpiration, water chemistry, and root length and other metabolic pathways. All these in turn affect both pollutant uptake and elimination.[266–268] After successful removal of heavy metals, phytoremediation faces a major problem of the disposal of the heavy metals accumulating plant waste. The amounts of heavy metals in the plant used for phytoremediation determine whether the plant needs to be landfilled or pyrolysis of biomass, smelting, or extraction. The ash made by the incineration of plant used for phytoremediation should be thrown in hazardous waste landfills and plants containing radioactive metals must be treated like radioactive waste. One more problem is the handling of heavy metals contaminated plants after harvesting to prevent secondary emissions. For example, *B. juncea* becomes brittle, dry, and cracking after harvesting.[269,270]

6.4 MICROBES-ASSISTED REMEDIATION

6.4.1 MICROBES WITH HEAVY METALS REMEDIATION CAPACITY

The microbes with heavy metal remediation potential are aerobic or anaerobic, while the aerobic ones are utilized more than the anaerobic ones for bioremediation process in different environments.[271] Microbes classified as algae, bacteria, and fungi possess heavy metals remediation efficiency. A single microbe can remediate single to multiple heavy metals (Table 6.2).

Algae are a type of photosynthetic organisms and occur in freshwater as well as in seawater. Algae can remediate heavy metals by adsorption, accumulation followed by intracellular localization and deposition. Moreover, algae can block the entry of heavy metals inside the cell and activate various defense mechanisms to get rid of heavy metal toxicity.[297–300] Algae adsorb heavy metals over a short period. Nonliving algae adsorb much better than living algae. Heavy metal adsorption ability of Pheophyta is the best. Alkaline environments are not favorable for heavy metal adsorption by algae.[301] Heavy metals enter the algal cells through membrane transport systems. Multiple heavy metal ions can enter through the membrane transporters and there occurs uptake competition among nutrients and heavy metal ions. After uptake these heavy metals may be compartmentalized and/or transformed to less toxic forms by binding or precipitation. Metal transport can occur intracellularly and localized at intraprotoplast sites such as vacuoles or chloroplasts.[300,302–304] Diatoms, the microalage, is the most abundant and diverse group of phytoplankton accounting for about 45% oceanic primary productivity. They play a key role in the biogeochemistry of heavy metals in fresh as well as marine habitats. To maintain optimal concentrations of heavy metals, diatoms have various mechanisms such as formation of organic ligands and exudation, alterations in permeability of membrane, expression of stress related proteins or antioxidants, release of metals into solution, or metal binding at metabolically inactive positions.[305] The diatoms participate in degeneration, speciation, and decontamination of chemical wastes and toxic heavy metals by adsorption and assimilation.[304]

TABLE 6.2 Microbes Having Remediation Potential Against Heavy Metals.

Heavy metal(s)	Category of microbes	Microbes	References
Ag, Cd, Cr, Cu, Fe, Mn, Methyl Hg, Pb, and Zn	Algae	*Chlamydomonas reinhardtii*, *Chlorella* sp., *Cladophora fascicularis*, *Cladophora* sp., *Cystoseira crinitophylla*, *Euglena gracilis*, *Fucus vesiculosus*, *Hydrodictylon*, *Oedogonium* sp., *Nostoc* sp., *Rhizoclonium* sp., *Sargassum filipendula*, *S.* sp., *Scenedesmus* sp., *Spirogyra* sp., *Spirulina* sp., and *Ulva lactuca*	19, 272–278
As, B, Cd, Cr, Cu, Hg, Ni, Pb and Zn	Bacteria	*Acinetobacter lwoffii*, *Arthrobacter viscosus*, *Bacillus aerius*, *B. amyloliquefaciens*, *B. cereus*, *B. licheniformis*, *B. megaterium*, *Bacillus* sp., *B. subtilis*, *Bacillus thuringiensis*, *Bradyrhizobium* sp., *Brevibacterium* sp., *Chryseobacterium* sp., *Cupriavidus metallidurans*, *C.* sp. *Desulfovibrio desulfuricans*, *Enterobacter asburiae*, *E. ludwigii*, *Escherichia coli*, *Kocuria flava*, *Geobacter metallireducens*, *Klebsiella pneumonia*, *K.* sp., *Ochrobactrum* sp., *Rhizobium metallidurans*, *Pseudomonas aeruginosa*, *P. fluorescens*, *P. veronii*, *P. putida*, *P.* sp., *Rhodobacter capsulatus*, *Sporosarcina ginsengisoli*, *Staphylococcus epidermidis*, *Sulfurospirillum barnesii*, and *Vitreoscilla* sp.	19, 277, 279–289
As, Cd, Co, Cr, Cu, Fe, Hg, Mn, Ni, Pb, and Zn	Fungi	*Absidia cylindroslora*, *Aspergillus A. niger*, *A. fumigates*, *A. versicolor*, *Aspergillus* sp., *Candidapara psilosis*, *Cephalotheca foveolata*, *Coniothyrium* sp., *Coprinellus* sp., *Fomitopsis meliae*, *Fusarium* sp., *Gloeophyllum sepiarium*, *Penicillium brevicompactum*, *P. chrysogenum P. simplicissimum*, *Pleurotus ostreatus*, *Rhizopus microsporus*, *R. oryzae*, *Saccharomyces cerevisiae*, *Termitomyces clypeatus Trichoderma brevicompactum*, and *T. ghanense*	19, 277, 290–296

They adsorb heavy metals that then enter the cells through the membrane transporters. Inside the cell, various kinds of organic ligands, glutathione, phytochelatins, and metallothioneins play role in metal complexation and accumulation. Heavy metals are accumulated mainly in the vacuole. An increase in size and number of vacuoles has been found in microalgae because of accumulation of heavy metals. However, heavy metal deposition can also occur in chloroplasts and mitochondria. In this way, diatoms participate in heavy metal removal and survive metal toxicity.[306] The use of algae and their derivatives for heavy metal remediation of adulterated environments is called phycoremediation. In addition to algae, bacteria also hold the potential to remediate heavy metal polluted ecosystems. Bacteria are small, occur abundantly, grow and multiply rapidly, easy to cultivate and are capable to survive in a wide spectrum of environments. Bacteria and fungi assist in bioremediation of heavy metals polluted environments through adsorption by functional groups of EPS, accumulation, and redox state change.[307,308] Mycoremediation is highly advantageous of being a cost-effective and ecofriendly approach to reduce soil and water pollution. Various characteristics such as rapid growth, vast hyphal network, synthesis, and release of ligninolytic enzymes, large surface area to volume ratio, high heavy metal resistance, higher resistance to pH and temperature flucctuations, and synthesis of metal-chelating proteins make fungi suitable for use in bioremediation.[309–312] Fungi adopt adorption, bioaccumulation, and biovolatilization to remove heavy metals. Biovolatilization is the enzymatic transformation of inorganic and organic compounds into volatile forms inside the cells.[313–316] The mycorrhizal fungus, *Funneliformis geosporum* improves soil quality and plant growth in Zn polluted soil by decreasing Zn accumulation in the plant.[317] *Pleurotus ostreatus* removes Mn from polluted water with the help of surfactants. This is because the latter increase the surface area and sites for metal-binding on hyphae which leads to accumulation and effective remediation of Mn.[318] *Aspergillus* sp. removed Cr and Ni from industrial waste water by biosorption.[319] *Trametes pubescence* is capable of removing Ni and Pd by absorption from polluted soil and water due to the activity of laccase.[320] *Rhizomucor*, *Fusarium*, and *Emericella* species play role in remediation of As-contaminated agricultural soils by enhancing the enzyme activity and physio-chemical properties of soil.[310]

6.4.2 MECHANISMS ADOPTED BY MICROBES FOR HEAVY METALS REMEDIATION

Microbes-assisted bioremediation is an economic and ecofriendly approach.[321] Heavy metal polluted environments have diverse microbial phylotypes and it leads to multifold remediation in contrast to pure microbial cultures. Microbes also exhibit diversity in metabolism and myriad of biomolecules to transform heavy metals into nonhazardous forms.[322] The uptake, transformation, and removal of heavy metals depend upon microbial cell wall composition, physiology of the cell, physicochemical parameters such as pH, temperature, contact time, concentration of ions, and initial concentration of metal ions.[323] Microbes (algae, bacteria, and fungi) have various resistance mechanisms to survive in the heavy metal contaminated environments. Beause of these resistance mechanisms, microbes are capable of assisting in heavy metal remediation of polluted environments (Figure 6.3).

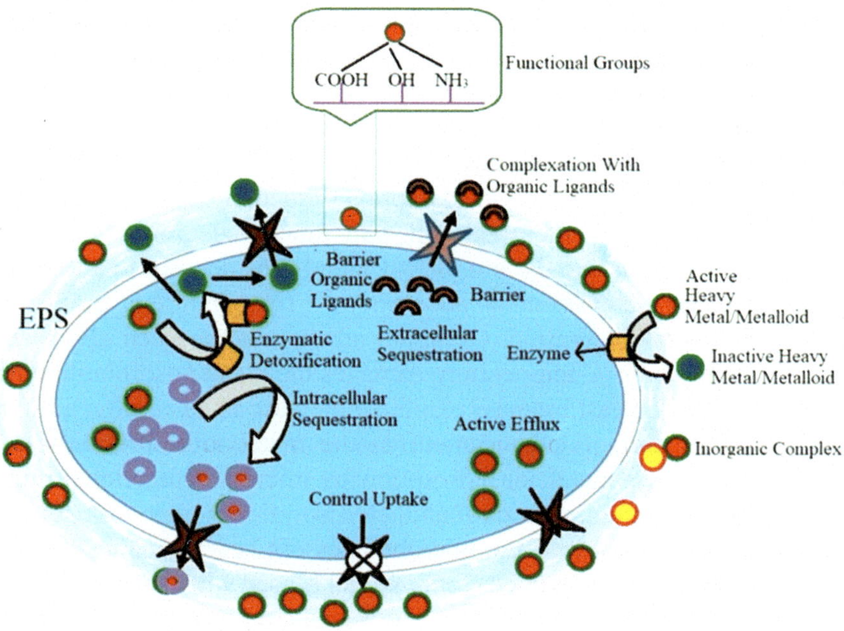

FIGURE 6.3 Mechanisms of microbial-assisted remediation of heavy metals.

6.4.2.1 ADSORPTION

The heavy metal adsorption on the surface of microbes occurs by various biological structures like EPS. EPS act as adsorbents to many heavy metals like Ag, Cr, Cu, Pb, Zn, etc.[315,324–328] EPS are made up of lipids, proteins, nucleic acids, and carbohydrates and exhibit numerous binding sites (amino, carboxylic, hydroxyl, phosphate, and sulphydryl functional groups) for metal ions.[301,329,330] The heavy metal adsorption with EPS such as exopolysaccharides is a nonmetabolic process, does not need energy, and occurs because of electrostatic interaction between oppositely charged metal ions and EPS functional groups.[331] Several factors such as concentration of biomass and its contact time, pH, and temperature affect heavy metal adsorption.[301] Higher biomass concentration increases electrostatic attraction between the adsorbent and the adsorbate, formatting agglomerates and creating hindrance in the effective metal binding. At lower concentrations of biomass, the metal-to-adsorbent ratio improves and consequently the adsorption increases. Adsorption is high when the initial metal ions concentrations are high on the microbial biomass since this offers a positive driving force at high initial concentrations of metal ions that increases the adsorption rate.[332] At greater initial levels of metal ions, competition occurs for the lesser metal ion binding sites of the biomass.[333] High pH of the sorption system enhances adsorption. This is because at low pH, the binding sites of the microbial biomass gain a positive charge because of greater concentration of H^+ ions.[334] The metal cations and H^+ ions then compete for binding sites, causing decrease in adsorption of metal ions on the adsorbent. Moreover, at high pH, microbial cell wall functional groups because of deprotonation of sites for the binding of metal ions acquire total negative charge. This also increases metal adsorption.[335] Temperature is another key factor in the metal adsorption. Optimum adsorption has been observed at 20–35 °C, while >45 °C temperature damages the metal adsorbing biomass proteins. However, yeast biomass exhibited increase in adsorption at higher temperature because of enhancement in the metal ions affinity for the binding sites.[332] The metal ion adsorption by microbial functional groups generally follows Langmuir and Freundlich adsorption isotherm models. The latter model is applicable if the microbes possess heterogeneous surfaces. Moreover, the Brunauer-Emmett-Teller model is used when microbes adsorb heavy metals in multiple layers.[336–339] The Scatchard plots are also employed to illustrate adsorption of heavy metal ions.[340]

6.4.2.2 EXTRACELLULAR SEQUESTRATION

Microbes are equipped with various biochemicals to protect themselves from heavy metal toxicity through extracellular sequestration. Bacteria and fungi produce siderophores e.g. pyoverdine.[341–343] These siderophores form Fe-complexes and assist microbes to overcome Fe deficiency. Besides Fe, siderophores are capable of forming complexes with Cu, Ni, and Zn and defend microbes from heavy metal toxicity.[344–346] Microbes also secrete glutathione that forms complexes with heavy metal ions and play role in their adsorption but not in their entry into the living cells. Hydrogen sulfide produced by microbes also helps in heavy metal detoxification. *Desulfovibrio desulfuricans* has been observed unaffected at heavy metal ions high concentrations due to hydrogen sulfide production. The excreted hydrogen sulfide precipitates the heavy metal ions and prevents cellular toxocity.[347–349] Biosurfactants are amphiphilic molecules that reduce surface and interfacial tension between surfaces. These are used in nutraceutical, pharmaceutical, and cosmetic industry as well as for remediation of heavy metal polluted environments.[350] Biosurfactants have characterists making them more beneficial than the synthetic surfactants.[351,352] Heavy metal excess increases production of biosurfactants.[353] A biosurfactant removes heavy metals by complexing or accumulation at a solid solution interface leading to an interaction between metal and biosurfactant.[354] The biosurfactant–metal complex leaves the soil surface by desorption and makes micelles. The separation of biosurfactant from the metal complex can also occur by precipitation.[355] or by bringing the pH less than 5. The recovered heavy metals can be reused and this decreases the requirement of ore mining or production of heavy metals.[356] The anionic biosurfactants from the yeast, *Candida sphaerica*, removed Fe (95%), Zn (90%), and Pb (79%) from metal-contaminated soil. These biosurfactants also recovered Cd (87%) and Pb (75%) from aqueous solution.[357] An anionic biosurfactant from *C. tropicalis* grown in distilled water with 2.5% molasses, 2.5% frying oil, and 4% corn steep liquor has been found effective in removing Cu, Pb, and Zn.[358] A marine sponge-associated bacteria (MSI 54), found as *Bacillus* sp., produced a biosurfactant (an anionic lipopeptide) with high affinity for Cd, Hg, Mn, and Pb. This biosurfactant at a 2.0× critical micelle concentration (CMC) removed Cd (99.93%), Hg (75.5%), Mn (89.5%), and Pb (97.73%) from 1000 ppm of the respective metal solution. The fresh cabbage, carrot, and lettuce after treartment with 2.0× CMC of lipopeptide exhibited effecient reduction of heavy metals from the surface.[359] These biosurfactants of microbial origin also play a role in bioleaching of heavy metals. For example, the microbial biosurfactants (acidic- and bolaform glycolipids) from various strains of *Starmerella bombicola* yeast have exhibited the potential of Cu and Zn leaching from low-grade secondary materials.[360] Bioleaching with mixed bacterial culture of oxidizing bacteria (*Acidithiobacillus thiooxidans* and *A. ferrooxidans*) and rhamnolipids biosurfactants help in the removal of Fe and Zn. The use of oxidizing bacteria and rhamnolipids in combination futher improved the metal removal percentage.[361] Microbes also secrete organic compounds that bind to heavy metal ions through chelation. For example, citrate (chelating with Al^{3+}), oxalic acid (binding with Cu and Cd), glomalin protein (binding with Cu, Pb, and Cd), chitin (binding with Cd), and polyphenols (binding with Cd) are secreted by microbes in the extracellular region for metal chelation.[362]

6.4.2.3 BIOACCUMULATION

The uptake of heavy metals inside the microbes takes place through membrane transporters. A single-membrane transport system can take up multiple metal ions.[300–305,363] The import of heavy metals occurs through membrane channels, secondary carriers, and primary active transporters. These occur in the inner lipid membrane, such as GlpF, Fps, and Mer T/P in channel transporters; NixA, Hxt7, and Pho84 in secondary carriers; and MntA, cdtB/Ip_3327, TcHMA3, and CopA in primary active transporters. The outer lipid membrane consists of some membrane transporters like the porin channel transporters.[364] Channel transporters do not need energy for the import of compounds while secondary carriers and primary active transporters are energy dependent. Inside microbes, proteins and peptide ligands (i.e., storage system) play a role in heavy metal sequestration. When the rate of absorption exceeds the rate of desorption, the storage or accumulation of heavy metals starts. This process of accumulation is called bioaccumulation. To overcome the limitations of natural microbial bioaccumulation of heavy metals such as continuous nutrient supply for the continuous growth and multiplication of microbes, maintenance of aeration to accommodate aerobic/anaerobic requirements, accidental emergence of genetically modified microbes, and genetically engineered microbes might prove better alternatives.[365] The genetically engineered yeast has been observed exhibiting plant-like characteristics and acting as a hyperaccumulator of various heavy metals present in the contaminated aquatic environment.[366]

6.4.2.4 INTRACELLULAR SEQUESTRATION

To prevent the heavy metal ions form reaching their elevated toxic levels, microbes exhibit intracellular sequestration. Many microbes form intracellular metal complexes with ligands such as sulfides, cysteine-rich metallothioneins, and cytosolic polyphosphates. The *Cryptococcus humicola* cells store Cd(II) and Co(II) polyphosphate to remediate heavy metals.[367,368] Some other bacteria and cyanobacteria also deposit the cytosolic polyphosphates of heavy metal ions.[369] The metallothionein-like proteins possess cysteine residues acting as a sink for heavy metals present in excess. A significant correlation has been found between Hg and sulfhydryl groups of metallothioneins.[370] Under heavy metal excess, metallothioneins overexpress to overcome the heavy metal excess. Metallothioneins in *Synechococcus* sp. bind to Cd(II) and Zn(II) to decrease their toxicity.[371] Some organisms upregulate their enzymes, producing amino acids and organic acids used to make complexes with heavy metals naturally. *Acidithiobacillus ferrooxidans* ATCC 23270 has been observed to enhance the histidine biosynthesis under 40 mM $CuSO_4$ application. This cytoplasmic histidine chelates Cu ions and provides protection from oxidative damage.[372]

6.4.2.5 ENZYMATIC DETOXIFICATION (REDOX STATE CHANGE)

Microbes are efficient in changing the ionic state of metal ions through oxidation, reduction, methylation, and demethylation that in turn influence the availability, solubility, and mobility of metal ions in aquatic and terrestrial ecosystems.[322, 373] Heavy metals occur in more than one ionic form and more commonly at higher oxidation states. The heavy metal ions are less soluble at higher oxidation state compared to the lower oxidation state. The enzymes found in microbes change the oxidation state of metal ions from higher to lower and enhance their solubility, for example, oxidation of U caused by *Thiobacillus ferrooxidans* and *T. thiooxidans* enzymatic.[374] The enzyme catalyzed transformation of Cr(VI) to Cr(III) results in the formation of insoluble hydroxides. Upflow anaerobic packed bed reactors, based on sulfur reducing bacteria (SRB), are useful to remediate Al(III), Cu(II), Fe(III), Mg(II), Ni(II), and Zn(II) from contaminated waters.[375] The SRB also function with Cu–Fe bimetallic particles to promote remediation and to reduce toxicity potential of heavy metals.[376] Under aerobic conditions, *Bacillus amyloliquefaciens* uses glucose to reduce Cr(VI).[378] The oxidation of phenols can also be coupled to the reduction of Cr(VI).[377] The reduction of Hg(II) to a less toxic and volatile Hg0 form is catalyzed by merA reductase, for example in *Pseudomonas putida* FB1 and *Pseudomonas* sp. B50A.[379–382] The genetically engineered bacterial strains consisting of merA are more efficient to convert Hg(II) into Hg0.[383] Some microbes are capable of converting As to less toxic methylated arsenic compounds. Moreover, As(III) can be oxidized to As(V) to reduce its toxigenic potential. The As(III) oxidation occurs slowly in natural conditions but microbes such as *Thermus* sp. enhance As(III) oxidation faster than abiotic processes.[384] Microbial enzyme catalyzed heavy metal ion reduction has been noticed in the laboratory scale paper pulp deinking method.[385] Microbial enzymes like arylsulfatase, β-glucosidase, and dehydrogenase improve soil quality in situ by removing heavy metals.[386] The biotransformation percentage of the heavy metal ion disintegrating microorganisms has been found to be 31%, 20%, 27%, 7%, 22%, 5%, and 18% for Cd, Co, Cr, Ni, Pb, Zn, and (As and Hg), respectively.[388] The benefit to utilize enzymes for bioremediation is that these can be easily detected before, during, and after remediation and act as biological sensors for heavy metal concentrations.[277]

6.4.2.6 ACTIVE EFFLUX

In addition to the presence of membrane transporters for the exit of heavy metals, microbes have the potential to alter the biochemical pathways to stop metal uptake. Microbial cell membrane possesses abundant metal-exporting proteins like P-type efflux ATPase, proton–cation antiporters, ABC transporters, and cation diffusion facilitator. Arsenite efflux occurs by a cell membrane-exporting protein accompanied assistance of ATPase.[388,389] ABC transporters participate to remove metal ions and thereby save microbes from toxic heavy metals.[390–392] Membrane transporters are present in both prokaryotes and eukaryotes. For example, P-type ATPases (ZntA and CadA) in bacteria offering Zn(II), Cd(II), and Pb(II) resistance; an *Escherichia coli* pump, CopA, for efflux of Cu(I); several different families of transporters for arsenicals and antimonials (such as *E. coli* ArsAB ATPase offering As(III) and Sb(III) resistance and *S. cerevisiae* transporters (Acr3p and Ycf1p) confering arsenite[As(III)] resistance). The resistance to arsenate.[As(V)] occurs by the reduction of As(V) to As(III) by the enzyme arsenate reductase. The latter enzyme has three families, two in bacteria and one in *S. cerevisiae.*[393]

6.4.3 FACTORS AFFECTING MICROBIAL-ASSISTED REMEDIATION

In microbial-assisted bioremediation, the selection of microbe is the most important thing for successful removal of pollutants from the adulterated ecosystem. The size, growth and multiplication rate, heavy metal uptake, and accumulation efficiency of the microbes are the factors accounting for the success of microbial-assisted bioremediation. Microbes with small size, rapid multiplication, multiple metal uptake efficiency, high metal accumulation, and metal detoxification potential are the best suited for bioremediation. The remediation efficacy of the microbes is affected adversely with unwanted waste substances, inadequate supply of nutrients, unfavorable environmental conditions (aeration, moisture, pH, temperature), and low availability of the pollutant. In other words, microbes perform remediation of heavy metals effectively only under optimum nutrient supply, pH, and favorable environmental conditions that in turn increase the synthesis of biomolecules, protecting the microbes from abiotic stress and improving their activity.[394–396]

Plant community composition and seasonal fluctuations affect the microbial respiration, biomass, and activity.[397] Soil microbial diversity of mixed forests is greater than the pure forests and hence bioremediation is higher in mixed forests.[398] Microbial-assisted bioremediation should be a safe technology with no risk to the environment and people as it involves the use of microbes found naturally in the ecosystem. Microbes have the potential to accumulate heavy metals and can cause further contamination. Therefore, after heavy metal uptake these must be eliminated from the contaminated place.

6.5 FUTURE SCOPE AND CHALLENGES OF BIOREMEDIATION

To save environment from the hazardous effects of heavy metal pollution, various measures have already been taken.[399] Further research should be done on the physical, chemical, and physiological behavior of microbes in heavy metal-contaminated soil, water, and gaseous environments to develop techniques leading to increased heavy metal tolerance in microbes.[322] Since algal microbes are potent microbes for heavy metal sorption from the soil, future research should focus on algae in context to improve their bioremediation efficiency. However, more applied research is required to be done in context to microbes mediated heavy metal remediation. Techniques based on CRISPER can be employed on various mycobacteria and fungi.[400] The omics-based biotechnology develops industrial microbial strains resilient to the prevailing environmental conditions because of their improved genetic capabilities and heavy metal tolerance.[401,402] Advancement of new biotechnologies is the need of the hour to promote the use of non-model organisms for heavy metal uptake, for example, removal/uptake of heavy metals from agricultural soils.[403] The combination of microbes with advanced materials can also bring highly efficient heavy metal ion remediation.[404] However, this needs further investigations. Advances in phytoremediation-integrated approaches such as phytoremediation-chemical, phytoremediation-microbial, and phytoremediation-genetic engineering, changing heavy metals availability, enhancing plant tolerance, improving plant growth, increasing phytoextraction and phytostabilization etc. may bring imrovements in phytoremediation.[405] In addition, plant breeding and plant biotechnology programs must be employed to improve the remediation properties of plant root exudates which in turn improve the heavy metal remediation efficacy of rhizobacteria.[406] Although the microbes (algae, bacteria, and fungi) and plants used for bioremediation are easily accessible, biological remediation alone is not enough to treat heavy metals and restore the contaminated sites. To sum up, conventional approaches merged with modern technologies might prove a promising approach to accelerate the heavy metal bioremediation process. Moreover, a comprehensive search of an area-wise and pollutant-type database is much needed to decide which area and pollutant needs preferential and urgent remediation.

6.6 CONCLUSION

Bioremedition is a effective, less costly, and ecofriendly approach to combat heavy metal pollution in contarst to physical and chemical methods. Plants and microbes possess various heavy metal resistance mechanisms and are capable of withstanding heavy metal toxicity. Because of this inherent tendency of metal tolerance, plants and microbes are utilized for the restoration of polluted environments (with organic pollutants and heavy metals). Phytoremediation involves the mechanisms of phytoextraction/phytoaccumulation, phytostabilization, phytovolatilization, phytodegradation, and phyto-microbial remediation. Microbial-assisted remediation utilizes algae, bacteria, and fungi consecrated with various structural and physiological characteristics useful for the restoration of heavy metals adulterated ecosystems. These microbes use mechanisms namely adsorption, bioaccumulation, extracellular sequestration, intracellular sequestration, enzymatic detoxification (redox state change), and active efflux to remove heavy metals. Phytoremediation and microbial bioremediation in combination, that is, phyto-microbial remediation system is a more effective biremediation strategy. Bioremediation can be further improved by using various modern-day technologies.

KEYWORDS

- **heavy metal pollution**
- **bioremediation**
- **microbes**
- **plants**
- **hyperaccumulation**
- **environment**
- **toxicity**

REFERENCES

1. El-Kady, A. A.; Abdel-Wahhab, M. A. Occurrence of Trace Metals in Foodstuffs and their Health Impact. *Trends Food Sci. Technol.* **2018,** *75*, 36–45.
2. Bakshi, S.; Banik, C.; He, Z. The Impact of Heavy Metal Contamination on Soil Health. In *Managing Soil Health for Sustainable Agriculture Volume 2: Monitoring and Management*; Reicosky, D., Ed.; Burleigh Dodds Science Publishing: Cambridge, UK, 2018.
3. Hayyat, M. S.; Adnan, M.; Awais, M.; Toor, M. D.; Bilal, H. M. Effect of Heavy Metal (Ni) on Plants and Soil: A Review. *IJAR* **2020,** *6* (7), 313–318.
4. Madhu, P. M.; Sadagopan, R. S. Effect of Heavy Metals on Growth and Development of Cultivated Plants with Reference to Cadmium, Chromium and Lead-A Review. *J. Stress Physiol. Biochem.* **2020,** *16* (3), 84–102.
5. Zhang, H.; Xu, Z.; Guo, K.; Huo, Y.; He, G.; Sun, H.; Guan, Y.; Xu, N.; Yang, W.; Sun, G. Toxic Effects of Heavy Metal Cd and Zn on Chlorophyll, Carotenoid Metabolism and Photosynthetic Function in Tobacco Leaves Revealed by Physiological and Proteomics Analysis. *Ecotoxicol. Environ. Saf.* **2020,** *202*, 110856.
6. Wang, X.; Zhang, C.; Qiu, B.; Ashraf, U.; Azad, R.; Wu, J.; Ali. S.; Biotransfer of Cd along a Soil-Plant-Mealybug-Ladybird Food Chain: A Comparison with Host Plants. *Chemosphere* **2017,** *168,* 699–706.
7. Bornhorst, J.; Kipp, A. P.; Haase, H.; Meyer, S.; Schwerdtle, T. The Crux of Inept Biomarkers for Risks and Benefits of Trace Elements. *Trac. Trends Anal. Chem.* **2018,** *104*, 183–190.
8. Khalid, S.; Shahid, M.; Niazi, K. N.; Murtaza, B.; Bibi, I.; Dumat, C. A Comparison of Technologies for Remediation of Heavy Metal Contaminated Soils. *J. Geochem. Explor.* **2016,** *182*, 247–268.
9. Azimi, A.; Azari, A.; Rezakazemi, M.; Ansarpour, M. Removal of Heavy Metals from Industrial Wastewaters: A Review. *Chem. Bio. Eng. Rev.* **2017,** *4*, 37–59.
10. Fang, X.; Li, J.; Li, X.; Pan, S.; Zhang, X.; Sun, X.; Shen, J.; Han, W.; Wang, L. Internal Pore Decoration with Polydopamine Nanoparticle on Polymeric Ultrafiltration Membrane for Enhanced Heavy Metal Removal. *Chem. Eng. J.* **2017,** *314*, 38–49.
11. Zamri, M. F. M. A.; Kamaruddin, M. A.; Yusoff, M. S.; Aziz, H. A.; Foo, K. Y. Semi-Aerobic Stabilized Landfill Leachate Treatment by Ion Exchange Resin: Isotherm and Kinetic Study. *Appl. Water Sci.* **2017,** *7*, 581–590.
12. d'Halluin, M.; Rull-Barrull, J.; Bretel, G.; Labrugere, C.; Le Grognec, E.; Felpin. F. X. Chemically Modified Cellulose Filter Paper for Heavy Metal Remediation in Water. *ACS Sustain. Chem. Eng.* **2017,** *5* (2), 1965–1973.
13. Emamverdian, A.; Ding, Y.; Mokhberdoran, F.; Xie, Y. Heavy Metal Stress and Some Mechanisms of Plant Defense Response. *Sci. World J.* **2015,** *2015*, 756120.
14. Rastgoo, L.; Alemzadeh, A.; Afsharifar, A. Isolation of Two Novel Isoforms Encoding Zinc-and Copper-Transporting P1b-Atpase from Gouan (*Aeluropus Littoralis*). *Plant Omics J.* **2011,** *4*, 377–383.
15. Sharma, P.; Jha, A. B.; Dubey, R. S.; Pessarakli, M. Reactive Oxygen Species, Oxidative Damage, and Antioxidative Defense Mechanism in Plants under Stressful Conditions. *J. Bot.* **2012,** 2012.
16. Singh, A.; Prasad, S. M.; Singh, S.; Singh, M. Phytoremediation Potential of Weed Plants' Oxidative Biomarker and Antioxidant Responses. *Chem. Ecol.* **2016,** *32*, 684–706.

17. Sharma, R.; Bhardwaj, R.; Handa, N.; Gautam, V.; Kohli, S. K.; Bali, S.; Kaur, P.; Thukral, A. K.; Arora, S.; Ohri, P.; Vig, A. P. Responses of Phytochelatins and Metallothioneins in Alleviation of Heavy Metal Stress in Plants: An Overview. *Plant Met. Interact.* **2016,** 263–283.
18. Dhalaria, R.; Kumar, D.; Kumar, H.; Nepovimova, E.; Kuča, K.; Torequl, I. M.; Verma, R. Arbuscular Mycorrhizal Fungi as Potential Agents in Ameliorating Heavy Metal Stress in Plants. *Agronpmy* **2020,** *10* (6), 815.
19. Yin, K.; Wang, Q.; Lv, M.; Chen, L. Microorganism Remediation Strategies Towards Heavy Metals. *Chem. Eng. J.* **2019,** *360*, 1553–1563.
20. Haldar, S.; Ghosh, A. Microbial and Plant-Assisted Heavy Metal Remediation in Aquatic Ecosystems: A Comprehensive Review. *3 Biotech.* **2020,** *10* (5), 1–3.
21. Alvarez, A.; Saez, J. M.; Costa, J. S. D.; Colin, V. L.; Fuentes, M. S.; Cuozzo, S. A.; Benimeli, C. S.; Polti M. A.; Amoroso, M. J. Actinobacteria: Current Research and Perspectives of Bioremediation of Pesticides and Heavy Metals. *Chemosphere* **2017,** *166*, 41–62.
22. Bhattacharya, S. The Role of *Spirulina* (*Arthrospira*) in the Mitigation of Heavy Metal Toxicity: An Appraisal. *J. Environ. Pathol. Toxicol. Oncol.* **2020,** *39* (2), 149–157.
23. Pourrut, B.; Jean, S.; Silvestre, J.; Pinelli, E. Lead-induced DNA Damage in *Vicia Faba* Root Cells: Potential Involvement of Oxidative Stress. *Mutat. Res.* **2011,** *726*, 123–128.
24. Pourrut, B.; Shahid, M.; Dumat, C.; Winterton, P.; Pinelli, E. Lead Uptake, Toxicity, and Detoxification in Plants. *Rev. Environ. Contam. Toxicol.* **2011,** *213*, 113–136.
25. Shahid, M.; Khalid, S.; Abbas, G.; Shahid, N.; Nadeem, M.; Sabir, M.; Aslam, M.; Dumat, C. Heavy Metal Stress and Crop Productivity; in *Crop Production and Global Environmental Issues*; Hakeem, K. R., Ed.; Springer International Publishing, 2015; pp. 1–25.
26. United Nations Environmental Protection/Global Program of Action (UNEP/GPA). Why the Marine Environment Needs Protection from Heavy Metals. *Heavy Metals.* UNEP/GPA, 2004.
27. Shakoor, M.; Niazi, N.; Bibi, I.; Rahman, M.; Naidu, R.; Dong, Z.; Shahid, M.; Arshad, M. Unraveling Health Risk and Speciation of Arsenic from Ground Water in Rural Areas of Punjab, Pakistan. *Int. J. Environ. Res. Public Health* **2015,** *12*, 12371–12390.
28. Shankar, B. S. A Critical Assay of Heavy Metal Pollution Index for the Ground Waters of Peenya Industrial Area, Bangalore, India. *Environ. Monitor. Assess.* **2019,** *191* (5), 289.
29. Muhammad, S.; Ullah, R.; Jadoon, I. A. K. Heavy Metals Contamination in Soil and Food and their Evaluation for Risk Assessment in the Zhob and Loralai Valleys, Baluchistan Province. *Pak. Microchem. J.* **2019,** *149*, 103971.
30. Nandakumar, P. B. A.; Dushenkov, V.; Motto, H.; Raskin, I. Phytoextraction: The Use of Plants to Remove Heavy Metals from Soils. *Environ. Sci. Technol.* **1995,** *29*, 1232–1238.
31. Khalid, S.; Shahid, M.; Niazi, N. K.; Murtaza, B.; Bibi, I.; Dumat, C. A Comparison of Technologies for Remediation of Heavy Metal Contaminated Soils. *J. Geochem. Explor.* **2017,** *182*, 247–268.
32. Singh, J.; Kalamdhad, A. S. Effects of Heavy Metals on Soil, Plants, Human Health and Aquatic Life. *Int. Res. Chem. Environ.* **2011,** *1*, 15–21.
33. Jaishanker, M.; Tseten, T.; Anbalagan, N.; Mathew, B. B.; Beeregowda, K. N. Toxicity, Mechanism and Health Effects of Some Heavy Metals. *Interdiscip. Toxicol.* **2014,** *7*, 60–72.

34. Bhatt, K.; Maheshwari, D. K. Decoding Multifarious Role of Cow Dung Bacteria in Mobilization of Zinc Fractions along with Growth Promotion of *C. Annuum* L. *Sci. Rep.* **2019,** *9* (1), 14232.
35. Goyal, D.; Yadav, A.; Prasad, M.; Singh, T. B.; Shrivastav, P.; Ali, A.; Dantu, P. K., Mishra, S. Effect of Heavy Metals on Plant Growth: An Overview. In: *Contaminants in Agriculture*; Naeem, M., Ansari, A., Gill, S., Eds.; Springer: Cham. 2020; pp. 79–101.
36. Hong, Y. J.; Liao, W.; Yan, Z. F.; Bai, Y. C.; Feng, C. L.; Xu, Z. X.; Xu, D. Y. Progress in the Research of the Toxicity Effect Mechanisms of Heavy Metals on Freshwater Organisms and their Water Quality Criteria in China. *J. Chem.* **2020,** (2020).
37. Mombo, S.; Foucault, Y.; Deola, F.; Gaillard, I.; Goix, S.; Shahid, M.; Schreck, E.; Pierart, A.; Dumat, C. Management of Human Health Risk in the Context of Kitchen Gardens Polluted by Lead and Cadmium Near a Lead Recycling Company. *J. Soils Sediments* **2015,** *16*, 1214–1224.
38. Xiong, T.; Dumat, C.; Pierart, A.; Shahid, M.; Kang,Y.; Li, N.; Bertoni, G.; Laplanche, C. Measurement of Metal Bioaccessibility in Vegetables to Improve Human Exposure Assessments: Field Study of Soil-Plant-Atmosphere Transfers in Urban Areas, South China. *Environ. Geochem. Health* **2016,** *38*, 1283–1301.
39. Awa, S. H.; Hadibarata, T. Removal of Heavy Metals in Contaminated Soil by Phytoremediation Mechanism: A Review. *Water Air Soil Pollut.* **2020,** *231* (2), 1–5.
40. Jaiswal, A.; Verma, A.; Jaiswal, P. Detrimental Effects of Heavy Metals in Soil, Plants, and Aquatic Ecosystems and in Humans. *J. Environ. Pathol. Toxicol. Oncol.* **2018,** *37* (3), 183–197.
41. Wei, Z.; Van L. Q.; Peng, W.; Yang, Y.; Yang, H.; Gu, H.; Lam, S. S.; Sonne, C. A Review on Phytoremediation of Contaminants in Air, Water and Soil. *J. Hazard. Mater.* **2021,** *403*, 123658.
42. Kurade, M. B.; Ha, Y. H.; Xiong, J. Q.; Govindwar, S. P.; Jang, M.; Jeon, B. H. Phytoremediation as a Green Biotechnology Tool for Emerging Environmental Pollution: A Step forward Towards Sustainable Rehabilitation of the Environment. *Chem. Eng. J.* **2021,** *19*, 129040.
43. Narayanan, M.; Natarajan, D.; Kandasamy, G.; Kandasamy, S.; Shanmuganathan, R.; Pugazhendhi, A. Phytoremediation Competence of Short Term Crops on Magnesite Mine Tailing. *Chemosphere* **2021,** *270*, 128641.
44. Parihar, J. K.; Parihar, P. K.; Pakade, Y. B.; Katnoria, J. K. Bioaccumulation Potential of Indigenous Plants for Heavy Metal Phytoremediation in Rural Areas of Shaheed Bhagat Singh Nagar, Punjab (India). *Environ. Sci. Pollut. Res.* **2021,** *28* (2), 2426–2442.
45. Zulfahmi, I.; Kandi, R. N.; Huslina, F.; Rahmawati, L.; Muliari, M.; Sumon, K. A.; Rahman, M. M. Phytoremediation of Palm Oil Mill Effluent (POME) Using Water Spinach (*Ipomoea Aquatica* Forsk). *Environ. Technol. Innov.* **2021,** *21*, 101260.
46. Yan, L.; Van, L. Q.; Sonne, C.; Yang, Y.; Yang, H.; Gu, H.; Ma, N. L.; Lam, S. S.; Peng, W. Phytoremediation of Radionuclides in Soil, Sediments and Water. *J. Hazard. Mater.* **2021,** *407*, 124771.
47. Lee, B. X.; Hadibarata, T.; Yuniarto, A. Phytoremediation Mechanisms in Air Pollution Control: A Review. *Water Air Soil Pollut.* **2020,** *231* (8), 1–3.
48. Gacia, E.; Soto, D. X.; Roig, R.; Catalan, J. Phragmites Australis as A Dual Indicator (Air and Sediment) of Trace Metal Pollution in Wetlands-the Key Case of Flix Reservoir (Ebro River). *Sci. Total Environ.* **2021,** *765*, 142789.

49. Alkorta, I.; Becerril, J. M.; Garbisu, C. Phytostabilization of Metal Contaminated Soils. *Rev. Environ. Health* **2010,** *25*, 135–146.
50. Guo, P.; Wang, T.; Liu, Y.; Xia, Y.; Wang, G.; Shen, Z.; Chen, Y. Phytostabilization Potential of Evening Primrose (*Oenothera Glazioviana*) for Copper-Contaminated Sites. *Environ. Sci. Pollut. Res.* **2014,** *21*, 631–640.
51. Fatnassi, I. C., Chiboub, M.; Saadani, O.; Jebara, M.; Jebara, S. H. Phytostabilization of Moderate Copper Contaminated Soils Using Co-inoculation of *Vicia Faba* with Plant Growth Promoting Bacteria. *J. Basic Microbiol.* **2015,** *55*, 303–311.
52. Garcia-Sanchez, M.; Kosnar, Z.; Mercl, F.; Aranda, E.; Tlustos, P. A Comparative Study to Evaluate Natural Attenuation, Mycoaugmentation, Phytoremediation and Microbial Assisted Phytoremediation Strategies for the Bioremediation of an Aged PAH-Polluted Soil. *Ecotoxicol. Environ. Saf.* **2018,** *147*, 165–174.
53. Sun, Y. B.; Zhou, Q. X.; Liu, W. T.; An, J.; Xu, Z. Q.; Wang, L. Joint Effects of Arsenic and Cadmium on Plant Growth and Metal Bioaccumulation: A Potential Cd Hyperaccumulator Cd Hyperaccumulator and As-Excluder *Bidens Pilosa* L. *J. Hazard. Mater.* **2009,** *165*, 1023–1028.
54. Pignattelli, S.; Colzi, I.; Buccianti, A.; Cecchi, L.; Arnetoli, M.; Monnanni, R.; Gabbrielli, R.; Gonnelli, C. Exploring Element Accumulation Patterns of a Metal Excluder Plant Naturally Colonizing a Highly Contaminated Soil. *J. Hazard. Mater.* **2012,** *227–228*, 362–369.
55. Chamba, I.; Rosado, D.; Kalinhoff, C.; Thangaswamy, S.; Sánchez-Rodríguez, A.; Gazquez, M. J. Erato Polymnioides-A Novel Hg Hyperaccumulator Plant in Ecuadorian Rainforest Acid Soils with Potential of Microbe Associated Phytoremediation. *Chemosphere* **2017,** *188*, 633–641.
56. Çelik, Ö.; Akdaş, E. Y. Tissue-Specific Transcriptional Regulation of Seven Heavy Metal Stress-Responsive Mirnas and their Putative Targets in Nickel Indicator Castor Bean (*R. Communis* L.) Plants. *Ecotoxicol. Environ. Saf.* **2019,** *170*, 682–690.
57. Ashraf, M.; Ahmad, M. S. A.; Ozturk, M. *Plant Adaptation and Phytoremediation.* Springer: Berlin, 2010.
58. Ahlfeld, D. P.; Heidari, M. Applications of Optimal Hydraulic Control to Ground-Water Systems. *J. Water Resour. Plan Manag.* **1994,** *120*, 350–365.
59. Mays, L. W.; Todd, D. K. *Groundwater Hydrology*; 3rd Edn.; John Wily and Sons, Inc., Arizona State University, 2005.
60. Nwoko, C. O. Trends in Phytoremediation of Toxic Elemental and Organic Pollutants. *Afr. J. Biotechnol.* **2010,** *9*, 6010–6016.
61. Nyer, E. K.; Gatliff, E. G. Phytoremediation. *Groundw. Monit. Remediat.* **1996,** *16*, 58–62.
62. Ferro, A.; Gefell, M.; Kjelgren, R.; Lipson, D. S.; Zollinger, N.; Jackson, S. Maintaining Hydraulic Control Using Deep Rooted Tree Systems. In *Advances in Biochemical Engineering/Biotechnology*; Springer, 2003; pp. 125–156.
63. Hirsh, S. R.; Compton, H. R.; Matey, D. H.; Wrobel, J. G.; Schneider, W. H. Five-Year Pilot Study: Aberdeen Proving Ground, Maryland. *Phytoremediat: Transform. Control Contam.* **2003,** 635–659
64. Negri, M. C.; Gatliff, E. G.; Quinn, J. J.; Hinchman, R. R. Root Development and Rooting at Depths. *Phytoremediat: Transform. Control Contam.* **2004,** *1*, 233–262.
65. Muthusaravanan, S.; Sivarajasekar, N.; Vivek, J. S.; Paramasivan, T.; Naushad, M.; Prakashmaran, J.; Gayathri, V.; Al-Duaij, O. K. Phytoremediation of Heavy Metals: Mechanisms, Methods and Enhancements. *Environ. Chem. Lett.* **2018,** *11* (4), 1339–1359.

66. Ali, S.; Abbas, Z.; Rizwan, M.; Zaheer, I. E.; Yavaş, İ.; Ünay, A.; Abdel-Daim, M. M.; Bin-Jumah, M.; Hasanuzzaman, M.; Kalderis, D. Application of Floating Aquatic Plants in Phytoremediation of Heavy Metals Polluted Water: A Review. *Sustainability* **2020,** *12* (5), 1927.
67. Rezania, S.; Ponraj, M.; Talaiekhozani, A.; Mohamad, S. E.; Md Din, M. F.; Taib, S. M.; Sabbagh, F.; Sairan, F. M. Perspectives of Phytoremediation Using Water Hyacinth for Removal of Heavy Metals, Organic and Inorganic Pollutants in Wastewater. *J. Environ. Manag.* **2015,** *163*, 125–133.
68. Eze, V. C.; Harvey, A. P. Extractive Recovery and Valorization of Arsenic from Contaminated Soil Through Phytoremediation Using *Pteris Cretica. Chemosphere* **2018,** *208*, 484–492.
69. Bai, L.; Liu, X. L.; Hu, J.; Li, J.; Wang, Z. L.; Han, G.; Li, S. L.; Liu, C. Q. Heavy Metal Accumulation in Common Aquatic Plants in Rivers and Lakes in the Taihu Basin. *Int. J. Environ. Res. Public Health* **2018,** *15* (12), E2857.
70. Sricoth, T.; Meeinkuirt, W.; Saengwilai, P.; Pichtel, J.; Taeprayoon, P. Aquatic Plants for Phytostabilization of Cadmium and Zinc in Hydroponic Experiments. *Environ. Sci. Pollut. Res. Int.* **2018,** *25* (15), 14964–14976.
71. Sharma, S.; Singh, B.; Manchanda, V. K. Phytoremediation: Role of Terrestrial Plants and Aquatic Macrophytes in the Remediation of Radionuclides and Heavy Metal Contaminated Soil and Water. *Environ. Sci. Pollut. Res. Int.* **2015,** *22* (2), 946–962.
72. Mkandawire, M. Biogeochemical Behaviour and Bioremediation of Uranium in Waters of Abandoned Mines. *Environ. Sci. Pollut. Res. Int.* **2013,** *20* (11), 7740–7767.
73. Sood, A.; Uniyal, P. L.; Prasanna, R.; Ahluwalia, A. S. Phytoremediation Potential of Aquatic Macrophyte, *Azolla. Ambio* **2012,** *41* (2), 122–137
74. Zhao, F. J.; Dunham, S. J.; Mcgrath, S. P. Arsenic Hyperaccumulation by Different Fern Species. *New Phytol.* **2002,** *156*, 27–31.
75. Mains, D.; Craw, D.; Rufaut, C.; Smith, C. Phytostabilization of Gold Mine Tailings, New Zealand. Part 1: Plant Establishment in Alkaline Saline Substrate. *Int. J. Phytoremediat.* **2006,** *8*, 131–147.
76. Mains, D.; Craw, D.; Rufaut, C.; Smith, C. Phytostabilization of Gold Mine Tailings from New Zealand. Part 2: Experimental Evaluation of Arsenic Mobilization during Revegetation. *Int. J. Phytoremediat.* **2006,** *8*, 163–183.
77. Kalve, S.; Sarangi, B. K.; Pandey, R. A.; Chakrabarti, T. Arsenic and Chromium Hyperaccumulation by an Ecotype of *Pteris Vittata*-Prospective for Phytoextraction from Contaminated Water and Soil. *Curr. Sci.* **2011,** *100*, 888–894.
78. Sakakibara, M.; Ohmori, Y.; Ha, N. T.; Sano, S.; Sera, K. Phytoremediation of Heavy Metal-Contaminated Water and Sediment by *Eleocharis Acicularis*. In *CLEAN Soil Air Water* **2011,** *31* (8), 735–741.
79. García-Salgado, S.; García-Casillas, D.; Quijano-Nieto, M. A.; Bonilla-Simón, M. M. Arsenic and Heavy Metal Uptake and Accumulation in Native Plant Species from Soils Polluted by Mining Activities. *Water Air Soil Pollut.* **2012,** *223*, 559–572.
80. Su, Y.; Han, F. X.; Chen, J.; Sridhar, B. M.; Monts, D. L. Phytoextraction and Accumulation of Mercury in Three Plant Species: Indian Mustard (*Brassica Juncea*), Beard Grass (*Polypogon Monospeliensis*) and Chinese Brake Fern (*Pteris Vittata*). *Int. J. Phytoremed.* **2008,** *10*, 547–560.
81. Xie, Q. E.; Yan, X. L.; Liao, X. Y.; Li, X. The Arsenic Hyperaccumulator Fern *Pteris Vittata* L. *Environ. Sci. Technol.* **2009,** *43*, 8488–8495.

82. Rathinasabapathi, B. Arsenic Hyperaccumulator Fern *Pteris Vittata*: Utilities for Arsenic Phytoremediation and Plant Biotechnology. In *Working with Ferns*; Fernández, H.; Kumar, A.; Revilla, M. A. Eds.; Springer: New York, NY, USA, 2011; pp. 261–269.
83. Datta, R.; Das, P.; Tappero, R.; Punamiya, P.; Elzinga, E.; Sahi, S.; Feng, H.; Kiiskila, J.; Sarkar, D. Evidence for Exocellular Arsenic in Fronds of *Pteris Vittata*. *Sci. Rep.* **2017,** *7*, 2839.
84. Saraswat, S.; Rai, J. P. N. Chemistry and Ecology Phytoextraction Potential of Six Plant Species Grown in Multimetal Contaminated Soil Phytoextraction Potential of Six Plant Species Grown in Multimetal Contaminated Soil. *Chem. Ecol.* **2009,** *25*, 1–11.
85. Bani, A.; Pavlova, D.; Echevarria, G.; Mullaj, A.; Reeves, R. D.; Morel, J. L.; Sulçe, S. Nickel Hyperaccumulation by the Species of *Alyssum* and *Thlaspi* (Brassicaceae) from the Ultramafic Soils of the Balkans. *Bot. Serbica.* **2010,** *31* (1), 3–14.
86. Altinozlu, H.; Karagoz, A.; Polat, T.; Unver, I. Nickel Hyperaccumulation by Natural Plants in Turkish Serpentine Soils. *Turk. J. Bot.* **2012,** *36*, 269–280.
87. Mengoni, A.; Cecchi, L.; Gonnelli, C. Nickel Hyperaccumulating Plants and Alyssum Bertolonii: Model Systems for Studying Biogeochemical Interactions in Serpentine Soils. In Bio-Geo Interactions in Metal-Contaminated Soils; Kothe, E., Varma, A., Eds.; Volume 31, Springer: Berlin/Heidelberg, Germany, **2012,** pp. 279–296.
88. Slatter, K. A. Nickel Accumulation and Tolerance in *Berkheya Codii* and its Application in Phytoremediation. Master's Thesis, University of Kwazulu: Natal, South Africa, 2013.
89. Broadhurst, C. L.; Chaney, R. L. Growth and Metal Accumulation of an Alyssum Murale Nickel Hyperaccumulator Ecotype Co-cropped with Alyssum Montanum and Perennial Ryegrass in Serpentine Soil. *Front. Plant Sci.* **2016,** *7*, 451.
90. Lombi, E.; Zhao, F. J.; Dunham, S. J.; Mcgrath, S. P. Phytoremediation of Heavy Metal-Contaminated Soils. *J. Environ. Qual.* **2001,** *30*, 1919.
91. Bolan, N. S.; Adriano, D. C.; Naidu, R. Role of Phosphorus in (Im) Mobilization and Bioavailability of Heavy Metals in the Soil Plant System. In *Reviews of Environmental Contamination and Toxicology*. Springer, 2003; pp. 1–44.
92. Zhang, Y. F.; He, L. Y.; Chen, Z. J.; Wang, Q. Y.; Qian, M.; Sheng, X. F. Characterization of ACC Deaminase-Producing Endophytic Bacteria Isolated from Copper-Tolerant Plants and their Potential in Promoting the Growth and Copper Accumulation of *Brassica Napus*. *Chemosphere* **2011,** *81* (1), 57–62.
93. Ramana, S.; Tripathi, A. K.; Kumar, A.; Dey, P.; Saha, J. K.; Patra, A. K. Evaluation of *Furcraea Foetida* (L.) Haw. For Phytoremediation of Cadmium Contaminated Soils. *Environ. Sci. Pollut. Res.* **2021,** *25*, 1–5.
94. Ramana, S.; Tripathi, A. K.; Kumar, A.; Dey, P.; Saha, J. K.; Patra, A. K. Phytoremediation of Soils Contaminated with Cadmium by *Agave Americana*. *J. Nat. Fibers*. **2021,** *15*, 1–9.
95. Yu, B.; Peng, Y.; Xu, J.; Qin, D.; Gao, T.; Zhu, H.; Zuo, S.; Song, H.; Dong, J. Phytoremediation Potential of *Youngia Japonica* (L.) DC: A Newly Discovered Cadmium Hyperaccumulator. *Environ. Sci. Pollut. Res.* **2021,** *21* (5), 6044–6057.
96. Boros-Lajszner, E.; Wyszkowska, J.; Kucharski, J. Phytoremediation of Soil Contaminated with Nickel, Cadmium and Cobalt. *Int. J. Phytoremed.* **2021,** *21* (3), 252–262.
97. Jiang, W.; Liu, D.; Hou, W. Hyperaccumulation of Lead by Roots, Hypocotyls and Shoots of *Brassica Juncea*. *Biol Planta* **2000,** *41* (4), 603–606.
98. Angelova, V. R.; Perifanova-Nemska, M.; Uzunova, G.; Ivanov, K.; Lee, H. Potential of Sunflower (*Helianthus Annuus* L.) for Phytoremediation of Soils Contaminated with Heavy Metals. *World J. Sci. Eng. Technol.* **2016,** *10*, 1–11.

99. Fulekar, M. H.; Sharma, J.; Tendulkar, A. Bioremediation of Heavy Metals Using Biostimulation in Laboratory Bioreactor. *Environ. Monit. Assess.* **2012,** *184*, 7299–7307.
100. Claire-Lise, M.; Nathalie, V. The Use of the Model Species *Arabidopsis Halleri* Towards Phytoextraction of Cadmium Polluted Soils. *New Biotechnol.* **2012,** 30, 9–14.
101. Zhang, Z.; Wen, X.; Huang, Y.; Inoue, C.; Liang, Y. Higher Accumulation Capacity of Cadmium Than Zinc by *Arabidopsis Halleri* ssp. *Germmifera* in the Field Using Different Sowing Strategies. *Plant Soil* **2017,** *418*, 1–12.
102. Tang, Y. T.; Qiu, R. L.; Zeng, X. W.; Ying, R. R.; Yu, F. M.; Zhou, X. Y. Lead, Zinc, Cadmium Hyperaccumulation and Growth Stimulation in *Arabis Paniculata* Franch. *Environ. Exp. Bot.* **2009,** *66*, 126–134.
103. Nematian, M. A.; Kazemeini, F. Accumulation of Pb, Zn, C and Fe in Plants and Hyperaccumulator Choice in Galali Iron Mine Area, Iran. *Int. J. Agric. Crop Sci.* **2013,** *5*, 426.
104. Conesa, H. M.; Faz. Á.; Arnaldos, R. initial Studies for the Phytostabilization of a Mine Tailing from the Cartagena-La Union Mining District (SE Spain). *Chemosphere* **2007,** *66*, 38–44.
105. Sharma, P.; Tripathi, S.; Chandra, R. Highly Efficient Phytoremediation Potential of Metal and Metalloids from the Pulp Paper Industry Waste Employing *Eclipta Alba* (L) and *Alternanthera Philoxeroide* (L): Biosorption and Pollution Reduction. *Bioresour. Technol.* **2021,** *319*, 124147.
106. Nero, BF. Phytoremediation of Petroleum Hydrocarbon-Contaminated Soils with Two Plant Species: *Jatropha Curcas* and *Vetiveria Zizanioides* at Ghana Manganese Company Ltd. *Int. J. Phytoremed.* **2021,** *21* (2), 171-80.
107. Yang, S. X.; Deng, H.; Li, M. S. Manganese Uptake and Accumulation in a Woody Hyperaccumulator, *Schima Superba*. *Plant Soil Env.* **2008,** *54*, 441–446.
108. Rizzi, L.; Petruzzelli, G.; Poggio, G.; Guidi, G. V. Soil Physical Changes and Plant Availability of Zn and Pb in a Treatability Test of Phytostabilization. *Chemosphere* **2004,** *57*, 1039–1046.
109. Chen, B.; Ai, W.; Gong, H.; Gao, X.; Qiu, B. Cleaning up of Heavy Metals-Polluted Water by a Terrestrial Hyperaccumulator *Sedum Alfredii* Hance. *Front. Biol.* **2013,** *8*, 599–605.
110. Chibuike, G.; Obiora, S. Heavy Metal Polluted Soils: Effect on Plants and Bioremediation Methods. *Appl. Environ. Soil Sci.* **2014,** 2014.
111. Ebbs, S. D.; Kochian, L. V. Toxicity of Zinc and Copper to *Brassica* Species: Implications for Phytoremediation. *J. Environ. Qual.* **1997,** *26*, 776–781.
112. Chehregani, A.; Malayeri, B. E. Removal of Heavy Metals by Native Accumulator Plants. *Int. J. Agric. Biol.* **2007,** *9*, 462–465.
113. Jin, X. F.; Liu, D.; Islam, E.; Mahmood, Q.; Yang, X. E.; He, Z. L.; Stoffella, P. J. Effects of Zinc on Root Morphology and Antioxidant Adaptations of Cadmium-Treated *Sedum Alfredii* H. *J. Plant. Nutr.* **2009,** *31* (10), 1642–1656.
114. Alford, É. R.; Pilon-Smits, E. A.; Fakra, S. C.; Paschke, M. W. Selenium Hyperaccumulation by *Astragalus* (Fabaceae) Does Not Inhibit Root Nodule Symbiosis. *Am. J. Bot.* **2012,** *99*, 1930–1941.
115. Wu, Z. Z.; Yang, J. Y.; Zhang, Y. X.; Wang, C. Q.; Guo, S. S.; Yu, Y. Q. Growth Responses, Accumulation, Translocation and Distribution of Vanadium in Tobacco and its Potential in Phytoremediation. *Ecotoxicol. Environ. Saf.* **2021,** *207*, 111297.
116. Brewer, C. Industrial Hemp (*Cannabis Sativa*) for Phytoremediation of Uranium Contaminated Soil. *Science* 2020.

117. Jomjun, N.; Siripen, T.; Maliwan, S., et al. Phytoremediation of Arsenic in Submerged Soil by Wetland Plants. *Int. J. Phytoremediat.* **2010,** *13*, 35–46.
118. Sakakibara, M.; Watanabe, A.; Inoue, M.; Sano, S.; Kaise, T. *Phytoextraction and Phytovolatilization of Arsenic from As-Contaminated Soils by Pteris Vittata.* In Proceedings of the Annual International Conference on Soils, Sediments, Water and Energy, 2010, 11 (1), pp. 26.
119. Mirza, N.; Pervez, A.; Mahmood, Q.; Shah, M. M.; Shafqat, M. N. Ecological Restoration of Arsenic Contaminated Soil by *Arundo Donax* L. *Ecol. Eng.* **2011,** *31* (12), 1949–1956.
120. Bennicelli, R.; Stępniewska, Z.; Banach, A., et al. The Ability of *Azolla Caroliniana to* Remove Heavy Metals (Hg(II), Cr(III), Cr(VI)) from Municipal Waste Water. *Chemosphere* **2004,** *55*, 141–146.
121. Moreno, F. N.; anderson, C. W. N.; Stewart, R. B., et al. Mercury Phytoextraction and Phytovolatilisation from Hg-Contaminated Artisanal Mine Sites. *Phytoremediat. Mercur. Mine Wastes* **2004,** 147–159.
122. Moreno, F. N.; Anderson, C. W. N.; Stewart, R. B.; Robinson, B. H. Phytoremediation of Mercury Contaminated Mine Tailings by Induced Plant-Mercury Accumulation. *Environ. Pract.* **2004,** *6*, 165–175.
123. Leung, H.; Ye, Z.; Wong, M. Survival Strategies of Plants Associated with Arbuscular Mycorrhizal Fungi on Toxic Mine Tailings. *Chemosphere* **2007,** *66*, 905–915.
124. Baghour, M.; Moreno, D. A.; Hernández, J., et al. Influence of Root Temperature on Phytoaccumulation of As, Ag, Cr, and Sb in Potato Plants (*Solanum Tuberosum* L. Var. *Spunta*). *J. Environ. Sci. Heal. A Toxic Hazard Subst. Environ. Eng.* **2001,** *36*, 1389–1401.
125. Gorinova, N.; Nedkovska, M.; Todorovska, E.; Simova-Stoilova, L.; Stoyanova, Z.; Georgieva, K.; Demirevska-Kepova, K.; Atanassov, A.; Herzig, R. Improved Phytoaccumulation of Cadmium by Genetically Modified Tobacco Plants (*Nicotiana Tabacum* L.). Physiological and Biochemical Response of the Transformants to Cadmium Toxicity. *Environ. Pollut.* **2007,** *141* (1), 161–170.
126. Rashid, A.; Ayub, N.; Ahmad, T.; Gul, J.; Khan, A. G. Phytoaccumulation Prospects of Cadmium and Zinc by Mycorrhizal Plant Species Growing in Industrially Polluted Soils. *Environ. Geochem. Health* **2009,** *31*, 91–98.
127. Redjala, T.; Zelko, I.; Sterckeman, T.; Legué, V.; Lux, A. Relationship Between Root Structure and Root Cadmium Uptake in Maize. *Environ. Exp. Bot.* **2011,** *71* (2), 241–248.
128. Udoka, O. C.; Ekanem, E. O.; Harami, M. D.; Tafawa, A. Phytoaccumulation Potentials of *Tamarindus Indica*. *Int. J. Innov. Sci. Res.* **2014,** *11*, 72–78.
129. Chinmayee, M. D.; Mahesh, B.; Pradesh, S.; Mini, I.; Swapna, T. S. The Assessment of Phytoremediation Potential of Invasive Weed *Amaranthus Spinosus* L. *Appl. Biochem. Biotechnol.* **2012,** *161* (6), 1550–1559.
130. Zayed, A.; Gowthaman, S.; Terry, N. Phytoaccumulation of Trace Elements by Wetland Plants: I. Duckweed. *J. Environ. Qual.* **1998,** *27*, 715–721.
131. Qian, J. H.; Zayed, A.; Zhu, Y. L. Yu, M.; Terry, N. Phytoaccumulation of Trace Elements by Wetlands Plants: Uptake and Accumulation of Ten Trace Elements by Twelve Plant Species. *J. Environ. Qual.* **1999,** *5*, 1448–1455.
132. Boonyapookana, B.; Upatham, E. S.; Kruatrachue, M.; Pokethitiyook, P.; Singhakaew, S. Phytoaccumulation and Phytotoxicity of Cadmium and Chromium in Duckweed *Wolffia Globosa*. *Int. J. Phytoremediat.* **2002,** *1* (2), 87–100.
133. Arora, A.; Saxena, S.; Sharma, D. K. Tolerance and Phytoaccumulation of Chromium by Three *Azolla* Species. *World J. Microbiol. Biotechnol.* **2006,** *22*, 97–100.

134. Chen, G. C.; Liu, Z.; Zhang, J.; Owens, G. Phytoaccumulation of Copper in Willow Seedlings Under Different Hydrological Regimes. *Ecol. Eng.* **2012,** *44*, 285–289.
135. Mukherjee, A.; Bandyopadhyay, A.; Dutta, S.; Basu, S. Phytoaccumulation of Iron by Callus Tissue of *Clerodendrum indicum* (L). *Chem. Ecol.* **2013,** *29*, 564–571.
136. Kamal, M.; Ghaly, A. E.; Mahmoud, N.; Cotecôté, R. Phytoaccumulation of Heavy Metals by Aquatic Plants. *Environ. Int.* **2004,** *21* (8), 1029–1039.
137. Abioye, O. P.; Agamuthu, P.; Abdul Aziz A. Phytoaccumulation of Zinc and Iron by Jatropha Curcas Grown in Used Lubricating Oilcontaminated Soil. *Malays. J. Sci.* **2010,** *29*, 207–213.
138. Zurayk, R.; Sukkariyah, B.; Baalbaki, R.; Ghanem, D. A. Ni Phytoaccumulation in *Mentha Aquatica* L. and *Mentha Sylvestris* L. *Water Air Soil Pollut.* **2002,** *139*, 355–364.
139. Adhikari, T.; Kumar, A. Phytoaccumulation and Tolerance of *Riccinus Communis* L. to Nickel. *Int. J. Phytoremediat.* **2012,** *14*, 481–492.
140. Boonyapookana, B.; Parkpian, P.; Techapinyawat, S., et al. Phytoaccumulation of Lead by Sunflower (*Helianthus Annuus*), Tobacco (*Nicotiana Tabacum*), and Vetiver (*Vetiveria Zizanioides*). *J. Environ. Sci. Heal. A Toxic Hazard Subst. Environ. Eng.* **2005,** *40*, 117–137.
141. Dinesh, M.; Kumar, M. V.; Neeraj, P.; Shiv, B. Phytoaccumulation of Heavy Metals in Contaminated Soil Using Makoy (*Solenum Nigrum* L.) and Spinach (*Spinacia Oleracea* L.) Plant. *Sciences (New York)* **2014,** *2*, 350–354.
142. Tzvetkova, C.; Bozhkov, O. Study of Rhenium Phytoaccumulation in White Clover (*Trifolium Repens*) and Water Fern (*Salvinia Natans* L.). In: *7th WSEAS International Conference on Environment, Ecosystems and Development*, 2009; pp. 123–126.
143. Shahandeh, H.; Hossner, L. R. Role of Soil Properties in Phytoaccumulation of Uranium. *Water Air Soil Pollut.* **2002,** *141*, 165–180.
144. Marbaniang, D.; Chaturvedi, S. S. Phytoaccumulation of Zinc by *Scirpus Mucronatus* (L.) Palla Ex Kerner. *Keanean J. Sci.* **1994,** *1*, 69–75.
145. Khellaf, N.; Zerdaoui, M. Phytoaccumulation of Zinc by the Aquatic Plant, *Lemna Gibba* L. *Bioresour. Technol.* **2009,** *100*, 6137–6140.
146. Vázquez, S.; Agha, R.; Granado, A., et al. Use of White Lupin Plant for Phytostabilization of Cd and as Polluted Acid Soil. *Water Air Soil Pollut.* **2006,** *177*, 349–365.
147. Varun, M.; Souza, R. D.; Pratas, J., et al. Evaluation of Phytostabilization, a Green Technology to Remove Heavy Metals from Industrial Sludge Using *Typha Latifolia* L. Experimental Design. *Biotechnol. Bioinf. Bioeng.* **2011,** *1*, 137–145.
148. King, D. J.; Doronila, A. I.; Feenstra, C., et al. Phytostabilisation of Arsenical Gold Mine Tailings Using Four *Eucalyptus* Species: Growth, Arsenic Uptake and Availability after Five Years. *Sci. Total Environ.* **2008,** *406*, 35–42.
149. Ehsan, M.; Santamaría-Delgado, K.; Vásquez-Alarcón, A., et al. Phytostabilization of Cadmium Contaminated Soils by *Lupinus Uncinatus* Schldl. *Span J. Agric. Res.* **2009,** *7*, 390–397.
150. Domínguez, M. T.; Madrid, F.; Marañón, T.; Murillo, J. M. Cadmium Availability in Soil and Retention in Oak Roots: Potential for Phytostabilization. *Chemosphere* **2009,** *76*, 480–486.
151. Zhang, X.; Xia, H.; Li, Z., et al. Potential of Four forage Grasses in Remediation of Cd and Zn Contaminated Soils. *Bioresour. Technol.* **2010,** *101*, 2063–2066.
152. Shackira, A. M.; Puthur, J. T. Enhanced Phytostabilization of Cadmium by a Halophyte-*Acanthus Ilicifolius* L. *Int. J. Phytoremediat.* **2017,** *19*, 319–326.

153. Brunner, I.; Luster, J.; Günthardt-Goerg, M. S.; Frey, B. Heavy Metal Accumulation and Phytostabilisation Potential of Tree Fine Roots in a Contaminated Soil. *Environ. Pollut.* **2008,** *152*, 559–568.
154. Yang, S.; Liang, S.; Yi, L., et al. Heavy Metal Accumulation and Phytostabilization Potential of Dominant Plant Species Growing on Manganese Mine Tailings. *Front. Environ. Sci. Eng.* **2014,** *8*, 394–404.
155. Bidar, G.; Garçon, G.; Pruvot, C.; Dewaele, D.; Cazier, F.; Douay, F.; Shirali, P. Behavior of *Trifolium Repens* and *Lolium Perenne* Growing in a Heavy Metal Contaminated Field: Plant Metal Concentration and Phytotoxicity. *Environ. Pollut.* **2007,** *141* (3), 546–553.
156. Soudek, P.; Petrová, Š.; Vaněk, T. Phytostabilization or Accumulation of Heavy Metals by Using of Energy Crop *Sorghum* sp. In 3rd International Conference on Biology, Environment and Chemistry IPCBEE. IACSIT Press: Singapore, 2012.
157. Ali, N. A.; Bernal, M. P.; Ater, M. Tolerance and Bioaccumulation of Copper in *Phragmites Australis* and *Zea Mays*. *Plant Soil* **2002,** *239*, 103–111.
158. Radziemska, M.; Vaverková, M. D.; Baryła, A. Phytostabilization Management Strategy for Stabilizing Trace Elements in Contaminated Soils. *Int. J. Environ. Res. Public Health.* **2017,** *14*, 958.
159. Heaton, A. C.; Rugh, C. L.; Wang, N. J.; Meagher, R. B. Phytoremediation of Mercury and Methylmercury Polluted Soils Using Genetically Engineered Plants. *J. Soil Contam.* **1998,** *1* (4), 497–509.
160. Rugh, C. L.; Senecoff, J. F.; Meagher, R. B.; Merkle, S. A. Development of Transgenic Yellow Poplar for Mercury Phytoremediation. *Nat. Biotechnol.* **1998,** *16*, 925–928.
161. Rotkittikhun, P.; Chaiyarat, R.; Kruatrachue, M.; Pokethitiyook, P.; Baker, A. J. Growth and Lead Accumulation by the Grasses *Vetiveria Zizanioides* and *Thysanolaena Maxima* in Lead-Contaminated Soil Amended with Pig Manure and Fertilizer: A Glasshouse Study. *Chemosphere* **2007,** *61* (1), 45–53.
162. Zhao, L.; Li, T.; Zhang, X., et al. Pb Uptake and Phytostabilization Potential of the Mining Ecotype of *Athyrium Wardii* (Hook.) Grown in Pb-Contaminated Soil. *CLEAN Soil Air Water* **2016,** *44*, 1184–1190.
163. Dongmei, L.; Changqun, D. Restoration Potential of Pioneer Plants Growing on Lead-Zinc Mine Tailings in Lanping, Southwest China. *J. Environ. Sci.* **2008,** *21* (10), 1202–1209.
164. Beath, O. A.; Eppson, H. F.; Gillbert, C. S. *Bulletin*; Vol. 206; University of Wyoming, Agricultural Experiment Station Laramie: Wyoming, 1935, pp. 1–55.
165. Terry, N.; Carlson, C.; Raab, T. K.; Zayed, A. Rates of Selenium Volatilization Among Crop Species. *J. Environ. Qual.* **1992,** *21*, 341–344.
166. Terry, N.; Zayed, A. M.; De Souza, M. P.; Tarun, A. S. Selenium in Higher Plants. *Ann. Rev. Plant Biol.* **2000,** *51* (1), 401–432.
167. Leduc, D. L.; Terry, N. Phytoremediation of Toxic Trace Elements in Soil and Water. *J. Ind. Microbiol. Biotechnol.* **2005,** *32*, 514–520.
168. Van Der Heijden, M. G. A. *Mycorrhizal Ecology*. Springer: Berlin, 2003; pp. 243–265.
169. Ghassemzadeh, F.; Yousefzadeh, H.; Arbab-Zavar, M. H. Removing Arsenic and Antimony by *Phragmites Australis*: Rhizofiltration Technology. *J. Appl. Sci.* **2008,** *1* (9), 1668–1675.
170. Galal, T. M.; Eid, E. M.; Dakhil, M. A.; Hassan, L. M. Bioaccumulation and Rhizofiltration Potential of *Pistia Stratiotes* L. for Mitigating Water Pollution in the Egyptian Wetlands. *Int. J. Phytoremediat.* **2018,** *21* (5), 440–447.

171. Yadav, A. K.; Pathak, B.; Fulekar, M. H. Rhizofiltration of Heavy Metals (Cadmium, Lead and Zinc) from Fly Ash Leachates Using Water Hyacinth (*Eichhornia Crassipes*). *Int. J. Environ.* **2015,** *1* (1), 179–196.
172. Düreŝová, Z.; Šuňovská, A.; Horník, M.; Pipíška, M.; Gubišová, M.; Gubiš, J.; Hostin, S. Rhizofiltration Potential of *Arundo Donax* for Cadmium and Zinc Removal from Contaminated Wastewater. *Chem. Pap.* **2014,** *61* (11), 1452–1462.
173. Abubakar, M. M.; Ahmad, M. M.; Getso, B. U. Rhizofiltration of Heavy Metals from Eutrophic Water Using *Pistia Stratiotes* in a Controlled Environment. *IOSR J. Environ. Sci. Toxicol. Food Technol.* **2014,** *1* (6), 27–34.
174. Leblebici, Z.; Dalmiş, E.; Andeden, E. E. Determination of the Potential of Pistia Stratiotes L. In *Removing Nickel from the Environment by Utilizing its Rhizofiltration Capacity*; Vol. 62. Braz. Arch. Biol. Technol. 2019.
175. Yang, M.; Jawitz, J. W.; Lee, M. Uranium and Cesium Accumulation in Bean (*Phaseolus Vulgaris* L. Var. *Vulgaris*) and its Potential for Uranium Rhizofiltration. *J. Environ. Radioactiv.* **2015,** *140*, 42–49.
176. Mohamed, S.; Mahrous, A.; Elshahat, R.; Kassem, M. Accumulation of Iron, Zinc and Lead by *Azolla pinnata* and *Lemna minor* and activity in contaminated water. *Egypt. J. Chem.* **2021,** *64*(9), 5017–5030.
177. Thayaparan, M.; Iqbal, S. S.; Chathuranga, P. K.; Iqbal, M. C. Rhizofiltration of Pb by *Azolla Pinnata. Int. J. Environ. Sci.* **2013,** *1* (6), 1811.
178. Ignatius, A.; Arunbabu, V.; Neethu, J.; Ramasamy, E. V. Rhizofiltration of Lead Using an Aromatic Medicinal Plant *Plectranthus Amboinicus* Cultured in a Hydroponic Nutrient Film Technique (NFT) System. *Environ. Sci. Pollut. Res.* **2014,** *21* (22), 13007–13016.
179. Tomé, F. V.; Rodríguez, P. B.; Lozano, J. C. Elimination of Natural Uranium and 226Ra from Contaminated Waters by Rhizofiltration Using *Helianthus Annuus* L. *Sci. Total Environ.* **2008,** *391* (2–3), 351–357.
180. Lee, M.; Yang, M. Rhizofiltration Using Sunflower (*Helianthus Annuus* L.) and Bean (*Phaseolus Vulgaris* L. Var. *Vulgaris*) to Remediate Uranium Contaminated Groundwater. *J. Hazard. Mater.* **2010,** *171* (1–3). 589–596.
181. Haq, S.; Bhatti, A. A.; Dar, Z. A.; Bhat, S. A. Phytoremediation of Heavy Metals: An Eco-Friendly and Sustainable Approach. In *Bioremediation and Biotechnology*; Springer: Cham, 2020; pp. 215–231.
182. Newman, L., Eds. *Phytoremediation: Management of Environmental Contaminants*; Vol. 3; Springer: Cham, pp. 113–114.
183. Jutsz, A. M.; Gnida, A. Mechanisms of Stress Avoidance and Tolerance by Plants Used in Phytoremediation of Heavy Metals. *Arch. Environ. Prot.* **2015,** *41*, 104–114.
184. Jabeen, R.; Ahmad, A.; Iqbal, M. Phytoremediation of Heavy Metals: Physiological and Molecular Mechanisms. *Bot. Rev.* **2009,** *75*, 339–364.
185. Rascio, N.; Navari-Izzo, F. Heavy Metal Hyperaccumulating Plants: How and Why Do They Do It? and What Makes Them So Interesting? *Plant Sci.* **2011,** *180*, 169–181.
186. Muszynska, E.; Hanus-Fajerska, E. Why are Heavy Metal Hyperaccumulating Plants So Amazing? *Biotechnol. J. Biotechnol. Comput. Biol. Bionanotechnol.* **2015,** *91* (4), 265–271.
187. Dai, H.; Wei, S.; Twardowska, I.; Han, R.; Xu, L. Hyperaccumulating Potential of *Bidens Pilosa* L. for Cd and Elucidation of its Translocation Behavior Based on Cell Membrane Permeability. *Environ. Sci. Pollut. Res. Int.* **2017,** *21* (29), 23161–23167.

188. Mourato, M. P.; Moreira, I. N.; Leitão, I.; Pinto, F. R.; Sales, J. R.; Martins, L. L. Effect of Heavy Metals in Plants of the Genus *Brassica*. *Int. J. Mol. Sci.* **2015,** *11* (8), 17975–17998.
189. Bothe, H.; Słomka, A. Divergent Biology of Facultative Heavy Metal Plants. *J. Plant Physiol.* **2017,** *219*, 45–61.
190. Kidd, P. S.; Monterroso, C. Metal Extraction by *Alyssum Serpyllifolium* Ssp. Lusitanicum on Mine-Spoil Soils from Spain. *Sci. Total Environ.* **2005,** *331* (1–3), 1–11.
191. Rabhi, M.; Ferchichi, S.; Jouini, J.; Hamrouni, M. H.; Koyro, H. W.; Ranieri, A.; Abdelly, C.; Smaoui, A. Phytodesalination of a Salt-Affected Soil with the Halophyte *Sesuvium Portulacastrum* L. to Arrange in Advance the Requirements for the Successful Growth of a Glycophytic Crop. *Bioresour. Technol. 101*, 2010; 6822–6828.
192. Seyyed G. M.; Mohamd J. Seghatoleslami Phytoremediation: A Review. *Adv. Agric. Biol.* **2013,** *1* (1), 5–11.
193. Ayyappan, D.; Sathiyaraj, G.; Ravindran, K. C. Phytoextraction of Heavy Metals by *Sesuvium Portulacastrum* L. A Salt Marsh Halophyte from Tannery Effluent. *Int. J. Phytoremediat.* **2016,** *11* (5), 453–459.
194. Verma, N.; Singh, M. Biosensors for Heavy Metals. *Biometals* **2005,** *18*, 121–129.
195. Padmavathiamma, P. K.; Li, L. Y. Phytoremediation Technology, Hyper-Accumulation Metals in Plants. *Water Air Soil Pollut.* **2007,** *184*, 105–126.
196. United States Environmental Protection Agency (USEPA). Introduction to Phytoremediation. EPA 600/R-99/107. U. S. Environmental Protection Agency, office of Research and Development, Cincinnati, 2000.
197. López-Chuken, U. J. Hydroponics and Environmental Clean-Up. In *Hydroponics-A Standard Methodology for Plant Biological Researches*; Toshiki, A., Ed.; Intech: Shanghai, China, 2012.
198. Raskin, I.; Ensley, B. D. *Phytoremediation of Toxic Metals: Using Plants to Clean up the Environment*. Wiley: New York, 2000.
199. Salt, D. E.; Pickering, I. J.; Prince, R. C.; Gleba, D.; Dushenkov, S.; Smith, R. D.; Raskin, I. Metal Accumulation by Aquacultured Seedlings of Indian Mustard. *Environ. Sci. Technol.* **1997,** *31* (6), 1636–1644.
200. Sylvain, B.; Mikael, M. H.; Florie, M.; Emmanuel, J.; Marilyne, S.; Sylvain, B.; Domenico, M. Phytostabilization of As, Sb and Pb by Two Willow Species (*S. Viminalis* and *S. Purpurea*) on former Mine Technosols. *Catena* **2016,** *136*, 44–52.
201. Wong, M. H. Ecological Restoration of Mine Degraded Soils, with Emphasis on Metal Contaminated Soils. *Chemosphere* **2003,** *50*, 775–780.
202. Mendez, M. O.; Maier, R. M. Phytostabilization of Mine Tailings in Arid and Semiarid Environments-An Emerging Remediation Technology. *Environ. Health Perspect.* **2008,** *116*, 278–283.
203. Meharg, A. A. The Mechanistic Basis of Interactions Between Mycorrhizal Associations and Toxic Metal Cations. *Mycol. Res.* **2003,** *107*, 1253–1265.
204. Bucking, H.; Heyser, W. Elemental Composition and Function of Polyphosphates in Ectomycorrhizal Fungi-An X-Ray Microanalytical Study. *Mycol. Res.* **1999,** *103*, 31–39.
205. Yang, X.; Feng, Y.; He, Z.; Stoffella, P. J. Molecular Mechanisms of Heavy Metal Hyperaccumulation and Phytoremediation. *J. Trace Elem. Med. Biol.* **2005,** *18*, 339–353.
206. Khanday, M.; Bhat, R. A.; Haq, S.; Dervash, M. A.; Bhatti, A. A.; Nissa, M.; Mir, M. R. Arbuscular Mycorrhizal Fungi Boon for Plant Nutrition and Soil Health. In *Soil Science: Agricultural and Environmental Prospectives*; Hakeem, K. R.; Akhtar, J.; Sabir, M., Eds.; Springer International: Berlin, 2016; pp. 317–332.

207. Bhat, R. A.; Dervash, M. A.; Mehmood, M. A.; Bhat, M. S.; Rashid, A.; Bhat, J. I. A.; Singh, D. V.; Lone, Rn. (Mycorrhizae): A Sustainable Industry for Plant and Soil Environment. In *Mycorrhiza-Nutrient Uptake, Biocontrol, Ecorestoration*; Varma, A., et al., Eds.; Springer International: Berlin, 2017; pp. 473–502.
208. Sofi, N. A.; Bhat, R. A.; Rashid, A.; Mir, N. A.; Mir, S. A.; Lone, R. Rhizosphere Mycorrhizae Communities an Input for Organic Agriculture. In *Mycorrhiza-Nutrient Uptake, Biocontrol, Ecorestoration*; Varma A, et al., Eds.; Springer international: Berlin, 2017; pp. 387–413.
209. Rengel, Z.; Zhang, W. H. Role of Dynamics of Intracellular Calcium in Aluminium-Toxicity Syndrome. *New Phytol.* **2003,** *159*, 295–314.
210. Bluskov, S.; Arocena, J.; Omotoso, O.; Young, J. Uptake, Distribution, and Speciation of Chromium in *Brassica Juncea*. *Int. J. Phytoremediat.* **2005,** *7*, 153–165.
211. Wei, S.; Zhou, Q.; Wang, X. Identification of Weed Plants Excluding the Uptake of Heavy Metals. *Environ. Int.* **2005,** *31*, 829–834.
212. Kramer, U. Metal Hyperaccumulation in Plants. *Annu. Rev. Plant Biol.* **2010,** *61*, 517–534.
213. Tam, P. C. Heavy Metal Tolerance by Ectomycorrhizal Fungi and Metal Amelioration by *Pisolithus Tinctorius*. *Mycorrhiza* **1995,** *5*, 181–187.
214. Hall, J. Cellular Mechanisms for Heavy Metal Detoxification and Tolerance. *J. Exp. Bot.* **2002,** *53*, 1–11.
215. Taylor, G. J. Exclusion of Metals from the Symplasm: A Possible Mechanism of Metal Tolerance in Higher Plants. *J. Plant Nutr.* **1987,** *10*, 1213–1222.
216. Bringezu, K.; Lichtenberger, O.; Leopold, I.; Neumann, D. Heavy Metal Tolerance of *Silene Vulgaris*. *J. Plant Physiol.* **1999,** *154*, 536–546.
217. Delhaize, E.; Ryan, P. R. Aluminum Toxicity and Tolerance in Plants. *Plant Physiol.* **1995,** *107*, 315.
218. Larsen, P. B.; Degenhardt, J. T. C. Y.; Stenzler, L. M.; Howell, S. H.; Kochian, L. V. Aluminum-Resistant *Arabidopsis* Mutants that Exhibit Altered Patterns of Aluminum Accumulation and Organic Acid Release from Roots. *Plant Physiol.* **1998,** *117*, 9–17.
219. Islam, M. S.; Ueno, Y.; Sikder, M. T.; Kurasaki, M. Phytofiltration of Arsenic and Cadmium from the Water Environment Using Micranthemum Umbrosum (Jf GMEL) sf Blake as a Hyperaccumulator. *Int. J. Phytoremed.* **2013,** *15*, 1010–1021.
220. Bolan, N. S.; Park, J. H.; Robinson, B.; Naidu, R.; Huh, K. Y. Phytostabilization: A Green Approach to Contaminant Containment. *Adv. Agron.* **2011,** *112*, 145–204.
221. Marques, A.; Rangel, A. O. S. S.; Castro, P. M. L. Remediation of Heavy Metal Contaminated Soils: Phytoremediation as a Potentially Promising Clean-up Technology. *Crit. Rev. Environ. Sci. Technol.* **2009,** *39*, 622–654.
222. Malik, N.; Biswas, A. Role of Higher Plants in Remediation of Metal Contaminated Sites. *Sci. Rev. Chem. Commun.* **2012,** *2*, 141–146.
223. Rahman, M. A.; Reichman, S. M.; De Filippis, L.; Sany, S. B. T.; Hasegawa, H. Phytoremediation of Toxic Metals in Soils and Wetlands: Concepts and Applications. In *Environmental Remediation Technologies for Metal-Contaminated Soils*; Hasegawa, H., Rahman, M. M., Rahman, I., Eds.; Springer: Tokyo, Japan, 2016; pp. 161–195.
224. Cristaldi, A.; Conti, G. O.; Jho, E. H.; Zuccarello, P.; Grasso, A.; Copat, C.; Ferrante, M. Phytoremediation of Contaminated Soils by Heavy Metals and Pahs. A Brief Review. *Environ. Technol. Innov.* **2017,** *8*, 309–326.
225. Karlson, U.; Frankenberger, W. T. Accelerated Rates of Selenium Volatilization from California Soils. *Soil Sci. Soc. Am. J.* **1989,** *53*, 749–753.

226. Parker, D. R.; Page, A. L.; Thomason, D. N. Salinity and Boron Tolerances of Candidate Plants for the Removal of Selenium from Soils. *J. Environ. Qual.* **1991,** *20*, 157–164.
227. Rugh, C. L.; Wilde, H. D.; Stack, N. M.; Thompson, D. M.; Summers, A. O.; Meagher, R. B. Mercuric Ion Reduction and Resistance in Transgenic *Arabidopsis Thaliana* Plants Expressing a Modified Bacterial Mera Gene. *Proc. Natl Acad. Sci. U S A.* **1996,** *93*, 3182–3187.
228. Wang, J.; Feng, X.; anderson, C. W. N.; Xing, Y.; Shang, L. Remediation of Mercury Contaminated Sites-A Review. *J. Hazard. Mater.* **2012,** *221*, 1–18.
229. Liphadzi, M. S.; Kirkham, M. B.; Musil, C. F. Phytoremediation of Soil Contaminated with Heavy Metals: A Technology for Rehabilitation of the Environment. *S. Afr. J. Bot.* **2005,** *71*, 24–37.
230. Ali, H.; Khan, E.; Sajad, M. A. Phytoremediation of Heavy Metals-Concepts and Applications. *Chemosphere* **2013,** *91*, 869–881
231. Meagher, R. B. Phytoremediation of Toxic Elements and Organic Pollutants. *Curr. Opin. Plant Biol.* **2000,** *3*, 153–162.
232. Rugh, C. L.; Bizily, S. P.; Meagher, R. B. Phytoremediation of Environmental Mercury Pollution. In *Phytoremediation of Toxic Metals Using Plants to Clean Up the Environment*. Wiley: Newyork, 2000; pp. 151–171.
233. Maheswari, K. U.; Rajeswari, K. 17 *Toxicity of Heavy Metals-Phytoremediation Techniques.* In Reshaping Agriculture and Nutrition Linkages for Food and Nutrition Security; 2016, *326*, 143.
234. Favas, P. J.; Pratas, J.; Varun, M.; D'Souza, R.; Paul, M. S. Phytoremediation of Soils Contaminated with Metals and Metalloids at Mining Areas: Potential of Native Flora. In *Environmental Risk Assessment of Soil Contamination*; Maria, C., Hernandez, S., Eds.; Intech: Shanghai, China, 2014.
235. Susarla, S.; Medina, V. F.; Mccutcheon, S. C. Phytoremediation: An Ecological Solution to Organic Chemical Contamination. *Ecol. Eng.* **2002,** *18*, 647–658.
236. Dixit, R.; Malaviya, D.; Pandiyan, K.; Singh, U. B.; Sahu, A.; Shukla, R.; Singh, B. P.; Rai, J. P.; Sharma, P. K.; Lade, H. Bioremediation of Heavy Metals from Soil and Aquatic Environment: An Overview of Principles and Criteria of Fundamental Processes. *Sustainability* **2015,** *7*, 2189–2212.
237. Jan, S.; Parray, J. A. *Approaches to Heavy Metal Tolerance in Plants*; Springer: New Delhi, India, 2016.
238. Yadav, S. K. Heavy Metals Toxicity in Plants: An Overview on the Role of Glutathione and Phytochelatins in Heavy Metal Stress Tolerance of Plants. *S. Afr. J. Bot.* **2010,** *71* (2), 167–179.
239. Rahman Z. an Overview on Heavy Metal Resistant Microorganisms for Simultaneous Treatment of Multiple Chemical Pollutants at Co-Contaminated Sites and their Multipurpose Application. *J. Hazard. Mater.* **2020,** 122682.
240. Martínez-Viveros, O.; Jorquera, M.; Crowley, D.; Gajardo, G.; Mora, M. Mechanisms and Practical Considerations Involved in Plant Growth Promotion by Rhizobacteria. *J. Soil Sci. Plant Nutr.* **2010,** *10*, 293–319.
241. Ramadan, E. M.; Abdelhafez, A. A.; Hassan, E. A.; Saber, F. M. Plant Growth Promoting Rhizobacteria and their Potential for Biocontrol of Phytopathogens. *Afr. J. Microbiol. Res.* **2016,** *10*, 486–504.
242. Bhattacharyya, P.; Jha, D. Plant Growth-Promoting Rhizobacteria (PGPR): Emergence in Agriculture. *World J. Microbiol. Biotechnol.* **2012,** *28*, 1327–1350.

243. Manoj, S. R.; Karthik, C.; Kadirvelu, K.; Arulselvi, P. I.; Shanmugasundaram, T.; Bruno, B.; Rajkumar, M. Understanding the Molecular Mechanisms for the Enhanced Phytoremediation of Heavy Metals Through Plant Growth Promoting Rhizobacteria: A Review. *J. Environ. Manag.* **2020,** *254*, 109779.
244. Vejan, P.; Abdullah, R.; Khadiran, T.; Ismail, S.; Nasrulhaq Boyce, A. Role of Plant Growth Promoting Rhizobacteria in Agricultural Sustainability-A Review. *Molecules* **2016,** *21*, 573.
245. Singh, J. S.; Pandey, V. C.; Singh, D. Efficient Soil Microorganisms: A New Dimension for Sustainable Agriculture and Environmental Development. *Agric. Ecosyst. Environ.* **2011,** *140*, 339–353.
246. Dell'mour, M.; Schenkeveld,W.; Oburger, E.; Fischer, L.; Kraemer, S.; Puschenreiter, M.; Lämmerhofer, M.; Koellensperger, G.; Hann, S. Analysis of Iron-Phytosiderophore Complexes in Soil Related Samples: LC-ESI-MS/MS versus CE-MS. *Electrophoresis* **2012,** *33*, 726–733.
247. Ahmed, E.; Holmström, S. J. Siderophores in Environmental Research: Roles and Applications. *Microb. Biotechnol.* **2014,** *7*, 196–208.
248. Ahemad, M.; Kibret, M. Mechanisms and Applications of Plant Growth Promoting Rhizobacteria: Current Perspective. *J. King Saud Univ. Sci.* **2014,** *26*, 1–20.
249. Chandra, P.; Singh, E. Applications and Mechanisms of Plant Growth-Stimulating Rhizobacteria. In *Plant-Microbe Interaction: An Approach to Sustainable Agriculture*; Choudhary, D.; Varma, A.; Tuteja, N., Eds.; Springer: Singapore, 2016; pp. 37–62.
250. Choudhary, D. K.; Varma, A.; Tuteja, N. *Plant-Microbe Interaction: An Approach to Sustainable Agriculture*; Springer: New Delhi, India, 2017.
251. Milošević, N. A.; Marinković, J. B.; Tintor, B. B. Mitigating Abiotic Stress in Crop Plants by Microorganisms. *Zbornik Matice Srpske Za Prirodne Nauke* **2012,** *123*, 17–26.
252. Nadeem, S. M.; Zahir, Z. A.; Naveed, M.; Nawaz, S. Mitigation of Salinity-Induced Negative Impact on the Growth and Yield of Wheat by Plant Growth-Promoting Rhizobacteria in Naturally Saline Conditions. *Ann. Microbiol.* **2013,** *63*, 225–232.
253. Kumar, P. R.; Raina, S. K.; Kumar, S.; Bhagat, K. P.; Singh, Y.; Bal, S. K. Adaptation and Mitigation Strategies of Plant Under Drought and High-Temperature Stress. *Clim. Chang. Plant Abiot. Stress Toler.* **2013,** 421–436.
254. Glick, B. R. Bacteria with acc Deaminase Can Promote Plant Growth and Help to Feed the World. *Microbiol. Res.* **2014,** *169*, 30–39.
255. Spaepen, S.; Vanderleyden, J. Auxin and Plant-Microbe Interactions. *Cold Spring Harb. Perspect. Biol.* **2011,** *1* (4), A001438.
256. Ahemad, M. Implications of Bacterial Resistance Against Heavy Metals in Bioremediation: A Review. *J. Inst. Integr. Omics Appl. Biotechnol.* **2012,** *3*, 3.
257. Tangahu, B. V.; Sheikh Abdullah, S. R.; Basri, H.; Idris, M.; Anuar, N.; Mukhlisin, M. A Review on Heavy Metals (As, Pb and Hg) Uptake by Plants through Phytoremediation. *Int. J. Chem. Eng.* 2011.
258. Wang, A. S.; Angle, J. S.; Chaney, R. L.; Delorme, T. A.; Reeves, R. D. Soil pH Effects on Uptake of Cd and Zn by *Thlaspi Caerulescens*. *Plant Soil.* **2006,** *281*, 325–337.
259. Chaney, R. L.; Angle, J. S.; Broadhurst, C. L.; Peters, C. A.; Tappero, R. V.; Sparks, D. L. Improved Understanding of Hyperaccumulation Yields Commercial Phytoextraction and Phytomining Technologies. *J. Environ. Qual.* **2007,** *36*, 1429–1443.
260. Zeng, F.; Ali, S.; Zhang, H.; Ouyang, Y.; Qiu, B.; Wu, F.; Zhang, G. The Influence of pH and Organic Matter Content in Paddy Soil on Heavy Metal Availability and their Uptake by Rice Plants. *Environ. Pollut.* **2011,** *159*, 84–91.

261. Pourghasemian, N.; Ehsanzadeh, P.; Greger, M. Genotypic Variation in Safflower (*Carthamus* Spp.) Cadmium Accumulation and Tolerance Affected by Temperature and Cadmium Levels. *Environ. Exp. Bot.* **2013,** 87, 218–226.
262. Zhou, H.; Zhou, X.; Zeng, M.; Liao, B. H.; Liu, L.; Yang, W. T.; Wu, Y. M.; Qiu, Q. Y.; Wang, Y. J. Effects of Combined Amendments on Heavy Metal Accumulation in Rice (*Oryza sativa* L.) Planted on Contaminated Paddy Soil. *Ecotoxicol. Environ. Saf.* **2014,** *101*, 226–232.
263. Rosenfeld, C. E.; Chaney, R. L.; Martínez, C. E. Soil Geochemical Factors Regulate Cd Accumulation by Metal Hyperaccumulating *Noccaea Caerulescens* (J. Presl & C. Presl) F. K. Mey in Field-Contaminated Soils. *Sci. Total Environ.* **2018,** *616*, 279–287.
264. Ma, Y.; Oliveira, R. S.; Freitas, H.; Zhang, C. Biochemical and Molecular Mechanisms of Plant-Microbe-Metal interactions: Relevance for Phytoremediation. *Front. Plant Sci.* **2016,** *7*, 918.
265. Sheoran, V.; Sheoran, A. S.; Poonia, P. Factors Affecting Phytoextraction: A Review. *Pedosphere* **2016,** *26*, 148–166.
266. Yu, X. Z.; Peng, X. Y.; Xing, L. Q. Effect of Temperature on Phytoextraction of Hexavalent and Trivalent Chromium by Hybrid Willows. *Ecotoxicology* **2010,** *19*, 61–68.
267. Bhargava, A.; Carmona, F. F.; Bhargava, M.; Srivastava, S. Approaches for Enhanced Phytoextraction of Heavy Metals. *J. Environ. Manag.* **2012,** *105*, 103–120.
268. Mccormack, M. L.; Guo, D. Impacts of Environmental Factors on Fine Root Lifespan. *Front. Plant Sci.* **2014,** *5*, 205.
269. Szczygłowska, M.; Piekarska, A.; Konieczka, P.; Namieśnik, J. Use of Brassica Plants in the Phytoremediation and Biofumigation Processes. *Int. J. Mol. Sci.* **2011,** *12*, 7760–7771.
270. Gomes, M. A. D. C.; Hauser-Davis, R. A.; De Souza, A. N.; Vitória, A. P. Metal Phytoremediation: General Strategies, Genetically Modified Plants and Applications in Metal Nanoparticle Contamination. *Ecotoxicol. Environ. Saf.* **2016,** *134*, 133–147.
271. Azubuike, C. C.; Chikere, C. B.; Okpokwasili, G. C. Bioremediation Techniques-Classification Based on Site of Application: Principles, Advantages, Limitations and Prospects. *World J. Microbiol. Ayangbenro Biotechnol.* **2016,** *31* (11), 180.
272. Howe, G.; Merchant, S. Heavy Metal-Activated Synthesis of Peptides in *Chlamydomonas Reinhardtii*. *Plant Physiol.* **1992,** *98*, 127–136.
273. Mendoza-Cozatl, D.; Devars, S.; Loza-Tavera, H.; Moreno-Sanchez, R. Cadmium Accumulation in the Chloroplast of *Euglena Gracilis*. *Physiol. Plant.* **2002,** *115*, 276–283.
274. Aviles, C.; Loza-Tavera, H.; Terry, N.; Moreno-Sanchez, R. Mercury Pretreatment Selects an Enhanced Cadmium-Accumulating Phenotype in *Euglena Gracilis*. *Arch. Microbiol.* **2003,** *180*, 1–10.
275. Kumaran, N. S.; Sundaramanicam, A.; Bragadeeswaran, S. Adsorption Studies on Heavy Metals by Isolated Cyano Bacterial Strain (*Nostoc* Sp.) from Uppanar Estuarine Water, Southeast Coast of India. *J. Appl. Sci. Res.* **2011,** *1* (11), 1609–1615.
276. Mane, P. C.; Bhosle, A. B. Bioremoval of Some Metals by Living Algae *Spirogyra* sp. *and Spirullina* sp. from Aqueous Solution. *Int. J. Environ. Res.* **2012,** *6* (6), 571–576.
277. Ojuederie, O. B.; Babalola, O. O. Microbial and Plant-Assisted Bioremediation of Heavy Metal Polluted Environments: A Review. *Int. J. Environ. Res. Public Health.* **2017,** *11* (12), 1504.
278. Quiroga-Flores, R.; Guédron, S.; Achá, D. High Methylmercury Uptake by Green Algae in Lake Titicaca: Potential Implications for Remediation. *Ecotoxicol. Environ. Saf.* **2021,** *207*, 111256.

279. Esringu, A.; Tarun. M.; Gunes, A.; Karaman, M. R. Roles of *Bacillus Megaterium* in Remediation of Boron, Lead, and Cadmium from Contaminated Soil. *Commun. Soil Sci. Plan. Anal.* **2014,** *45*, 1741–1759.
280. Syed, S.; Chinthala, P. Heavy Metal Detoxification by Different *Bacillus* Species Isolated from Solar Salterns. *Scientifica* **2015,** *2015*, 319760.
281. Su, L. S. Isolation and Identification of Heavy Metal-Tolerant Bacteria from an Industrial Site as a Possible Source for Bioremediation of Cadmium, Lead and Nickel. *Adv. Environ. Biol.* **2016,** *10*, 10–15.
282. Wei, W.; Wang, Q.; Li, A.; Yang, J.; Ma, F.; Pi, S.; Wu, D. Biosorption of Pb (II) from Aqueous Solution by Extracellular Polymeric Substances Extracted from *Klebsiella* sp. J1: Adsorption Behavior and Mechanism Assessment. *Sci. Rep.* **2016,** *6*, 31575.
283. Mosharaf, M. K.; Tanvir, M. Z. H.; Haque, M. M.; Haque, M. A.; Khan, M. A. A.; Molla, A. H.; Alam A. Z.; Islam, M. S.; Talukder, M. R. Metaladapted Bacteria Isolated from Wastewaters Produce Biofilms by Expressing Proteinaceous Curli Fimbriae and Cellulose Nanofibers. *Front. Microbiol.* **2018,** *9*, 1334.
284. Sujkowska-Rybkowska, M.; Ważny, R. Metal Resistant Rhizobia and Ultrastructure of *Anthyllis Vulneraria* Nodules from Zinc and Lead Contaminated Tailing in Poland. *Int. J. Phytorem.* **2018,** *20*, 709–720.
285. Alviz-Gazitua, P.; Fuentes-Alburquenque, S.; Rojas, L. A.; Turner, R. J.; Guiliani, N.; Seeger, M. The Response of *Cupriavidus Metallidurans* CH34 to Cadmium Involves Inhibition of the Initiation of Biofilm formation, Decrease in Intracellular C-Di-GMP Levels, and a Novel Metal Regulated Phosphodiesterase. *Front. Microbiol.* **2019,** *10*, 1499.
286. Li, F.; Zheng, Y.; Tian, J.; Ge, F.; Liu, X. W.; Tang, Y.; Feng, C. *Cupriavidus* sp. Strain Cd02-Mediated pH Increase Favoring Bioprecipitation of Cd^{2+} in Medium and Reduction of Cadmium Bioavailability in Paddy Soil. *Ecotoxicol. Environ. Saf.* **2019,** *184*, 109655.
287. Qiao, W.; Zhang, Y.; Xia, H.; Luo, Y.; Liu, S.; Wang, S.; Wang, W. Bioimmobilization of Lead by *Bacillus subtilis* X3 Biomass Isolated from Lead Mine Soil Under Promotion of Multiple Adsorption Mechanisms. *R. Soc. Open Sci.* **2019,** *6*, 181701.
288. Biswas, J. K.; Banerjee, A.; Sarkar, B.; Sarkar, D.; Sarkar, S. K.; Rai, M. Vithanage M. Exploration of an Extracellular Polymeric Substance from Earthworm Gut Bacterium (*Bacillus Licheniformis*) for Bioflocculation and Heavy Metal Removal Potential. *Appl. Sci.* **2020,** *11* (1), 349.
289. Sun, W.; Cheng, K.; Sun. K. Y.; Ma, X. Microbially Mediated Remediation of Contaminated Sediments by Heavy Metals: A Critical Review. *Curr. Pollut. Rep.* **2021,** 1–12.
290. Taştan, B. E.; Ertuğrul, S.; Donmez, G. Effective Bioremoval of Reactive Dye and Heavy Metals by *Aspergillus Versicolor*. *Bioresour. Technol.* **2010,** *101* (3), 870–876.
291. Muneer, B.; Iqbal, M. J.; Shakoori, F. R.; Shakoori, A. R. Tolerance and Biosorption of Mercury by Microbial Consortia: Potential Use in Bioremediation of Wastewater. *Pak. J. Zool.* **2013,** *41* (1), 247–254.
292. Vaseem, H.; Singh V. K.; Singh M. P. Heavy Metal Pollution due to Coal Washery Effluent and its Decontamination Using a Macrofungus, *Pleurotus Ostreatus*, *Ecotoxicol. Environ. Saf.* **2017,** *145*, 42–49.
293. Oladipo, O. G.; Awotoye, O. O.; Olayinka, A.; Bezuidenhout, C. C.; Maboeta, M. S. Heavy Metal Tolerance Traits of Filamentous Fungi Isolated from Gold and Gemstone Mining Sites, *Braz. J. Microbiol.* **2018,** *41* (1), 29–37.

294. Lin, Y.; Xiao, W.; Ye, Y.; Wu, C.; Hu, Y.; Shi, H. Adaptation of Soil Fungi to Heavy Metal Contamination in Paddy Fields-A Case Study in Eastern China. *Environ. Sci. Pollut. Res.* **2020,** *27*, 27819–27830.
295. Zhang, D.; Yin, C.; Abbas, N.; Mao, Z.; Zhang, Y. Multiple Heavy Metal Tolerance and Removal by an Earthworm Gut Fungus *Trichoderma Brevicompactum* QYCD-6. *Sci. Rep.* **2020,** *11* (1), 1–9.
296. Ali, E. A.; Sayed, M. A.; Abdel-Rahman, T. M., Hussein, R. Fungal Remediation of Cd (ii) from Wastewater Using Immobilization Techniques. *RSC Adv.* **2021,** *11* (8), 4853–4863.
297. Bilal, M.; Rasheed, T.; Sosa-Hernandez, J. E.; Raza, A.; Nabeel, F.; Iqbal H. Biosorption: An Interplay Between Marine Algae and Potentially Toxic Elements-A Review. *Mar. Drugs* **2018,** *11* (2), 65.
298. Salama, E. S.; Roh, H. S.; Dev, S.; Khan, M. A.; Abou-Shanab, R. A. I.; Chang, S. W.; Jeon, B. H. Algae as a Green Technology for Heavy Metals Removal from Various Wastewater. *World J. Microbiol. Biotechnol.* **2019,** *35*, 75.
299. Ahmad, S.; Pandey, A.; Pathak, V. V.; Tyagi, V. V.; Kothari, R. Phycoremediation: Algae as Eco-Friendly Tools for the Removal of Heavy Metals from Wastewaters. In *Bioremediation of Industrial Waste for Environmental Safety*; Bharagava, R. N., Saxena, G., Eds.; Springer: Singapore, **2020;** pp. 53–76.
300. Sen Gupta, G.; Yadav, G.; Tiwari, S. Bioremediation of Heavy Metals: A New Approach to Sustainable Agriculture. In *Restoration of Wetland Ecosystem: A Trajectory Towards a Sustainable Environment*; Upadhyay, A. K., Singh, R., Singh, D. P., Eds.; Springer, Singapore, 2020; pp. 195–226.
301. Lin, Z.; Li, J.; Luan, Y.; Dai, W. Application of Algae for Heavy Metal Adsorption: A 20-Year Meta-Analysis. *Ecotoxicol. Environmen. Safe.* **2020,** *190*, 110089.
302. Fortin, C.; Campbell, P. G. Thiosulfate Enhances Silver Uptake by a Green Alga: Role of Anion Transporters in Metal Uptake. *Environ. Sci. Technol.* **2001,** *35*, 2214–2218.
303. Perales-Vela, H. V.; Peña-Castro, J. M.; Cañizares-Villanueva, R. O. Heavy Metal Detoxification in Eukaryotic Microalgae. *Chemosphere* **2006,** *64*, 1–10.
304. Marella, T. K.; Saxena, A.; Tiwari, A. Diatom Mediated Heavy Metal Remediation: A Review. *Bioresour. Technol.* **2020,** *305*, 123068.
305. Luoma, S. N.; Rainbow, P. S. Why is Metal Bioaccumulation So Variable? Biodynamics as a Unifying Concept. *Environ. Sci. Technol.* **2005,** *39*, 1921–1931.
306. Balzano, S.; Sardo, A.; Blasio, M.; Chahine, T. B.; Dell'Anno, F.; Sansone, C.; Brunet, C. Microalgae Metallothioneins and Phytochelatins and their Potential Use in Bioremediation. *Front. Microbiol.* **2020,** *11*, 517.
307. Purchase, D.; Scholes, L. N.; Revitt, D. M.; Shutes, R. B. E. Effects of Temperature on Metal Tolerance and the Accumulation of Zn and Pb by Metal-Tolerant Fungi Isolated from Urban Runoff Treatment Wetlands, *J. Appl. Microbiol.* **2009,** *106*, 1163–1174.
308. Liu, S. H.; Zeng, G. M.; Niu, Q. Y.; Liu, Y.; Zhou, L.; Jiang, L. H.; Tan, X. F.; Xu, P.; Zhang, C.; Cheng, M. Bioremediation Mechanisms of Combined Pollution of PAHs and Heavy Metals by Bacteria and Fungi: A Mini Review. *Bioresour. Technol.* **2017,** *224*, 25–33.
309. Bhattacharya, S.; Das, A. Mycoremediation of Congo Red Dye by Filamentous Fungi. *Braz. J. Microbiol.* **2011,** *41* (4), 1526–1536.
310. Singh, M.; Srivastava, P. K.; Verma, P. C.; Kharwar, R. N.; Singh, N.; Tripathi, R. D. Soil Fungi for Mycoremediation of Arsenic Pollution in Agriculture Soils. *J. Appl. Microbiol.* **2015,** *119*, 1278–1290.

311. Kapahi, M.; Sachdeva, S. Mycoremediation Potential of *Pleurotus* Species for Heavy Metals: A Review. *Bioresour. Bioprocess.* **2017,** *4,* 32.
312. Khan, I.; Aftab, M.; Shakir, S.; Ali, M.; Qayyum, S.; Rehman, M. U.; Haleem, K. S.; Touseef, I. Mycoremediation of Heavy Metal (Cd and Cr)-Polluted Soil Through Indigenous Metallotolerant Fungal Isolates. *Environ. Monit. Assess.* **2019,** *191.*
313. Urík, M.; Cernanský, S.; Ševc, J.; Šimonovičová, A.; Littera, P. Biovolatilization of Arsenic by Different Fungal Strains. *Water Air Soil Pollut.* **2007,** *181* (1), 337–342.
314. Boriová, K.; Cernanský, S.; Matúš, P.; Bujdoš, M.; Šimonovičová, A. Bioaccumulation and Biovolatilization of Various Elements Using Filamentous Fungus *Scopulariopsis Brevicaulis*. *Lett. Appl. Microbiol.* **2014,** *59,* 217–223.
315. Fomina, M.; Gadd, G. M. Biosorption: Current Perspectives on Concept, Definition and Application. *Bioresour. Technol.* **2014,** *160,* 3–14.
316. Igiri, B. E.; Okoduwa, S. I. R.; Idoko, G. O.; Akabuogu, E. P.; Adeyi, A. O.; Ejiogu, I. K. Toxicity and Bioremediation of Heavy Metals Contaminated Ecosystem from Tannery Wastewater: A Review. *J. Toxicol.* **2018,** *2018,* 2568038..
317. Abu-Elsaoud, A. M.; Nafady, N. A.; Abdel-Azeem, A. M. Arbuscular Mycorrhizal Strategy for Zinc Mycoremediation and Diminished Translocation to Shoots and Grains in Wheat. *PLoS One* **2017,** *11* (11), E0188220.
318. Wu, M.; Xu, Y.; Ding, W.; Li, Y.; Xu, H. Mycoremediation of Manganese and Phenanthrene by *Pleurotus Eryngii* Mycelium Enhanced by Tween 80 and Saponin, *Appl. Microbiol. Biotechnol.* **2016,** *100,* 7249–7261.
319. Congeevaram, S.; Dhanarani, S.; Park, J.; Dexilin, M.; Thamaraiselvi, K. Biosorption of Chromium and Nickel by Heavy Metal Resistant Fungal and Bacterial Isolates. *J. Hazard. Mater.* **2007,** *141* (1–2), 270–277.
320. Enayatizamir, N.; Liu, J.; Wang, L.; Lin, X.; Fu, P. Coupling Laccase Production from *Trametes Pubescence* with Heavy Metal Removal for Economic Waste Water Treatment. *J. Water Process Eng.* **2020,** *37,* 101357.
321. Ayangbenro, A. S.; Babalola, O. O. A New Strategy for Heavy Metal Polluted Environments: A Review of Microbial Biosorbents, *Int. J. Environ. Res. Public Health.* **2017,** *14,* 94.
322. Njoku, K. L.; Asunmo, M. O.; Ude, E. O.; Adesuyi, A. A.; Oyelami, A. O. The Molecular Study of Microbial and Functional Diversity of Resistant Microbes in Heavy Metal Contaminated Soil. *Environ. Technol. Innov.* **2020,** *17,* 100606.
323. Tiwari, S.; Lata, C. Heavy Metal Stress, Signaling, and Tolerance Due to Plant-Associated Microbes: An Overview. *Front. Plant Sci.* **2018,** *1* (9), 452.
324. Loaec, M.; Olier, R.; Guezennec, J. Uptake of Lead, Cadmium and Zinc by a Novel Bacterial Exopolysaccharides. *Water Res.* **1997,** *31,* 1171–1179.
325. Teitzel, G. M.; Parsek, M. R. Heavy Metal Resistance of Biofilm and Planktonic *Pseudomonas Aeruginosa*. *Appl. Environ. Microbiol.* **2003,** *69,* 2313–2320.
326. Manivasagan, P.; Kim, S. K. Extracellular Polysaccharides Produced by Marine Bacteria. *Adv. Food Nutr. Res.* **2014,** *72,* 79–94.
327. Yue, Z. B.; Li, Q.; Li, C. C.; Chen, T. H.; Wang, J. Component Analysis and Heavy Metal Adsorption Ability of Extracellular Polymeric Substances (EPS) from Sulfate Reducing Bacteria. *Bioresour. Technol.* **2015,** *194,* 399–402.
328. Deschatre, M.; Ghillebaert, F.; Guezennec, J.; Colin, C. S. Sorption of Copper(II) and Silver(I) by Four Bacterial Exopolysaccharides. *Appl. Biochem. Biotechnol.* **2013,** *171,* 1313–1327.

329. Shi, Y.; Huang, J.; Zeng, G.; Gu, Y.; Chen, Y.; Hu, Y.; Tang, B.; Zhou, J.; Yang, Y.; Shi, L. Exploiting Extracellular Polymeric Substances (EPS) Controlling Strategies for Performance Enhancement of Biological Wastewater Treatments: An Overview. *Chemosphere* **2017,** *180,* 396–411.
330. Wei, L.; Li, Y.; Noguera, D. R.; Zhao, N.; Song, Y.; Ding, J.; Zhao, Q.; Cui, F. Adsorption of Cu^{2+} and Zn^{2+} by Extracellular Polymeric Substances (EPS) in Different Sludges: Effect of EPS Fractional Polarity on Binding Mechanism. *J. Hazard. Mater.* **2017,** *321,* 473–483.
331. Kim S.-Y., J.-H. Kim, C.-J. Kim, D.-K. Oh, Metal Adsorption of the Polysaccharide Produced from *Methylobacterium Organophilum. Biotechnol. Lett.* 18 (1996) 1161–1164.
332. Ayangbenro, A. S.; Babalola, O. O.; Aremu, O. S. Bioflocculant Production and Heavy Metal Sorption by Metal Resistant Bacterial Isolates from Gold Mining Soil. *Chemosphere* **2019,** *231,* 113–120.
333. Pokethitiyook, P.; Poolpak, T. Biosorption of Heavy Metal from Aqueous Solutions. In *Phytoremediation*; Springer: Cham. **2016,** pp. 113–141.
334. Deng, X.; Wang, P. Isolation of Marine Bacteria Highly Resistant to Mercury and their Bioaccumulation Process. *Biore. Tech.* **2012,** *121,* 342–347.
335. Abbas, S. H.; Ismail, I. M.; Mostafa, T. M.; Sulaymon, A. H. Biosorption of Heavy Metals: A Review. *J. Chem. Sci. Technol.* **2014,** *3,* 74–102.
336. Gautam, R. K.; Gautam, P. K.; Banerjee, S.; Rawat, V.; Soni, S.; Sharma, S. K.; Chattopadhyaya, M. C. Biomass-Derived Biosorbents for Metal Ions Sequestration: Adsorbent Modification and Activation Methods and Adsorbent Regeneration. *J. Environ. Chem. Eng.* **2014,** *1* (1), 239–259.
337. Wang, J.; Li, Q.; Li, M. M.; Chen, T. H.; Zhou, Y. F.; Yue, Z. B. Competitive Adsorption of Heavy Metal by Extracellular Polymeric Substances (EPS) Extracted from Sulfate Reducing Bacteria. *Bioresour. Technol.* **2014,** *16,* 3374–376.
338. San Keskin, N. O.; Celebioglu, A.; Sarioglu, O. F.; Uyar, T.; Tekinay, T. Encapsulation of Living Bacteria in Electrospun Cyclodextrin Ultrathin Fibers for Bioremediation of Heavy Metals and Reactive Dye from Wastewater. *Colloids Surf. B: Biointerfaces.* **2018,** *161,* 169–176.
339. Sahmoune, M. N. Performance of Streptomyces Rimosus Biomass in Biosorption of Heavy Metals from Aqueous Solutions. *Microchem. J.* **2018,** 87–95.
340. Black, R.; Sartaj, M.; Mohammadian, A.; Qiblawey, H. A. Biosorption of Pb and Cu Using Fixed and Suspended Bacteria. *J. Environ. Chem. Eng.* **2014,** *2,* 1663–1671.
341. Chaturvedi, K. S.; Hung, C. S.; Crowley, J. R.; Stapleton, A. E.; Henderson, J. P. The Siderophore *Yersinia Bactin* Binds Copper to Protect Pathogens During Infection. *Nat. Chem. Biol.* **2012,** *8,* 731.
342. Yin, K.; Zhang, W.; Chen, L. Pyoverdine Secreted by *Pseudomonas Aeruginosa* as a Biological Recognition Element for the Fluorescent Detection of Furazolidone. *Biosens. Bioelectron.* **2014,** *51,* 90–96.
343. Yin, K.; Wu, Y.; Wang, S.; Chen, L. A Sensitive Fluorescent Biosensor for the Detection of Copper Ion Inspired by Biological Recognition Element Pyoverdine. *Sens. Actuators B: Chem.* **2016,** *232,* 257–263.
344. Oves, M.; Khan, M. S.; Qari, H. A. Ensifer Adhaerens for Heavy Metal Bioaccumulation, Biosorption, and Phosphate Solubilization Under Metal Stress Condition. *J. Taiwan Inst. Chem. Eng.* **2017,** *80,* 540–552.

345. Patel, P. S.; Shaikh, R. Sayyed, Modified Chrome Azurol S Method for Detection and Estimation of Siderophores Having Affinity for Metal Ions Other Than Iron. *Environ. Sustainability* **2018,** *1* (1), 81–87.
346. Sharma, R.; Bhardwaj, R.; Gautam, V.; Kohli, S. K.; Kaur, P.; Bali, R. S.; Saini, P.; Thukral, A. K.; Arora, S.; Vig, A. P. *Microbial Siderophores in Metal Detoxification and Therapeutics: Recent Prospective and Applications. Plant Microbiome: Stress Response*; Springer, 2018, pp. 337–350.
347. Voordouw, G. The Genus *Desulfovibrio*: The Centennial. *Appl. Environ. Microbiol.* **1995,** *61* (8), 2813.
348. Kieu, H. T. Q.; Muller, E.; Horn, H. Heavy Metal Removal in Anaerobic Semi-Continuous Stirred Tank Reactors by a Consortium of Sulfate-Reducing Bacteria. *Water Res.* **2011,** *45,* 3863–3870.
349. Ock Joo, J.; Choi, J. H.; Kim, I. H.; Kim, Y. K.; Oh, B. K. Effective Bioremediation of Cadmium (II), Nickel (II), and Chromium (VI) in a Marine Environment by Using *Desulfovibrio Desulfuricans. Biotechnol. Bioprocess Eng.* **2015,** *21* (5), 937–941.
350. Maneerat, S. Biosurfactants from Marine Microorganisms. *Songklanakarin J. Sci. Technol.* **2005,** *21* (6), 1263–1272.
351. Kiran, G. S.; Sabarathnam, B.; Selvin, J. Biofilm Disruption Potential of a Glycolipid Biosurfactant from Marine *Brevibacterium Casei. FEMS Immunol. Med. Microbiol.* **2010,** *59,* 432–438.
352. Franzetti, A.; Gandolfi, I.; Fracchia, L.; Van Hamme, J.; Gkorezis, P.; Marchant, R.; Banat, I. M. Biosurfactant Use in Heavy Metal Removal from Industrial Effluents and Contaminated Sites. *Biosurfactants Prod. Util. Technol. Econ.* **2014,** *159,* 361–369.
353. Cameotra, S. S.; Makkar, R. S. Biosurfact Yang Ant-Enhanced Bioremediation of Hydrophobic Pollutants. *Pure Appl. Chem.* **2010,** *82,* 97–116.
354. Miller, R. M. Biosurfactant-Facilitated Remediation of Metal-Contaminated Soils. *Environ. Health Perspect.* **1995,** *103,* 59–62.
355. Lee, Y. J.; Choi, J. K.; Kim, E. K.; Youn, S. H.; Yang, E. J. Field Experiments on Mitigation of Harmful Algal Blooms Using a Sophorolipid-Yellow Clay Mixture and Effects on Marine Plankton. *Harmful Algae* **2008,** *1* (2), 154–162.
356. Mulligan, C. N.; Yong, R. N.; Gibbs, B. F. Remediation Technologies for Metal-Contaminated Soils and Groundwater: An Evaluation. *Eng. Geol.* **2001,** *60,* 193–207.
357. Luna, J. M.; Rufino, R. D.; Sarubbo, L. A. Biosurfactant from *Candida Sphaerica* UCP0995 Exhibiting Heavy Metal Remediation Properties. *Process Saf. Environ. Prot.* **2016,** *102,* 558–566.
358. Da Rocha Junior, R. B.; Meira, H. M.; Almeida, D. G.; Rufino, R. D.; Luna, J. M.; Santos, V. A.; Sarubbo, LA. Application of a Low-Cost Biosurfactant in Heavy Metal Remediation Processes. *Biodegradation* **2019,** *31* (4), 215–233.
359. Ravindran, A.; Sajayan, A.; Priyadharshini, G. B.; Selvin, J.; Kiran, G. S. Revealing the Efficacy of Thermostable Biosurfactant in Heavy Metal Bioremediation and Surface Treatment in Vegetables. *Front. Microbiol.* **2020,** *11,* 222.
360. Castelein, M.; Verbruggen, F.; Van Renterghem, L.; Spooren, J.; Yurramendi, L.; Du Laing, G.; Boon, N.; Soetaert, W.; Hennebel, T.; Roelants, S.; Williamson, A. J. Bioleaching of Metals from Secondary Materials Using Glycolipid Biosurfactants. *Miner. Eng.* **2021,** *163,* 106665.
361. Diaz, M. A.; De Ranson, I. U.; Dorta, B.; Banat, I. M.; Blazquez, M. L.; Gonzalez, F.; Muñoz, J. A.; Ballester, A. Metal Removal from Contaminated Soils Through

Bioleaching with Oxidizing Bacteria and Rhamnolipid Biosurfactants. *Soil Sediment Contam. An. Int. J.* **2015,** *21* (1), 16–29.

362. Hrynkiewicz, K.; Złoch, M.; Kowalkowski, T.; Baum, C.; Niedojadło, K.; Buszewski, B. Strain-Specific Bioaccumulation and Intracellular Distribution of Cd^{2+} in Bacteria Isolated from the Rhizosphere, Ectomycorrhizae, and Fruit Bodies of Ectomycorrhizal Fungi. *Environ. Sci. Pollut. Res. Int.* **2015,** *22,* 3055–3067.
363. Yin, K.; Lv, M.; Wang, Q.; Wu, Y.; Liao, C.; Zhang, W.; Chen, L. Simultaneous Bioremediation and Biodetection of Mercury Ion through Surface Display of Carboxylesterase E2 from *Pseudomonas Aeruginosa* PA1. *Water Res.* **2016,** *103,* 383–390.
364. Jain, P. K.; Gupta, V. K.; Bajpai, V.; Lowry, M.; Jaroli, D. P. GMO's: Perspective of Bioremediation. In *Recent Advances in Environmental Biotechnology*; Jain, P. K., Gupta, V. K., Bajpai, V., Eds.; LAP Lambert Academic Publishing AG and Co.: KG, Germany, 2011; pp. 6–23.
365. Diep, P.; Mahadevan, R.; Yakunin, A. F. Heavy Metal Removal by Bioaccumulation Using Genetically Engineered Microorganisms. *Front. Bioeng. Biotechnol.* **2018,** *6,* 157.
366. Sun, G. L.; Reynolds, E. E.; Belcher, A. M. Using Yeast to Sustainably Remediate and Extract Heavy Metals from Waste Waters. *Nat. Sustain.* **2020,** *3,* 303–311.
367. Li, F.; Wang, W., Li, C., Zhu, R.; Ge, F.; Zheng, Y.; Tang, Y. Self-Mediated pH Changes in Culture Medium Affecting Biosorption and Biomineralization of Cd^{2+} by *Bacillus Cereus* Cd^{01}. *J. Hazard. Mater.* **2018,** *358,* 178–186.
368. Kulakovskaya, T.; Ryazanova, L.; Zvonarev, A.; Khokhlova, G.; Ostroumov, V.; Vainshtein, M. The Biosorption of Cadmium and Cobalt and Iron Ions by Yeast *Cryptococcus Humicola* At Nitrogen Starvation. *Folia Microbiol.* **2018,** *61* (4), 1–4.
369. Remonsellez, F.; Orell, A.; Jerez, C. A. Copper Tolerance of the Thermoacidophilic Archaeon *Sulfolobus metallicus*: Possible Role of Polyphosphate Metabolism. *Microbiol.* **2006,** *152,* 59–66.
370. Naik, M. M.; Dubey, S. Lead and Mercury-Resistant Marine Bacteria and their Application in Lead and Mercury Bioremediation. In *Marine Pollution and Microbial Remediation*; Springer, 2017; pp. 29–40.
371. Blindauer, C.; Harrison, M.; Parkinson, J.; Robinson, N.; Sadler, P. Isostructural Replacement of Zinc by Cadmium in Bacterial Metallothionein. *Met. Ions Biol. Med.* **2008,** *10,* 167–173.
372. Almárcegui, R. J.; Navarro, C. A.; Paradela, A.; Albar, J. P.; Von Bernath, D.; Jerez, C. A. New Copper Resistance Determinants in the Extremophile *Acidithiobacillus Ferrooxidans*: A Quantitative Proteomic Analysis. *J. Proteome Res.* **2014,** *13,* 946–960.
373. Kisielowska, E.; Hołda, A.; Niedoba, T. Removal of Heavy Metals from Coal Medium with Application of Biotechnological Methods. *Górnictwo I Geoinzynieria* **2010,** *34,* 93–104.
374. Cumberland, S. A.; Douglas, G.; Grice, K.; Moreau, J. W. Uranium Mobility in Organic Matter Rich Sediments: A Review of Geological and Geochemical Processes. *Earth Sci. Rev.* **2016,** *159,* 160–185.
375. Jong, T.; Parry, D. L. Removal of Sulfate and Heavy Metals by Sulfate Reducing Bacteria in Short-Term Bench Scale Upflow Anaerobic Packed Bed Reactor Runs. *Water Res.* **2003,** *37,* 3379–3389.
376. Zhou, Q.; Chen, Y.; Yang, M.; Li, W.; Deng, L. Enhanced Bioremediation of Heavy Metal from Effluent by Sulfate-Reducing Bacteria with Copper-Iron Bimetallic Particles Support. *Bioresour. Technol.* **2013,** *136,* 413–417.

377. Das, S.; Mishra, J.; Das, S. K.; Pandey, S.; Rao, D. S.; Chakraborty, A.; Sudarshan, M.; Das, N.; Thatoi, H. Investigation on Mechanism of Cr (VI) Reduction and Removal by *Bacillus Amyloliquefaciens*, a Novel Chromate Tolerant Bacterium Isolated from Chromite Mine Soil. *Chemosphere* **2014,** *96,* 112–121.
378. Horton, R. N.; Apel, W. A.; Thompson, V. S.; Sheridan, P. P. Low Temperature Reduction of Hexavalent Chromium by a Microbial Enrichment Consortium and a Novel Strain of *Arthrobacter Aurescens*. *BMC Microbiol.* **2006,** *6,* 1–8.
379. Moller, A. K.; Barkay, T.; Hansen, M. A.; Norman, A.; Hansen, L. H.; Sorensen, S. J.; Boyd, E.S,; Kroer, N. Mercuric Reductase Genes (Mera) and Mercury Resistance Plasmids in High Arctic Snow, Freshwater and Sea-Ice Brine. *FEMS Microbiol. Ecol.* **2014,** *87,* 52–63.
380. Noroozi, M.; Amoozegar, M.; Pourbabaei, A.; Naghavi, N.; Nourmohammadi, Z. Isolation and Characterization of Mercuric Reductase by Newly Isolated Halophilic Bacterium, *Bacillus Firmus* MN8. *Global J. Environ. Sci. Manag.* **2017,** *3,* 427–436.
381. Battistel, D.; Baldi, F.; Marchetto, D.; Gallo, M.; Daniele, S. A Rapid Electrochemical Procedure for the Detection of Hg(0) Produced by Mercuric-Reductase: Application for Monitoring Hg-Resistant Bacteria Activity. *Environ. Sci. Technol.* **2012,** *46,* 10675–10681.
382. Giovanella, P.; Cabral, L.; Bento, F.; Gianello, C.; Camargo, F. A. Mercury (II) Removal by Resistant Bacterial Isolates and Mercuric (II) Reductase Activity in a New Strain of *Pseudomonas* sp. B50A. *New Biotechnol.* **2016,** *33,* 216–223.
383. Jaysankar, N. R. D.; Vardanyan, L.; De J.; Ramaiah, N.; Vardanyan, L. Detoxification of Toxic Heavy Metals by Marine Bacteria Highly Resistant to Mercury. *Mar. Biotechnol.* **2008,** *11* (4), 471–477.
384. Gihring, T. M.; Banfield, J. F. Arsenite Oxidation and Arsenate Respiration by a New Thermus Isolate. *FEMS Microbiol. Lett.* **2001,** *204,* 335–340.
385. Nathan, V. K.; Rani, M. E.; Gunaseeli, R.; Kannan, N. D. Enhanced Biobleaching Efficacy and Heavy Metal Remediation Through Enzyme Mediated Lab Scale Paper Pulp Deinking Process. *J. Clean Prod.* **2018,** *203,* 926–932.
386. Mora, A. P. D.; Ortega-Calvo, J. J.; Cabrera, F.; Madejon, E. Changes in Enzyme Activities and Microbial Biomass After "in Situ" Remediation of a Heavy Metal-Contaminated Soil. *Appl. Soil Ecol.* **2005,** *21* (2), 125–137.
387. Pratush, A.; Kumar, A.; Hu, Z. Adverse Effect of Heavy Metals (As, Pb, Hg and Cr) on Health and their Bioremediation Strategies: A Review. *Int. Microbiol.* **2018,** *21,* 97–106.
388. Yang, H. C.; Fu, H. L.; Lin, Y. F.; Rosen, B. P. Pathways of Arsenic Uptake and Efflux, Current Topics in Membranes, Elsevier, **2012,** pp. 325–358.
389. Soto, D. F.; Recalde, A.; Orell, A.; Albers, S. V.; Paradela, A.; Navarro, C. A.; Jerez, C. A. Global Effect of the Lack of Inorganic Polyphosphate in the Extremophilic Archaeon *Sulfolobus Solfataricus*: A Proteomic Approach. *J. Proteomics* **2018,** *191,* 143–152.
390. Al-Gheethi, A. A.; Lalung, J.; Noman, E. A.; Bala, J.; Norli, I. Removal of Heavy Metals and Antibiotics from Treated Sewage Effluent by Bacteria. *Clean Technol. Environ. Policy* **2015,** *17,* 2101–2123.
391. Lerebours, A.; To, V. V.; Bourdineaud, J. P. Danio Rerio ABC Transporter Genes Abcb3 and Abcb7 Play a Protecting Role Against Metal Contamination. *J. Appl. Toxicol.* **2016,** *31* (12), 1551–1557.
392. Zammit, C. M.; Weiland, F.; Brugger, J.; Wade, B.; Winderbaum, L. J.; Nies, D. H.; Southam, G.; Hoffmann, P.; Reith, F. Proteomic Responses to Gold (III)-Toxicity in the Bacterium *Cupriavidus Metallidurans* CH34. *Metallomics* **2016,** *8,* 1204–1216.

393. Rosen, BP. Transport and Detoxification Systems for Transition Metals, Heavy Metals and Metalloids in Eukaryotic and Prokaryotic Microbes. *Comp Biochem. Physiol. A. Mol. Integ. Physiol.* **2002,** *131* (3), 689–693.
394. Liang, D.; He, W.; Li, C.; Wang, F.; Crittenden, J. C.; Feng, Y. Remediation of Nitrate Contamination by Membrane Hydrogenotrophic Denitrifying Biofilm Integrated in Microbial Electrolysis Cell. *Water Res.* **2021,** *188,* 116498.
395. Laroute, V.; Mazzoli, R.; Loubière, P.; Pessione. E.; Cocaign-Bousquet, M. Environmental Conditions Affecting Gaba Production in *Lactococcus Lactis* NCDO 2118. *Microorganisms* **2021,** *1* (1), 122.
396. Wan, W.; Hao, X.; Xing, Y.; Liu, S.; Zhang, X.; Li, X.; Chen, W.; Huang, Q. Spatial Differences in Soil Microbial Diversity Caused by Ph-Driven Organic Phosphorus Mineralization. *Land Degrad. Dev.* **2021,** *31* (2), 766–776.
397. Babur, E.; Dindaroğlu, T.; Solaiman, Z. M.; Battaglia, ML. Microbial Respiration, Microbial Biomass and Activity are Highly Sensitive to forest Tree Species and Seasonal Patterns in the Eastern Mediterranean Karst Ecosystems. *Sci. Total Environ.* **2021,** *775,* 145868.
398. Dong, X.; Gao, P.; Zhou, R.; Li, C.; Dun, X.; Niu, X. Changing Characteristics and Influencing Factors of the Soil Microbial Community During Litter Decomposition in a Mixed *Quercus Acutissima* Carruth. and *Robinia Pseudoacacia* L. forest in Northern China. *Catena* **2021,** *196,* 104811.
399. Rajkumar, H.; Naik, P. K.; Rishi, M. S. A New Indexing Approach for Evaluating Heavy Metal Contamination in Groundwater. *Chemosphere* **2020,** *245,* 125598.
400. Shapiro, R. S.; Chavez, A.; Collins, J. J. CRISPER Based Genomic Tools for the Manipulation of Genetically Intractable Microorganism. *Nat. Rev. Microbiol.* **2018,** *16,* 333–339.
401. Hemmat-Jou, M. M.; Safari-Sinegani, A. A.; Mirzaie-Asl, A.; Tahmourespour, A. Analysis of Microbial Communities in Heavy Metals Contaminated Soil Using the Metagenomic Approach. *Ecotoxicol.* **2018,** *21* (9), 1281–1291.
402. Sun, F. S.; Yu, G. H.; Ning, J. Y.; Zhu, X. D.; Goodman, B. A.; Wu, J. Biological Removal of Cadmium from Biogas Residues During Vermicomposting, and the Effect of Earthworm Hydrolysates on *Trichoderma Guizhouense* Sporulation. *Bioresour. Technol.* **2020,** *312,* 123635.
403. Qian, X.; Chen, L.; Sui, Y.; Chen, C.; Zhang, W.; Zhou, J.; Dong, W.; Jiang, M.; Xin, F.; Ochsenreither, K. Biotechnological Potential and Application of Microbial Consortia. *Biotechnol. Adv.* **2020,** *40,* 107500.
404. Xin, S.; Zeng, Z.; Zhou, X.; Luo, W.; Shi, X.; Wang, Q.; Deng, H.; Du, Y. Recyclable *Saccharomyces Cerevisiae* Loaded Nanofibrous Mats with Sandwich Structure Constructing Via Bio-Electrospraying for Heavy Metal Removal. *J. Hazard. Mater.* **2017,** *324,* 365–372.
405. Liu, S.; Yang, B.; Liang, Y.; Xiao, Y.; Fang, J. Prospect of Phytoremediation Combined with Other Approaches for Remediation of Heavy Metal Polluted Soils. *Environ. Sci. Pollut. Res.* **2020,** *21* (14), 16069–16085.
406. Mosa, K. A.; Saadoun, I.; Kumar, K.; Helmy, M.; Dhankher, O. P. Potential Biotechnological Strategies for the Cleanup of the Heavy Metals and Metalloids. *Front. Plan. Sci.* **2016,** *7,* 303.

CHAPTER 7

Bioinformatical and Biotechnological Advances for Bioremediation and Plant Pollution Control

KAYENAT SHEIKH[1*], KHALID RAZA[1], and SYED SALEEMULLAH[2]

[1]*Dept. of Computer Science, Jamia Millia Islamia, New Delhi, India*

[2]*Dept. of Bioscience, Jamia Millia Islamia, New Delhi, India*

[*]*Corresponding author. E-mail: kayenat996@gmail.com*

ABSTRACT

Environmental pollution is an unstoppable phenomenon that affects every living being. As the human population is increasing worldwide, natural ecosystems have declined in quality and an imbalance has created a negative impact. Most of the chemicals and pollutants are synthesized in industries and factories. The process of production has led to an increase in the unwanted by-products that are toxic to the environment. Most of the chemicals get absorbed by the water bodies and soil. These in return are transported back in the organisms such as plants and trees. The effect of pollution is the severe health damage to both the flora and fauna. As a result, the necessary scientific steps have been adopted and mechanisms employed for sustainability and survival.

The most acceptable solution to prevent the environment from getting sicker is to bring interdisciplinary sciences—bioinformatics and microbiology into action. Microorganisms such as bacteria, yeast, and fungi can integrate substrates within their systems and produce products. Some commonly known usages are bioremediation (waste treatment), biofiltrations,

Environmental Pollution Impact on Plants: Survival Strategies under Challenging Conditions. Tariq Aftab, PhD & Khalid Rehman Hakeem, PhD (Eds.)

biomining, biotreatment, biocatalysts, and more. Another application through which environment protection can be ensured are Biotools (biotechnological tools) and Bioinfotools (bioinformatics tools). The pollutants taken up by them are discarded back into the environment in the organic form. As a result, now the end products are nontoxic and suitable for reconsumption by the living resources present in the environment. All the conversions occur naturally and organically, without the involvement of synthetic processes. Pollution can be controlled by researching and understanding more about the microbial organisms using bioinformatical tools and applying biotechnology for immediate effects.

Many developed nations such as United States, Germany, Japan, and Sweden have been using these technologies for years to control pollution. Many economically unstable, highly populated, and polluted countries can integrate similar sustainable development plans.

7.1 INTRODUCTION

Environmental pollution since many decades has been a grave problem to both flora and fauna. The impact of air, water, and soil pollution has caused severe damage to plants and living forms. The availability of natural resources has been perturbed due to rise in industrial activities, excessive agriculture, and land mining that has also led to rapid urbanization. In addition to this, it has also led to severe contamination of different components of life.[1] One major concern in the scientific community is the contamination of food crops and other substantial plants by heavy metals and toxic effluents. This is a global issue that results in diseases and toxicity in both animals and human beings. It poses a great threat to the environment and life forms due to bioaccumulation via food chains. Uptake of toxic elements such as heavy metals by plants on absorption from soil increases the threat to human and animal health. Heavy metals such as Cd, As Hg, V, Ni, Zn, Cr, Ba, and Pb are quite common metals in the soil environment, incoming from different industrial sources.[2,3] Other contaminant includes petrol-based products and by-products. Chemicals containing oxygen–sulfur–nitrogen are the major constituents of petrol, also alkanes, aromatic compounds, asphaltene. Out of these, aromatic compounds are known to express carcinogenic and mutagenic properties. They are excessively toxic for environment as well.[3] For more than two decades, studies have been going on regarding development of sustainable methods to reduce the environmental pollution. Bioremediation is one such technique that can reach the goal of effective environmental damage control. Different bioremediation methods have been developed to combat the pollution issues. However, the diversity in pollutant types demands multiple methods to be integrated into one for effective cleansing. There is no single "golden tool" to restore the damaged environment.[4,5] Most common way for pollution control is the action of indigenous microorganisms on the polluted environment. This organic cleanup can only be executed when the environmental conditions are microbe-friendly and most suited for replication and metabolism.[5] The natural methods such as bioremediation save costs and remain nature friendly, thereby taking advantageous position over other physicochemical systems. Bioremediation is an overly broad term that includes several definitions. It is a process of biological degradation and mechanical breakdown of pollutants such as dyes, heavy metals, greenhouse gases, aromatic compounds, nuclear and industrial wastes, plastics, medical wastes, hydrocarbons, agricultural chemicals, chlorinated and fluorinated compounds into simpler forms.[4] Depending on the site of bioremediation techniques application, the entire setup can be divided into distinct categories namely, in situ and ex situ that would be discussed in detail in the upcoming section. However, the type of bioremediation technique applied to control pollution depends largely upon the type of pollutant, its nature and degree of causing pollution, investment, location, and environment. Special focus is paid on the abiotic factors such as pH, nutrients, and oxygen concentration in the affected environment to measure the success ratio of the bioremediation techniques. Most of the bioremediation is associated with reduction of hydrocarbons toxins from soil and water sources due to excess accumulation.[6-11] The increased demand and dependence on nonrenewable sources of energy such as coal and petroleum has contributed immensely to the environmental degradation, thereby increasing the quantity of pollutants.[12] In this chapter, we shall provide a brief overview on the applications of interdisciplinary sciences such as biotechnology, bioinformatics, and microbiology in successful application of bioremediation techniques on pollution treatment and environment restoration.

7.2 WHAT IS BIOREMEDIATION?

Let us try to understand what bioremediation means before diving into the different aspects of bioremediation. As discussed earlier, bioremediation is a broad term with several definitions, but the core idea is to eliminate the pollutants from the environment using sophisticated biological methods, varying depending upon the nature, size, and damaging intensity of the pollutants.[4,5] The term Bioremediation is referred to using life forms to destroy/degrade the contaminants. The most common strategy adopted is the Intrinsic Bioremediation or the natural attenuation. The most straightforward method is to allow the indigenous microbes to decay/degrade xenobiotic compounds from the natural environment through their metabolic systems. It also includes certain chemical and physical processes along with biological ones to eliminate the quantity of toxic components, concentration of contaminated compounds, volume, toxicity, and mass. Furthermore, biodegradation through aerobic and anaerobic processes, sometimes sorption, dilution, pollutant transformation, pollutant stabilization, and volatilization are accompanied too.[5]

7.2.1 TYPES OF BIOREMEDIATIONS

7.2.1.1 INTRINSIC BIOREMEDIATION

Natural attenuation is always preferred in environmental areas with slight/low contamination where other techniques are not applied. There is a limit to the process of intrinsic remediation. Certain microbes at the polluted site might lack the catabolic genes in their genetic material, thereby being inefficient in degrading the environmental contaminants.[13] The soil pollution is either taken care of by the intrinsic microbes. If the native organisms lack the genes, genetically modified microbes or wild type are introduced into the environment for accelerating the process of pollutant degradation or improving the transformation rate of xenobiotics. [14]

7.2.1.2 BIOAUGMENTATION

Bioinoculants are those microbes that are added to the polluted soil for improving the soil quality and plant health. They also act as Plant Growth Promoting Rhizobacteria. They improve the nutrients uptake by the plants, enhance resistance to pathogenic life forms, and control contaminants. They stay in a symbiotic relationship with the plants or sometimes free-living.[14,15] The plant roots (rhizospheric zone) are the key host in this relationship since they are the center for deposition of photosynthetic carbon. PGPR and PGPF are helpful as bioinoculants present at this zone for taking up nutrients and resisting diseases.[16] Soil is rich in toxic chemicals such as pesticides. The bioinoculants have the capability to degrade organic chemical compounds into simpler harmless forms. The Polychlorinated biphenyls, petroleum hydrocarbons, lead, mercury, zinc, arsenic, carbamates are degraded through the enzymatic activity by the microbes.[15] Some of the most used bioinoculants for bioremediation are *Pseudomonas* and its variants for removing polychlorinated biphenyls (PBH),[17,18] *Azospirillum* for crude oil removal,[19] *Azotobacter* for Zn and Pb,[20] *Bacillus subtilis* for Ni,[21] *Kluyvera* for Pb and Ni,[20] and many more.

7.2.1.3 PHYTOREMEDIATION/RHIZOREMEDIATION

Rhizoremediation is also a natural technique for environmental restoration at polluted sites. This method involves rhizodegradation and phytostimulation resulting in mutual benefits to both the components.[13] This technique for compound degradation was introduced to eradicate the pesticides and herbicides chemicals. However, latest research suggests that it can also be used to get rid of PAH. The various plant and microbial interactions in rhizosphere help in eliminating contaminants from the environment.[22] The process is natural since flavonoids and similar chemical compounds released by the plant roots trigger the growth of contaminant degrading bacteria.[23,24] Also, soil aeration due to constant growth of root and decay increases oxidative degradation of PAH compounds in soil. Some species of plants promote the rise in the number of degrading microbes that increase in huge volume, even beyond the rhizospheric zone.[25] The success of rhizoremediation by any plant species relies on multiple factors such as dense root branching that favors the habitation of large bacterial colonies, metabolism and primary and secondary plant products, interaction with the surrounding environment and organisms. [14] The plant species such as *Populus* sp. and *Salix* sp. are popularly used for rhizoremediation of PHC polluted soil. The aerenchyma tissues of these plants enhance the oxygen quantity into the deep layers of soil.[26] The nutrient in soil comes from the decay of root cap and loss of root cells by the mucigel secretions at the root tip regions. The rich rhizosphere zone enables bacterial cell proliferation.[14] Also, few organic compounds are released that are derived from photosynthesis processes. [27,28] These compounds are rich sources of carbon and nitrogen that promote survival of microbes and degradation of organic compounds for a long term. The root secretions consist of both water-soluble and water-insoluble compounds such as sugars, proteins, nucleotides, enzymes, acids, flavonoids, and more. Volatile compounds such as alcohols and phenols are also secreted. Catechin and coumarin present in phenolic compounds act as co-metabolites for degrading bacteria.[14,29,30]

The process of rhizoremediation depends on the age of the plant, timings of secretions, and the quantity of root secretions at the tips. Overall, the soil ecology consists majorly of rhizomicrobial populations; hence, this method is effective in catalyzing the polymerization of pollutants on the surfaces of roots and soil.[14,30,31] Most of the time degradation is the result of the sum of bacterial actions. Different strains populate the different regions in the soil and degrade simultaneously. [14,32]

7.3 INTERDISCIPLINARY APPROACHES

A single remediation technique is not enough to effectively minimize the pollutant amount from the soil and the nearby plant surroundings. A powerful combination of integrated technologies is a promising alternative. One integrated method is the quick oxidation by Fenton during pretreatment and later followed up by microbial degradation in the untreated soil. The efficiency for treating Persistent Organic Pollutants (POP) can reach up to 98% from 70% due to this combined therapy.[33] In addition, PAH removal is enhanced by the Fenton oxidation treatment and bioremediation.[33,34] The TCDD contaminated soils were cleaned using a similar strategy.[35] Major advantages are low levels of pollutants in soil making it less toxic, reduced incomplete mineralization, bioavailability of PAH, peroxide decomposition, and oxygen production after Fenton treatment to improve soil aeration and decomposition.[35,36]

Another approach utilized by research community is the metagenomics for extensive analysis and studies of microbes thriving in various habitats. This technique involves applications of interdisciplinary sciences such as bioinformatics and biotechnology.[37,38] The techniques are an amalgamation of both the sciences. Some biotechnological methods include isolation of genomic DNA from the bacterial samples present in the environment, amplification of 16s rRNA gene through PCR, RFLP, for the microbial diversity analyses. The bioinformatical method such as NGS has become popular in recent times. It has been used extensively to study the diversity of microbes and metabolic pathways. This is an effective method for accurate and efficient studies for the entire genome in microbial populations. NGS analysis of hot springs microbial genome has shown the presence of hydrocarbon degrading bacteria and their pathways existing in hot temperatures.[39] On compiling this information on NGS and genomic extraction and analyses, bacterial communities with pollution control potentials can be identified easily. Bioinformatics branch of study such as proteomics and genomics play a vital role in these studies. They are interlinked together. Genomics reveals the structural information from the environmentally extracted microbial data. Proteomics is used to study the protein components in individuals and communities, to analyze the protein expression from genes and adaptations by extremophiles.[40,41]

The presence of xenobiotics in the environment is extremely toxic to plants and other life forms. Polychlorinated biphenyls, heavy metals, hydrocarbons are carcinogenic. Accumulation of pollutants in environment causes health hazards. The toxicity of xenobiotics is persistent in the environment and by developing eco-friendly and effective measures, remediation can be done. The traditional and tedious methods of physical and chemical methods such as carbon adsorption, air stripping, are very time consuming and noneconomical. Bioremediation based on low expenses is an effective choice. The extremophiles' robust catalytic activities with genetic manipulations can be used as machinery for pollutant cleanup. The properties of extremophiles can be exploited for biodegradation in ample quantity.[37,42,43]

7.3.1 BIOTECHNOLOGICAL APPROACHES

The present study suggests that the degradation mechanism by extremophiles can be improved by applying recombinant DNA technology. It can be achieved on inserting highly degrading catabolic genes from the wild bacterial genome. Psychrophilic bacteria used for aromatic compounds bioremediation has shown positive results.[37] A recombinant bacterial strain *Pseudoaltermonas haloplanktis* TAC125 constructed by rDT has great potential to convert aromatic chemical compounds into catechols. Genetic engineering has improved the catabolic activities of *Pseudoaltermonas haloplanktis* TAC125. Production of aromatic oxidative activity in *Pseudoaltermonas haloplanktis* TAC125 has been encoded by toluene-o-xylene monooxygenase (isolated from Pseudomonas sp. OX1) also known for converting phenols, dimethylphenols, and aromatic compounds. Another bacterium called *Geobacter sp.* has been modified under biotechnology for better biodegradation activities. It reduces the quantity of uranium by enrichment in acetate concentration.[44-48]

Similarly, microorganisms have been known for the bioremediation of hydrocarbons. By enhancing the cell surface hydrophobicity, hydrocarbons uptake by them becomes feasible. The extreme saline conditions reduce the rates of biodegradation. Halophiles are potent in degrading petrol-based hydrocarbons through massive release of biosurfactants. Studies of microbial community based on their 16S rRNA gene by applying biotechnological tools have helped in gathering useful information. Bacteria such as *Chloroflexi* and *Chlorobi* are effective against arsenic pollutant. *Geobacillus* degrades organic compounds present in the environment. The hydrocarbon such as hexadecane is degraded by another specie of the same genera. *Pyrococcus* possess cyanide degrading enzymes. The enzymes produced by extremophiles have been used extensively for industrial and biotechnological purposes. More than 3000 enzymes have been extracted and studied. Extremozymes are highly stable. They are effective in degrading the polymers such as cellulases, proteases, amylases, and more. Marine extremozymes are also known for bioremediation activities, and also efficient in food, paper, pulp, chemicals, drugs industrial applications.[49,50]

One common and successful soil bioremediation technique is plasmid-mediated bioaugmentation. The inoculation of microorganisms can be done in two ways, cell bioaugmentation and genetic bioaugmentation. The plasmid-mediated or genetic bioaugmentation occurs through biotechnological processes.[51,52] This is a technology in which the self-transmissible plasmids of the donor bacteria are added to the soil by HGT to improve degradation capabilities of the existing microbial population. By employing Horizontal Gene Transfer (HGT), the naturally degrading genes can be transferred to the recipient bacteria via plasmid. The transformed bacteria adapts to the differing environmental conditions rapidly. Antibiotic and metal resistances are the developed characteristics of recombinant microbes. However, their effect on environmental pollutants degradation is still not fully explored. The core idea behind this process is to stimulate the contaminant degradation rate by enhancing the number of diverse native bacteria through HGT. Some recent examples are transfer of plasmids from *Pseudomonas pseudoalcaligenes* and *Pseudomonas sp.* to recipient bacteria in soil. It led to enhancement in the degradation of 3-phenoxybenzoic acid. Another study is where *P. putida* transferred its two catabolic plasmids for better rate of degradation of 2,4-D in the soil.[52,53] Using this tool, a contaminated field in Zhejiang, China was successfully bioremediated. Dichlorodiphenyltrichloroethane or DDT was degraded effectively after an *E.coli* strain TG I was introduced into the soil matrix.[54]

These were just a few notable examples of the broad applications of biotechnological advances into the bioremediation process. Similarly, another field of study that has contributed to the environmental sciences and plant pollution control is bioinformatics.

7.3.2 BIOINFORMATICAL APPROACHES

Increasing world population and food demands are met with innovative agricultural practices. This involves utilization of high yielding seeds, fertilizers, pesticides, herbicides, and weedicides. However, the excessive use of herbicides and different forms of chemicals increase the chemical concentration in soil.[55,56] The conventional bioremediation techniques are not fully capable of treating the poor soil. The advance scientific method of gene editing and systemic biology is applied to get rid of pesticide chemicals. Systems biology is helpful in collecting relevant information about the microbial systems. Systems biology helps in understanding the microbial interactions between different microbial communities. Genomics, proteomics, metabolomics, and transcriptomics are some of the Omics-based studies that aid system biology to fully grasp the dynamic systems of microbial life forms. The genetic level analysis of the bioremediation system is made possible with advancement in omics studies. High-throughput sequencing (HTS) and next-generation sequencing (NGS) techniques analyze the genes involved in pollutant degradation.[56]

Gene editing is also a remarkable technique for genetic manipulations. By applying recombination enzymes, DNA can be altered. Gene editing tools developed by the bioinformatics community help in improving the process of bioremediation.[56,57] Remarkable reduction in the concentration of xenobiotics, degradation of pesticides, and conversion of toxic chemicals into nontoxic chemicals have been noted. Tools such as CRISPR-Cas, Zinc Finger Nucleases, TALEN are popularly used for editing genes. These tools aim at providing best quality genes for optimum results.[56,58] The genetic makeup of wild type is changed at the root level for obtaining new microbes with desirable gene incorporation.[56,58]

Bioinformatics is also a source of biological database. It provides both curated, annotated and noncurated information. Many universities across the world have developed biodegradation databases for pesticide bioremediation. Some of them are University of Minnesota Biocatalysis/Biodegradation Database (UMBDD), which provides information on molecular mechanisms of biodegradation processes. It tells about the enzymes, genes, reactions that are involved in the xenopesticidal compounds microbial degradation. Another database called Bionemo is for the data on metabolic pathways, regulation, and factors involved in the degradation pathways. OxDBase is for gaining insights on oxygenases, ARHD, and ARCD for degrading pesticidal complex chemicals. BioCyc is a popular database for accessing the genetic and biochemical information of degradation in microbes.[59] MetaCyc is a prediction tool for metabolic pathways and developing catabolic pathways. PTID is dedicated to identifying and studying the interaction between pesticides and targets. MBGD is extensively used for comparative analysis of genomes of microbes. Metarouter contains huge data on biodegradation and its diversity. Pesticide Action Network (PAN) enables access to the toxicity list of different pesticides used in agricultural fields.[56,59]

7.4 FUTURE PROSPECTS

Bioremediation process is an enrichment process. Initial study and system diagnosis is important for the development of an execution plan. The soil matrix represents a multidynamic system. The idea behind investing time on advanced biotechnology and bioinformatics is to enhance the pre-existing bioremediation methods. It could either be carried out by genetic manipulations or inoculation. Amending the genome with desirable genes is necessary for carrying out the task.[60] Based on the plenty of research and reports it has been suggested that the bioremediation process occurs by various forms of organisms such as fungi, bacteria, and plants components. The natural attenuation and bioaugmentation are effective in reducing organic pollutants from the environment. However, it is not enough due to heavy and regular flow of diverse pollutants into the environment from industrial and nonindustrial sources. In-depth study and assessment of extremophiles, cellular plant components, fungi, bacteria, and other degrading organisms and the enzymes produced by them through biotechnological and bioinformatical methods can help in exploring them.[37] A lot is unknown about the extremophiles. Their ability to thrive in the extreme conditions and industrial applications are an advantage to the ecosystem for reestablishment. The different varieties of extremophiles, salt-tolerant, cold-tolerant, acid-tolerant, offer new advances to address the issue of pollution and climate change. Applying the advanced molecular tools and bioinformatical technologies to study them can help in identifying great microbial degrading potentials.[37] The Next Generation Sequencing (NGS), Proteomics, and Genomics can play an influential role in tackling the environment restoration process. In addition, the study of diverse microorganisms can lead to a way to identify more similar mechanisms. Evolutionary studies and better understanding of origins can play a role in developing methods for environmental sustainability. More innovative, scientifically relevant, cost-effective, and time-saving shall be discovered in the near future to combat the serious threat of rising pollution.[56]

KEYWORDS

- **bioinformatics**
- **bioremediation**
- **biotechnology**
- **microbiology**
- **environmental pollution**

REFERENCES

1. Emamverdian, A.; Ding, Y.; Mokhberdoran, F.; Xie, Y. Heavy Metal Stress and Some Mechanisms of Plant Defense Response. *Sci. World J.* **2015**, *2015*, 1–18.
2. Onakpa, M.; Njan, A.; Kalu, O. A Review of Heavy Metal Contamination of Food Crops in Nigeria. *Ann. Glob. Health* **2018**, *84* (3), 488–494.
3. Li, Q.; Liu, J.; Gadd, G. Fungal Bioremediation of Soil Co-Contaminated with Petroleum Hydrocarbons and Toxic Metals. *Appl. Microbiol. Biotechnol.* **2020**, *104* (21), 8999–9008.
4. Azubuike, C.; Chikere, C.; Okpokwasili, G. Bioremediation Techniques–Classification Based on Site of Application: Principles, Advantages, Limitations and Prospects. *World J. Microbiol. Biotechnol.* **2016**, *32* (11).
5. Verma, J.; Jaiswal, D. Book Review: Advances in Biodegradation and Bioremediation of Industrial Waste. *Front. Microbiol.* **2016**, *6*.
6. Frutos, F.; Escolano, O.; García, S.; Babín, M.; Fernández, M. Bioventing Remediation and Ecotoxicity Evaluation of Phenanthrene-Contaminated Soil. *J. Hazard. Mater.* **2010**, *183* (1-3), 806–813.
7. SUI, H.; LI, X. Modeling For Volatilization and Bioremediation of Toluene-Contaminated Soil by Bioventing. *Chin. J. Chem. Eng.* **2011**, *19* (2), 340–348.
8. Kim, S.; Krajmalnik-Brown, R.; Kim, J.; Chung, J. Remediation of Petroleum Hydrocarbon-Contaminated Sites by DNA Diagnosis-Based Bioslurping Technology. *Sci. Total Environ.* **2014**, *497–498*, 250–259.
9. Firmino, P.; Farias, R.; Barros, A.; Buarque, P.; Rodríguez, E.; Lopes, A.; dos Santos, A. Understanding The Anaerobic BTEX Removal in Continuous-Flow Bioreactors for Ex Situ Bioremediation Purposes. *Chem. Eng. J.* **2015**, *281*, 272–280.
10. Frutos, F.; Pérez, R.; Escolano, O.; Rubio, A.; Gimeno, A.; Fernandez, M.; Carbonell, G.; Perucha, C.; Laguna, J. Remediation Trials for Hydrocarbon-Contaminated Sludge from a Soil Washing Process: Evaluation of Bioremediation Technologies. *J. Hazard. Mater.* **2012**, *199–200*, 262–271.
11. Smith, E.; Thavamani, P.; Ramadass, K.; Naidu, R.; Srivastava, P.; Megharaj, M. Remediation Trials for Hydrocarbon-Contaminated Soils in Arid Environments: Evaluation of Bioslurry and Biopiling Techniques. *Int. Biodeterior. Biodegr.* **2015**, *101*, 56–65.

12. Khudur, L.; Shahsavari, E.; Miranda, A.; Morrison, P.; Nugegoda, D.; Ball, A. Evaluating the Efficacy of Bioremediating a Diesel-Contaminated Soil Using Ecotoxicological and Bacterial Community Indices. *Environ. Sci. Pollut. Res.* **2015**, *22* (19), 14809–14819.
13. Bisht, S.; Pandey, P.; Bhargava, B.; Sharma, S.; Kumar, V.; Sharma, K. Bioremediation of Polyaromatic Hydrocarbons (Pahs) Using Rhizosphere Technology. *Braz. J. Microbiol.* **2015**, *46* (1), 7–21.
14. Kuiper, I.; Lagendijk, E.; Bloemberg, G.; Lugtenberg, B. Rhizoremediation: A Beneficial Plant-Microbe Interaction. *Mol. Plant-Microbe Interact.* **2004**, *17* (1), 6–15.
15. Chaudhary, T.; Shukla, P. Bioinoculants For Bioremediation Applications and Disease Resistance: Innovative Perspectives. *Indian. J. Microbiol.* **2019**, *59* (2), 129–136.
16. Gouda, S.; Kerry, R.; Das, G.; Paramithiotis, S.; Shin, H.; Patra, J. Revitalization of Plant Growth Promoting Rhizobacteria for Sustainable Development in Agriculture. *Microb. Res.* **2018**, *206*, 131–140.
17. Narasimhan, K.; Basheer, C.; Bajic, V.; Swarup, S. Enhancement of Plant-Microbe Interactions Using a Rhizosphere Metabolomics-Driven Approach and its Application in the Removal of Polychlorinated Biphenyls. *Plant Physiol.* **2003**, *132* (1), 146–153.
18. Toussaint, J.; Pham, T.; Barriault, D.; Sylvestre, M. Plant Exudates Promote PCB Degradation by a Rhodococcal Rhizobacteria. *Appl. Microbiol. Biotechnol.* **2011**, *95* (6), 1589–1603.
19. Parewa, H.; Meena, V.; Jain, L.; Choudhary, A. Sustainable Crop Production and Soil Health Management Through Plant Growth-Promoting Rhizobacteria. *Role Rhizosph. Microb. Soil* **2018**, 299–329.
20. Ndeddy Aka, R.; Babalola, O. Effect of Bacterial Inoculation of Strains Ofpseudomonas Aeruginosa, Alcaligenes Feacalisandbacillus Subtilison Germination, Growth and Heavy Metal (Cd, Cr, and Ni) Uptake Ofbrassica Juncea. *Int. J. Phytoremed.* **2015**, *18* (2), 200–209.
21. Abou-Shanab, R.; El-Sheekh, M.; Sadowsky, M. Role of Rhizobacteria in Phytoremediation of Metal-Impacted Sites. *Emerg. Eco-Friend. Approach. Waste Manag.* **2018**, 299–328.
22. Chaudhry, Q.; Blom-Zandstra, M.; Gupta, S.; Joner, E. Utilising the Synergy between Plants and Rhizosphere Microorganisms to Enhance Breakdown of Organic Pollutants in the Environment (15 Pp). *Environ. Sci. Pollut. Res. Int.* **2004**, *12* (1), 34–48.
23. Thoma, G.; Lam, T.; Wolf, D. A Mathematical Model of Phytoremediation for Petroleum Contaminated Soil: Sensitivity Analysis. *Int. J. Phytoremed.* **2003**, *5* (2), 125–136.
24. Leigh, M.; Prouzová, P.; Macková, M.; Macek, T.; Nagle, D.; Fletcher, J. Polychlorinated Biphenyl (PCB)-Degrading Bacteria Associated with Trees in a PCB-Contaminated Site. *Appl. Environ. Microbiol.* **2006**, *72* (4), 2331–2342.
25. Leigh, M.; Fletcher, J.; Fu, X.; Schmitz, F. Root Turnover: An Important Source of Microbial Substrates in Rhizosphere Remediation of Recalcitrant Contaminants. *Environ. Sci. Technol.* **2002**, *36* (7), 1579–1583.
26. Zalesny, R.; Bauer, E.; Hall, R.; Zalesny, J.; Kunzman, J.; Rog, C.; Riemenschneider, D. Clonal Variation in Survival and Growth of Hybrid Poplar and Willow in Anin Situtrial on Soils Heavily Contaminated with Petroleum Hydrocarbons. *Int. J. Phytoremed.* **2005**, *7* (3), 177–197.
27. Salt, D.; Smith, R.; Raskin, I. Phytoremediation. *Ann. Rev. Plant Physiol. Plant Mol. Biol.* **1998**, *49* (1), 643–668.
28. Pilon-Smits, E. Phytoremediation. *Ann. Rev. Plant Biol.* **2005**, *56* (1), 15–39.

29. Yee, D.; Maynard, J.; Wood, T. Rhizoremediation of Trichloroethylene by a Recombinant, Root-Colonizing Pseudomonas Fluorescensstrain Expressing Toluene Ortho-Monooxygenase Constitutively. *Appl. Environ. Microbiol.* **1998**, *64* (1), 112–118.
30. Anderson, T.; Guthrie, E.; Walton, B. Bioremediation in the Rhizosphere. *Environ. Sci. Technol.* **1993**, *27* (13), 2630–2636.
31. Dekkers, L.; Mulders, I.; Phoelich, C.; Chin-A-Woeng, T.; Wijfjes, A.; Lugtenberg, B. The Sss Colonization Gene of the Tomato-Fusarium Oxysporum F. Sp. Radicis-Lycopersici Biocontrol Strain Pseudomonas Fluorescens WCS365 Can Improve Root Colonization of other Wild-Type Pseudomonas Spp. Bacteria. *Mol. Plant-Microbe Interact.* **2000**, *13* (11), 1177–1183.
32. Sriprang, R.; Hayashi, M.; Yamashita, M.; Ono, H.; Saeki, K.; Murooka, Y. A Novel Bioremediation System for Heavy Metals using the Symbiosis Between Leguminous Plant and Genetically Engineered Rhizobia. *J. Biotechnol.* **2002**, *99* (3), 279–293.
33. Venny; Gan, S.; Ng, H. Current Status and Prospects of Fenton Oxidation for the Decontamination of Persistent Organic Pollutants (Pops) in Soils. *Chem. Eng. J.* **2012**, *213*, 295–317.
34. Palmroth, M.; Langwaldt, J.; Aunola, T.; Goi, A.; Münster, U.; Puhakka, J.; Tuhkanen, T. Effect of Modified Fenton's Reaction on Microbial Activity and Removal of Pahs in Creosote Oil Contaminated Soil. *Biodegradation* **2006**, *17* (2), 29–39.
35. Kao, C.; Wu, M. Enhanced TCDD Degradation By Fenton's Reagent Preoxidation. *J. Hazard. Mater.* **2000**, *74* (3), 197–211.
36. Megharaj, M.; Naidu, R. Soil and Brownfield Bioremediation. *Microb. Biotechnol.* **2017**, *10* (5), 1244–1249.
37. Shukla, A.; Singh, A. Exploitation of Potential Extremophiles for Bioremediation of Xenobiotics Compounds: A Biotechnological Approach. *Curr. Genom.* **2020**, *21* (3), 161–167.
38. Rawat, N.; Joshi, G. Bacterial Community Structure Analysis of a Hot Spring Soil by Next Generation Sequencing of Ribosomal RNA. *Genomics* **2019**, *111* (5), 1053–1058.
39. Saxena, R.; Dhakan, D.; Mittal, P.; Waiker, P.; Chowdhury, A.; Ghatak, A.; Sharma, V. Metagenomic Analysis of Hot Springs in Central India Reveals Hydrocarbon Degrading Thermophiles and Pathways Essential for Survival in Extreme Environments. *Front. Microbiol.* **2017**, 7.
40. Burg, D.; Ng, C.; Ting, L.; Cavicchioli, R. Proteomics of Extremophiles. *Environ. Microbiol.* **2011**, *13* (8), 1934–1955.
41. Kumar, A.; Alam, A.; Tripathi, D.; Rani, M.; Khatoon, H.; Pandey, S.; Ehtesham, N.; Hasnain, S. Protein Adaptations in Extremophiles: An Insight into Extremophilic Connection of Mycobacterial Proteome. *Semin. Cell Dev. Biol.* **2018**, *84*, 147–157.
42. Yadav, M.; Shukla, A. K.; Srivastva, N.; Upadhyay, S. N.; & Dubey, S. K. Utilization of Microbial Community Potential for Removal of Chlorpyrifos: A Review. *Critic. Rev. Biotechnol.* **2016**, 36(4), 727–742.
43. Shukla, A. K.; Upadhyay, S. N.; & Dubey, S. K. Current Trends In Trichloroethylene Biodegradation: A Review. *Critic. Rev. Biotechnol.* **2014**, 34(2), 101–114.
44. Marques C. R. Extremophilic Microfactories: Applications in Metal and Radionuclide Bioremediation. *Front. Microbiol.* **2018**, 9, 1191.
45. Parrilli, E.; Papa, R.; Tutino, M. L.; & Sannia, G. Engineering of a Psychrophilic Bacterium for the Bioremediation of Aromatic Compounds. *Bioeng. Bugs* **2010**, 1(3), 213–216.

46. Bertoni, G.; Bolognese, F.; Galli, E.; & Barbieri, P. Cloning of the Genes for and Characterization of the Early Stages of Toluene and O-Xylene Catabolism in Pseudomonas Stutzeri OX1. *Appl. Environ. Microbiol.* **1996**, 62(10), 3704–3711.
47. Siani, L.; Papa, R.; Di Donato, A.; & Sannia, G. Recombinant Expression of Toluene o-Xylene Monooxygenase (ToMO) from Pseudomonas stutzeri OX1 in the Marine Antarctic Bacterium Pseudoalteromonas Haloplanktis TAC125. *J. Biotechnol.* **2006**, 126(3), 334–341.
48. Anderson, R. T.; Vrionis, H. A.; Ortiz-Bernad, I.; Resch, C. T.; Long, P. E.; Dayvault, R.; Karp, K.; Marutzky, S.; Metzler, D. R.; Peacock, A.; White, D. C.; Lowe, M.; & Lovley, D. R. Stimulating the In Situ Activity of Geobacter Species to Remove Uranium from the Groundwater of a Uranium-Contaminated Aquifer. *Appl. Environ. Microbiol.* **2003**, 69(10), 5884–5891.
49. Dumorné, K.; Córdova, D. C.; Astorga-Eló, M.; & Renganathan, P. Extremozymes: A Potential Source for Industrial Applications. *J. Microbiol. Biotechnol.* **2017**, 27(4), 649–659.
50. Adams, M.W.W.; Kelly, R.M. Enzymes Isolated from Microorganisms that Grow in Extreme Environments. *Chem. Eng. News*, **1995**, 73(51),32–42
51. Wiedenbeck, J.; & Cohan, F. M. Origins of Bacterial Diversity through Horizontal Genetic Transfer and Adaptation to New Ecological Niches. *FEMS Microbiol. Rev.* **2011**, 35(5), 957–976.
52. Garbisu, C.; Garaiyurrebaso, O.; Epelde, L.; Grohmann, E.; & Alkorta, I. Plasmid-Mediated Bioaugmentation for the Bioremediation of Contaminated Soils. *Front. Microbiol.* **2017**, 8, 1966.
53. Dejonghe, W.; Goris, J.; El Fantroussi, S.; Höfte, M.; De Vos, P.; Verstraete, W.; & Top, E. M. Effect of Dissemination of 2,4-Dichlorophenoxyacetic Acid (2,4-D) Degradation Plasmids on 2,4-D Degradation and on Bacterial Community Structure in Two Different Soil Horizons. *Appl. Environ. Microbiol.* **2000**, 66(8), 3297–3304.
54. Gao, C.; Jin, X.; Ren, J.; Fang, H.; & Yu, Y. Bioaugmentation of DDT-Contaminated Soil by Dissemination of the Catabolic Plasmid pDOD. *J. Environ. Sci.* **2015**, 27, 42–50.
55. Cazalis, V.; Loreau, M.; & Henderson, K. Do We Have to Choose between Feeding the Human Population and Conserving Nature? Modelling the Global Dependence of People on Ecosystem Services. *Sci. Total Environ.* **2018**, 634, 1463–1474.
56. Jaiswal, S.; Singh, D. K.; & Shukla, P. Gene Editing and Systems Biology Tools for Pesticide Bioremediation: A Review. *Front. Microbiol.* **2019**, *10*, 87.
57. Singh, V.; Gohil, N.; Ramírez García, R.; Braddick, D.; & Fofié, C. K. Recent Advances in CRISPR-Cas9 Genome Editing Technology for Biological and Biomedical Investigations. *J. Cell. Biochem.* **2018**, *119*(1), 81–94.
58. Waryah, C. B.; Moses, C.; Arooj, M.; & Blancafort, P. Zinc Fingers, TALEs, and CRISPR Systems: A Comparison of Tools for Epigenome Editing. *Methods Mol. Biol.* **2018**, *1767*, 19–63.
59. Arora, P. K.; & Bae, H. Integration of Bioinformatics to Biodegradation. *Biol. Proced. Online* **2014**, *16*, 8.
60. Lee, J. K.; & Kalia, V. C. Mapping Microbial Capacities for Bioremediation: Genes to Genomics. *Indian. J. Microbiol.* **2020**, *60*(1), 45–53.

CHAPTER 8

Ceratophyllum demersum (L.): An Aquatic Macrophyte for Phytoremediation

MUHAMMAD AASIM[1*], OZLEM AKGUR[1], and ZEMRAN MUSTAFA[2]

[1]*Department of Plant Protection, Faculty of Agricultural Sciences and Technologies, Sivas University of Science and Technology, Sivas, Turkey*

[2]*Department of Plant Prduction and Technology, Faculty of Agricultural Sciences and Technologies, Sivas University of Science and Technology, Sivas, Turkey*

[*]*Corresponding author. E-mail: mshazim@gmail.com*

ABSTRACT

Water bodies all over the world are continuously contaminated with variable types of pollutants due to urbanization, industrialization, and heavy use of pesticides. To clean these water bodies, different techniques have been employed including physical, chemical, and bioremediation techniques alone or in combination. *Ceratophyllum demersum* (L) is an important aquatic macrophyte used for the phytoremediation studies of water bodies containing pollutants like heavy metals. Besides that the plant has been reported as biological insecticide against stored grain pests, medicinal plant and exhibiting antimicrobial properties. In recent years, number of research studies enlighten the successful use of *C. demersum* against different heavy metals in water obtained from natural sources like rivers, lakes, ponds, or artificially induced polluted aquatic environment under lab conditions. *C. demersum* plant samples collected from natural aquatic bodies or regenerated under in vitro conditions have been used for the phytoremediation studies against different heavy metals like Cd, Cr, Ni, Pb, Zn, and Ni. This study

Environmental Pollution Impact on Plants: Survival Strategies under Challenging Conditions. Tariq Aftab, PhD & Khalid Rehman Hakeem, PhD (Eds.)

highlights the use and efficacy of *C. demersum* plants for phytoremediation studies along with mechanism and genes responsible for phytoremediation.

8.1 INTRODUCTION

Water contamination in recent years due to human-induced factors[1,2] and natural factors like climate change is increasing rapidly with ultimate result of water scarcity all over the world.[3] The discharge of variable pollutants ranging from inorganic to organic compounds and heavy metals from household or industries[4, 5] are exerting noxious impact on humans and aquatic ecosystem.[6] The pollutants in water bodies especially in underground water and drinking water resources have been present above toxic level,[7] causing deaths all over the World.[8] The use of wastewater as irrigation for urban agriculture may lead to other problems like low-quality agricultural crops[9] and accumulation of pollutants to the soil surface[10] and there is always a possibility of accumulation of nonbiodegradable pollutants in the biota.[11] To overcome the issue, wastewater reclamation is the only option for its use for industrial and agricultural sectors.[12]

Phytoremediation is the application of hyperaccumulator plants along with their specific rhizospheric microorganisms[13] to remediate the soil or water bodies from pollutants.[2] The idea of phytoremediation technique was planted 300 years ago with modern application started in 1983.[14] Phytoremediation is also known as agro-remediation, botano remediation, green remediation, green technology, or vegetative remediation.[15,16] This technique is gaining popularity due to its low cost, easy handling, eco-friendly, effectiveness, and applicable for a wide array of pollutants.[2,17] The main advantage of this approach is the in-situ application with reduced risk of multiplication of pollutants for both soil and aquatic sites[17] without generating hazardous byproducts.[18] The basic principle of phytoremediation is the removal of pollutants from polluted site by uptake through its root followed by accumulation on the body or discharge to the environment as a nontoxic pollutant. The overall process of phytoremediation includes the uptake, removal, immobilization, extraction, sequestration, remediation, and stabilization of pollutants.[19, 20] The mechanism of phytoremediation includes phyto-accumulation, phyto-extraction, phyto-degradation, phyto-stabilization, phyto-transformation, phyto-volatilization, rhizospheric bilological degradation, or rhizofiltration.[20, 21] However, the success of phytoremediation is dependent on combinations of factors like plant (type, growth, bioaccumulation potential), pollutants (type, concentration, pH of the medium), and environmental factors (humidity, temperature).[20]

The cost-effective and in-situ bioremediation (phytoremediation) of an aquatic ecosystem can be done by using aquatic plants and most of the aquatic plants are natural absorber of pollutants[22] known as hyperaccumulator plants. The hyperaccumulator plants are considered for phytoremediation studies as they can accumulate high contents of heavy metals (> 1000 ppm) in their body. However, these hyperaccumulator plants are rare in nature and approximately 400 hyperaccumulators have been documented.[23] However, fewer aquatic plants (macrophytes) have been reported as hyperaccumulator. These aquatic plants have the ability to extract and accumulate the pollutants in their body (roots and shoots) by developing an extensive root system.[24] However, the selection of proper aquatic plant in accordance with pollutant type is vital for successful phytoremediation studies. Hornwort (*Ceratophyllum demersum* L.) is an important macrophyte used for the phytoremediation of a wide array of pollutants.[21] It is a submerged macrophyte aquatic with high vegetative propagation and biomass production[25] under low illumination and in muddy water as eutrophic or oligotrophic.[26]

In recent years, extensive studies on phytoremediation potential of C. demersum against wide array of pollutants (mainly heavy meatals) have been documented. The present study enlightens the different aspects on phytoremediation ranging from collection or procurement of plants from type and concentration of pollutant, exposure time (phytoremediation time), and overall assessment of the system.

8.2 PHYTOREMEDIATION STUDIES

Phytoremediation studies for the remediation of water bodies by using aquatic or semiaquatic plant are a highly efficient and eco-friendly technique. The success of the whole process of phytoremediation is dependent on the number of variables and series of actions presented in Figure 8.1. The process comprises of selection and collection of plants to expose them to pollutants followed by analytical studies using variable techniques. During phytoremediation studies, certain factors like temperature, pH, and other culture conditions are highly significant.

8.2.1 COLLECTION OF C. DEMERSUM PLANTS

The collection of plants for phytoremediation studies is highly significant and these aquatic plants are generally collected from either natural fresh water bodies like lakes, rivers, ponds, etc. or contaminated sites (Table 8.1). The phytoremediation studies on *C. demersum* also revealed the collection of plants majorly from natural water bodies like lake,[28, 29, 30, 31, 32, 33, 34, 35] ponds[36, 37, 38, 39, 40] rivers, deltas and arbitury,[41, 42, 43, 44, 45, 46] and streams/irrigation canals. Procurement of plants from local aquariums is another source used for phytoremediation studies.[25, 47] However, it has certain disadvantages like high cost and low availibility. Most recently, in vitro micropropagated plants of *C. demersum* have also been applied for phytoremediation studies[48, 49, 50, 51] which provides an advantage of low-cost plants at desired amount and time. However, there is no study to date which enlightens the comparison of in vitro or ex vitro plants.

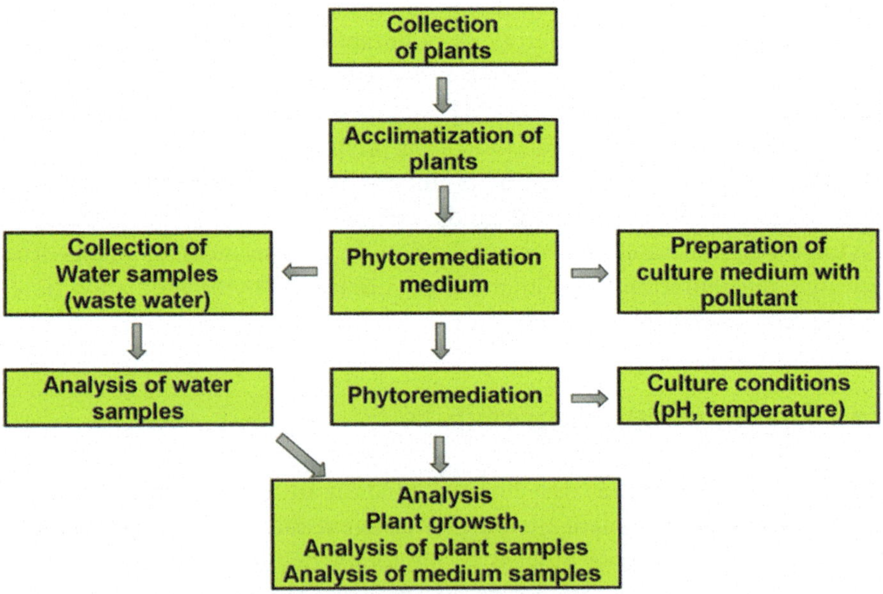

FIGURE 8.1 An overview of phytoremediation process.

8.2.2 PRETREATMENT OF COLLECTED PLANTS

After prourement of plants, the next step is the acclimatization of collected plant samples under specific conditions prior to exposing them to the pollutants. These plants are generally placed in containers of different sizes, greenhouse pond, or plastic tubs for a certain period of time. These containers, ponds, or tubs are filled with fresh water or nutrient medium like

Hoagland medium at different concentrations. Howver, most of the studies revealed the use of 10% Hoagland medium for acclimatization followed by exposing these plants to phytoremediation medium.

TABLE 8.1 An Overview of Collection or Procuremenet of *C. demersum* and Phytoremediation Medium for Phytoremediation Studies.

Plant sources	References
Lake	28, 29, 30, 31, 32, 33, 34, 35
Ponds	36, 37, 38, 39, 40
Rivers/ tributary	41, 42, 43, 44, 45, 46
Waste water	52
Unpolluted water bodies	53
In vitro	48, 49, 50, 51
Canals	25, 47
Aquariums	54, 55
Experimental medium	
10% Hoagland's solution	26, 27, 28, 29, 36, 38, 46, 48, 53, 56, 57, 58, 59, 60
0.1% Hoagland's solution	36, 61
½ Hoagland's nutrient solution	25
Aquarium culture	62
domestic waste water	54
Distilled water	51
Tanning home industry	55

8.2.3 PHYTOREMEDIATION MEDIUM

The phytoremediation medium using *C. demersum* plants comprised of collecting contaminated water from different sources or preparation of medium enriched with different concentrations of nutrient medium (Hoagland's solution) (Table 8.1). The most used culture medium is 10% Hoagland's solution.[26, 27, 28, 29, 36, 38, 46, 48, 53, 56, 57, 58, 59, 60] Some studies also revealed the use of ½ strength Hoagland's nutrient solution[25] or 0.1% Hoagland's solution.[36, 60] Aasim et al.[51] revealed the use of distilled water without any nutrient medium for phytoremediation studies of *C. demersum*. In all these studies, the phytoremediation medium was supplemented with pollutant used in the relative study. On the other hand, pytoremediation studies of *C. demersum* involve the direct use of collected water from different sources like contaminated water. These include the domesctic wastewater[54] or tanning home industry wastewater[54] or aquarium water or fishponds.[62] A study by Karataş et al.[63] revealed the use of drinking water, dam water, and tap water of laboratory using *C. demesrum* plants. In all these studies, collected water samples were tested prior to phytoremediation and tested again after placing plants for a certain period.

8.2.4 pH AND PHYTOREMEDIATION

The pH of the culture medium for phytoremediation is also an important factor and generally these experiments were performed at 7.0 pH.[25, 48, 51, 57, 64] Howver, some studies also revealed the use of slight saline medium of 7.1–7.9 pH[53, 59, 65, 66] and slight acidic medium like pH 6.0–6.2.[36, 43, 65] The analysis of a post-phytoremediaton medium revealed the change in pH of the medium that reached upto 10.[58] Similarly, slightly elevated pH of the phytoremediation medium have also been reported but dependant on the type and concentration of heavy metal and exposure time.[48] On the other hand, the pH of the water body from where plants are collected is also significant and change in pH has been documented when *C. demersum* plants were placed in that water.[63] Therefore, the determination of pre- and post-phytoremediaum medium is highly vital for phytoremediation studies and the change may vary depending on the type and conentration of pollutant and exposure time of *C. demersum* plants to the phytoremedium medium.

8.2.5 TEMPERATURE AND PHYTOREMEDIATION

Environmental factors like humidity and temperature can affect the phytoremediation process.[20] The studies using *C. demersum* are generally performed in open envireonment, lab, or greenhouse conditions and limited reports present experimental setup at specific temperautre without providing any impact of temperature. The optimization of temperature prior to phytoremediation study is highly significant. The phytoremediation study is carried out under controlled temperaure (25 ± 1 °C) conditions by using in vitro propagated *C. demersum*[48] in growthroom. But, some studies on phytoremediation using *C. demersum* reflect the experimental setup at different temperaureas like 26-30 °C,[36, 67] 23-25 °C,[62] 28 °C,[66] room temperature.[44, 51] Also, Poklonov[65] set up phytoremediation experiment at different temperatures (17, 20, 23, 26, 27 °C). To check the impact of temperature in open environement or in-situ application, there is need to check the deifferent temperatures ranging from 10 to 40 °C.

8.2.6 HEAVY METAL/POLLUTANTS TYPE AND CONCENTRATION

The success of phytoremediation of contaminated site or water body is dependant on the type of pollutant and their concentration. *C. demersum* is an important macrophyte used for phytoremediation studies and to date, it has been used for different types of pollutants ranging from heavy metals (Table 8.2) to radioactive elements.[40, 43] The plant is generally used for the phytoremediation studies of heavy metals like Cd, Pb, Zn, Ni, Cr, Cu, As, Mn, Co, V, and Fe (Table 8.2). In addition, some studies also reveal the phytoremediation of other elements like Ba, Li, Mo, Rb, Ti,[35] Sr,[34] and Se.[30] Besides that, the plant has been employed for the phytoremediation of nitrogenous compounds,[33, 34, 62, 68] dyes[66] or organic pollutants like SDS, phenol,[46] and EDTA.[64] In all these studies, metals, elements, or compounds were employed at different concentrations.

8.2.7 MASS AND EXPOSURE TIME OF C. DEMERSUM FOR PHYTOREMEDIATION

The mass of *C. demersum* plants used for phytoremediation studies varies in all experiments and researchers used different plant masses without mentioning any reason or optimization of plant mass per liter. Table 8.3 presents the information about plant mass used for phytoremediation studies. Results reveal the use of 1 g/L[60] to 12.5 g/L[48] plants for phytoremediation studies. But, some studies revealed the use of 5 plants (approximately 2 g wet plant samples) per replicate.[26, 28, 53, 55] There is need to optimize the plant mass for its use at commercial level for in situ phytoremediation in the future. On the other hand, time span or exposure time of plants to pollutant is one of the most important factors that affect the whole phytoremediation process. The studies revealed the culture of *C. demersum* plant for a short-term exposure of 6 and 12 h[56] to long-term exposure upto[56] days.[43] Table 8.3 presents the information about exposure time of different pollutants. In most of the studies, significant amount of metals were accumulaed in short time of less than 1 d and there is need to analyze experiment on hours basis depending on the type and concentration of the pollutant.

TABLE 8.2 An Overview About Different Pollutants Used for Phytoremedion Studies Using *C. demersum*.

Pollutant	References
Heavy metals	
Lead (Pb)	27, 28, 30, 31, 33, 35, 39, 42, 44, 45, 48, 52, 64, 67, 69
Cadmium (Cd)	30, 33, 34, 35, 36, 37, 42, 44, 45, 48, 53, 56, 62, 64, 67, 69
Zinc (Zn)	33, 34, 35, 36, 37, 41, 44, 52, 67, 69
Nickel (Ni)	25, 26, 33, 34, 35, 44, 67, 69
Chromium (Cr) (Cr III, Cr VI)	31, 34, 35, 38, 45, 52, 59, 67, 69
Copper (Cu)	33, 34, 35, 41, 44, 62, 67, 70
Arsenic (As)	26, 32, 45, 58, 60, 71, 72
Managanese (Mn)	34, 35, 44, 52, 67, 69
Cobalt (Co)	33, 34, 35, 69
Vanedium (V)	34, 35, 45, 67
Iron (Fe)	35, 44, 67
Aluminium (Al)	65, 69
Mercury (Hg)	45, 67
Caesium (^{133}Cs and ^{134}Cs)	40
Barium (Ba)	35
Lithium (Li)	35
Molybdenum (Mo)	35
Rubidium (Rb)	35
Titanium (Ti)	35
Strontium (Sr)	34
Uranium (U)	43
Selenium (Se) (Se VI)	30
Other pollutants	
NO2, NO3, NH3, PO4	68
SDS and phenol	46
EDTA	64
Urea	62
Methylene Blue	66
Water samples	
Water hardness with CaCO3	43
Water (Tap, drinking, dam)	63
Waste water	54

TABLE 8.3 An Overview About Types, Concentration, Exposure Time, and Analytical Methods Used for Phytoremedion Studies Using *C. demersum.*

Heavy metal/Salt with source	Plant	Concenetration/Salt	Exposure time	References
Cd ($CdCl_2.H_2O$)	n/a	0.1, 1, 10 mg/L	7 d	53
EDTA, Pb (Pb $[CH_3COO)_2$. $3H_2O$, Cd ($CdCl_2H_2O$)	-	0.5 mM (EDTA), 50 µg (Pb/mL), 0.5 µg (Cd/mL), 0.5 mM (EDTA) + 50 µg (Pb/mL), 0.5 mM (EDTA) + 0.5 µg (Cd/mL), 50 µg (Pb mL^{-1}) + 0.5 µg (Cd mL^{-1})	7 d	64
Cd	8 g/L	ST (0, 15, 20, 25, 50, 75 and 100 µM), MT (2.5, 5, 7.5 and 10 µM), LT (0.25, 0.50, 0.75 and 1.00 µM)	ST (6 and 12 h), MT (2, 4, 6, 8 d), LT (1, 2, 3 w)	56
Cd (CdCl2), Zn (ZnCl2)	10 g/L	Cd (10 µM) and Zn (10, 50, 100, and 200 µM)	1 w	36
Pb	5 Plant (2 g) /replicate	1, 10, 25, 50, 100 µM	1, 2, 4 and 7 d	28
As (Na2HAsO4)	5 Plant (2g)/replicate	0, 10, 50, 250 µM	1, 2, 4 or 7 d	26
Cd ($CdCl_2$)	5 Plant (2 g) /replicate	0, 1, 5, 10 µM	1, 2, 4 and 7 d	30
Cd ($CdCl_2$), Zn ($ZnCl_2$)	8 g/L	$CdCl_2$ (10 µmol/L), $ZnCl_2$ (10, 50, 100, 200 µmol/L)	1 w	37
Se (VI) Na_2SeO_4	-	20 µg Se/L, 10 mg Se/L	31 d	30
Ni	6.67 g/L	Ni (0.00, 1.0, 2.0, 4.0, 6.0 mg L^{-1})	1, 2, 3 and 14 d	25
Pb (Pb $(NO_3)_2$),	n/a	2, 4, 10, 15 mg/L	2, 4, 6, 9 and 12 d	31
Cr ($K_2Cr_2O_7 \cdot 5H_2O$)				
As [Arsenite (As(III)]	5 Plant (2 g)/replicate	0, 2, 5, 10, 20 and 40 µg	24 h and 48 h	58
As (arsenate or arsenite)	2 g/L	0, 1.25, 5, 10, 20, 40 µM	4 d	32
Cr III (CrCl3) Cr VI (K2Cr2O7)	4 g/replicate	1, 2, 5, and 10 MM	1, 2, 4, and 7 d	38
Cr(VI) ($K_2Cr_2O_7$)	5 g plant samples	Cr(VI) (0, 1, 5, 10 mM) NaCl (0, 125, 250, and 500 mM)	5 d	59

TABLE 8.3 *(Continued)*

Heavy metal/Salt with source	Plant	Concenetration/Salt	Exposure time	References
As(V) (Na2HAsO4)	-	0, 0.5, 1, 2, 5, 10, and 20 μM	4 w	72
As ($Na_2HAsO_4.7H_2O$)	1 g/L	2500 μg/L	30 d	60
Pb	2.67 g/L	5, 10, 20, 40, and 80 mM	7, 14 or 21 d	27
Pb ($Pb(NO_3)_2$	200 g/ 17 L	2, 4, 6 ppm	0, 15, 30, 45, and 60 d	39
Al (Al+H2SO4)	n/a	25 mg/L	0, 1, 6, 17, 28 d	65
Cd (Cd(NO3)2·4H2O) Pb (Pb(NO3)2	12.5 g/L	Cd (0.5, 1, 1.5 and 2 mg/L) Pb (25, 50, 75 and 100 mg/L)	1, 3, 5 d	48
Urea, Ni, Cu	-	Urea (2 mM), Ni (10 μM), Cu (10 μM), Urea + Ni (2 mM + 10 μM), Urea + Cu (2 mM + 10 μM)	48 h	62
Vn, As, Cd, Cr, Hg, Pb	-	300 μg/L (Vn), 80 μg/L (As), 10 μg/L (Cd), 80 μg/L (Cr), 0.08 μg/L (Hg) and 36 μg/L (Pb)	24 d	45
SDS - sodium dodecyl sulfate (C12H25SO4Na), Phenol (C6H5OH)	20 plants per m^2 pond	SDS and phenol (0, 0.1, 0.5, 1.0, 10, and 20 mg/L), SDS (0.1, 1.0 and 20 mg/L) +phenol (0.5 and 10 mg/L), phenol+SDS: 0 + 0 (control), 0.5 + 0.1 mg/L, 0.5+1.0 mg/L, 0.5+20 mg/L, 10+0.1 mg/L, 10+1.0, 10+20 mg/L	7 d	46
Cu ($CuSo4.5H_2O$)	3.5 g/L	40 mg/L	3, 7, 14, 21, and 28 d	70
^{133}Cs (CsCl), ^{134}Cs	-	0.008, 0.033, 0.133, 0.267, 0.533, 0.800, 1.067, and 1.333 mM)	^{133}Cs for 8, 16 and 24 d, ^{134}Cs for 8 d, ^{133}Cs for 8 d	40
Methylene Blue ($C_{16}H_{18}ClN_3 S$)	-	5 mg/L	5 d	66

8.2.8 ANALYSIS OF POLLUTANTS IN PLANTS AND WATER SAMPLES

At the end of the experiment, there is need to analyze the results of the phytoremediation study. The general parameters used for phytoremediation include its impact on plants (plant biomass, phytotoxicity), heavy metal or pollutants concentration in plants (mineral uptake, BCF), and remaining concentration in water. Besides that, number of other parameters have also been documented like enzymatic activities. For the estimation of bioaccumulation potential, different equipments have been used for plant and medium analysis. Atomic absorption spectrophotometer (AAS) or Flame AAS is the most commonly used equipment for the analysis of medium and plant samples,[25, 26, 27, 31, 48, 51, 53, 60] GF-AAS[62] ICP-MS,[32, 40, 58, 68] Fluorate-02–5 m fluorescence analyzer,[65] UV/Vis Microplate SP,[66] Plasma mass spectroscopy,[38, 59] HPLC,[28, 32] Spectroflourimeter,[36] AFS,[45] MP-AES,[52] ESR spectroscopy,[36] and UV/VIS SP.[72]

8.3 MOLECULAR MECHANISM OF PHYTOREMEDIATION IN *C. DEMERSUM*

After exposure to toxic materials, plant cells generate high levels of reactive oxygen species (ROS) like superoxide radicals, hydrogen peroxide, hydroxyl radical singlet oxygen, and lipid hydroperoxides, which are known to attack and degrade cell membranes. The reason of accumulation of high levels of ROS after toxic exposures may be due to imbalance caused in ROS-antioxidant production.[73] Although ROS is produced under normal circumstances and is used as a secondary messenger, higher concentration has detrimental effects. When the concentration of ROS exceeds the cell's capability to neutralize, the cell begins to experience oxidative stress. Oxidative stress causes peroxidation of lipids, mutation of nucleic acids, oxidation of proteins, and inhibition of enzymatic actions. If the cell is not able to cope with the oxidative stress, pathways leading to programmed cell death are activated. To combat the adverse effects of ROS, cells produce a variety of enzymatic and nonenzymatic antioxidants for the prevention of damages by converting the ROS to less harmful compounds. The antioxidative capacity of the organism also determines the tolerance to oxidative stress.[74, 75] The capability of plants to tolerate and accumulate the toxic materials is of great interest for economical removal of toxic wastes by phytoremediation. The detoxification mechanism in plants includes immobilization, exclusion, chelation, and compartmentalization. The genes responsible for the detoxification and accumulation mechanisms can be transformed and expressed in better adapted local plants to produce genetically modified plants with superior phytoremediation properties.[76] The possible molecular mechanism involved in detoxification of pollutants by *C. demersum* can be summarized as oxidative stress byproducts, antioxidant enzymes, antioxidant compounds, and related enzymes and phytohelatins (PCS).

8.3.1 OXIDATIVE STRESS BYPRODUCTS

Malondialdehyde (MDA) is a cytotoxic molecule produced as the end product during lipid peroxidation caused by ROS. It is thus a reliable indicator of the presence of the oxidative stress in the cell.[77] Previous studies reported that the MDA level increased significantly when *C. demersum* was exposed to lead,[27, 28, 48] chromium,[38] arsenate,[26] cadmium,[48] Copper,[78, 62] and nickel[62] stress.

Superoxide radical (O_2^-) and hydrogen peroxide (H_2O_2) are the most common ROS observed during toxic level of pollutants causing oxidative stress in cell. Both compounds were measured by Mishra et al.[26] in their assay and were found to have been increased notably after toxic arsenate concentrations were exceeded.

Electrical Conductivity (EC) is a measurement of ion leakage. If ROS concentrations go beyond a certain threshold, a programmed cell death (PCD) is activated. The amount of PCD can be quantified by measuring the ion leakage.[79] An increase in EC during chromium,[38] arsenate,[26] lead,[28] and copper[78] exposure was observed in parallel with the duration of the exposure.

8.3.2 ANTIOXIDATIVE ENZYMES

Superoxide dismutase (SOD) **(EC 1.15.1.1)** is the enzyme of first line of defense against oxidative stress in the cell. It catalyzes the reaction of transformation of superoxide radical O_2^- into hydrogen peroxide H_2O_2 and molecular oxygen O_2.[74] Previous reports revealed that SOD concentration increased gradually with increased concentration and duration of lead,[27, 28] chromium,[38] arsenate,[26] nickel,[62] and copper.[62, 78] The SOD concentration was observed to decrease after prolonged periods of time.

Catalase (CAT) **(EC 1.11.1.6)** is found in cytosol, peroxisome, and mitochondria acting as a scavenger of hydrogen peroxide, catalyzing the reaction of dismutating H_2O_2 to H_2O and O_2.[80] During exposure to lead,[27, 28] chromium,[38] nickel,[62] and copper.[62, 78] CAT showed synergistic effects with concentration and duration of exposure to stress. CAT concentration was increased to a peak value and decreasing after surpassing the specific concentration and duration. This suggests that after a certain threshold CAT may be not sufficient to protect the cell from oxidative stress.

Peroxidase (POD) or Guaiacol peroxidase **(EC 1.11.1.7)** is a highly important heme containing antioxidative enzyme that scavenges H_2O_2 by reacting it with guaiacol to produce oxidized glutathione and H_2O.[74] It is more efficient than CAT in protection against H_2O_2 damaging effects. POD concentration showed a significant increment with elevated concertation and exposure to lead[27, 28] nickel, and copper[62] in *C. demersum.*

Ascorbate peroxidase **(APX) (EC 1.11.1.11)** catalyzes the reaction of H_2O_2 with l-ascorbate to generate dihydroxy ascorbate and H_2O. There was a positive correlation between APX and lead concentration,[28] and arsenate concentration[26] while APX concentration decreased with time. Maleva et al.[62] found APX to decrease during nickel and copper toxicity treatment. APX level showed a considerable elevated concentration when treated cadmium supplemented with zinc[36] and copper.[78]

Glutathione reductase **(GR) (EC 1.8.1.7)** ***or glutathione-disulfide reductase*** is an enzyme that catalyzes the recycling of the oxidized glutathione disulfide to the reduced glutathione. GR showed an increase in concentration after treatment with arsenate[26] and lead[28] and decreased after longer exposure and higher concentration. When treated with cadmium, GR tended to increase.[29, 36]

Dehydroascorbate reductase **(DHAR) (EC 1.8.5.1)** also catalyzes the recycling of glutathione disulfide to glutathione. DHAR showed decreased levels when treated with cadmium, which was restored when supplemented with zinc.[36]

Monodehydroascorbate reductase **(MDHAR) (EC 1.6.5.4)** together with DHAR and GR maintain high concentration of reduced ascorbic acid and glutathione compared to oxidized forms.[75] MDHAR showed a similar pattern with DHAR, decreasing activity during cadmium toxicity and restored levels when supplemented with zinc.[36]

8.3.3 ANTIOXIDANT COMPOUNDS AND RELATED ENZYMES

Reduced glutathione **(GSH)** or γ-glutamyl-cysteinyl-glycine is produced in chloroplasts and the cytosol of plant cells by γ-glutamyl-cysteinyl synthetase and glutathione synthetase. It functions as an important nonprotein thiol compound scavenging ROS by oxidizing itself to oxidized glutathione.[74] GSH concentration was notably increased after exposure to arsenate,[26] cadmium,[29, 36] and decreased during exposure to lead[28] and first increasing up certain concentration after exposure to copper.[75]

Oxidized Glutathione **(GSSG)** or glutathione disulfide is the byproduct of oxidation of GSH. To replenish GSH, GR reduces the GSSG to GSS recycling the molecule in the antioxidant pathway. GSSG was increased compared to control during exposure to arsenate,[26] cadmium,[29] lead[28] and decreased during exposure to cadmium supplemented with zinc.[36]

γ-glutamate-cysteine synthase **(γGCS)** or glutamate-cysteine ligase catalyzes the reaction between glutamate and cysteine to generate gamma-glutamyl cysteine, as a precursor to gluthatione. Gamma-glutamyl cysteine is catalyzed to gluthatione by gluthatione synthase.[81] γGCS has been increased significantly during arsenate exposure[26] while remaining relatively unaffected during cadmium supplemented with zinc exposure.[36]

γ-glutamyl transpeptidase **(γGT) (EC 2.3.2.2)** is an enzyme that catalyzes the catabolic reaction of hydrolysis of GSH, GSSG, and glutathione synthase conjugates.[82] γGT increased significantly with increasing of concentration and exposure time to arsenate[26] while cadmium exposure did not change the γGT level significantly.

Glutathione-S-transferase **(GST) (EC 2.5.1.18)** is an important enzyme in cellular metabolism for detoxification by attaching GST to toxic compounds. GST showed increased levels compared to control during toxic effects of arsenate[26] and cadmium.[29,36]

Glutathione peroxidase **(GSH-PX) (EC 1.11.1.9)** is an important enzyme that catalyzes the hydrogen peroxide with GSH. When exposed to cadmium GSH-PX levels decreased significantly, which were reversed and even increased when supplemented with zinc.[36] Besides that, other important enzymes and compounds in detoxification metabolism in *C. demersum* have been summarized in Table 8.4.

8.4 PHYTOCHELATINS (PCs)

Phytochelatins (PCs) determine the tolerance of the plant to toxins and their capacity for phytoremediation. PCs are produced by oligomerization of gluthatione by phytochelatin synthase (PCS) (EC 2.3.2.15). PCs have the potential to react with many different toxic materials.[76, 83] During exposures to arsenate,[26] cadmium,[29] and lead,[28] PCs showed dramatic incensement with increased toxic concentrations. PCS showed also similar rise during arsenate[26] and cadmium[29] exposure. The gene coding for PCS in *C. demersum*, CdPCS1, was cloned into model organisms such as *Escherichia coli*, *Arabidopsis thaliana*,[84] and Tobacco (*Nicotiana tabacum*),[85] the transformed organisms developed higher tolerance to cadmium and arsenic toxicity. The transgenic organisms also accumulate more toxins in their body in a chelated form. These studies confirm the importance of PCs for detoxification and the possible use in phytoremediation.

TABLE 8.4 Enzymes and Compounds Measured in Assays of Toxicity and Phytoremediation Potential of *C. demersum*.

	Abbreviation	Compound	References
1	Asc	Ascorbate	36, 44, 62,
2	AGC	Enzymes of ascorbate–glutathione cycle	35
3	Car	Carotenoids	28, 38, 48, 62
4	Chl a, Chl b	Chlorophyll a and b	28, 38, 48, 55, 62, 78
5	Pro	Proline	38, 44, 62
6	CYS	Cysteine	26, 28, 29
7	CS	Cysteine synthase	26, 29
8	NP-SH	Nonprotein thiols	28, 29, 36, 78
9	PB-SH	Protein-bound thiols	36
10		Total thiols	36, 44,
11	PAL	Phenylalanine ammonia-lyase	27
12	PPO	Phenol oxidase	27
13	AR	Arsenate Reductase	26
14	NEA	Nonenzymatic antioxidants	44, 62

8.5 CONCLUSION

Detoxification of pollutants of different types, forms, and sources is highly important for the environmental sustanibility in the modern age of industrialization and urbanization. There are a number of techniques used for the remediation of these pollutants. Among these, phytoremediation using aquatic plants is a more sustainable and eco-friendly system for a wide array of pollutants. However, there is a need to optimize the plant rate and exposure time for an efficient phytoremediation for in situ studies. There is also need to establish genetic transformation in *C. demersum* to incorporate other genes related to phytoremediation or by incorporating the PCS1 gene from *C. demersum* or other sources to check the over expression of the said gene and phytoremediation.

KEYWORDS

- **aquatic ecosystem**
- **pollutants**
- **macrophyte**
- **phytoremediation**

REFERENCES

1. Rahman, M. A.; Hasegawa, H. Aquatic Arsenic: Phytoremediation Using Floating Macrophytes. *Chemosphere* **2011**, *83* (5), 633–646.
2. Nedjimi, B. Phytoremediation: A Sustainable Environmental Technology for Heavy Metals Decontamination. *SN Appl. Sci.* **2021**, *3* (3). https://doi.org/10.1007/s42452-021-04301-4.
3. Calzadilla, A.; Rehdanz, K.; Tol, R. S. J. Water Scarcity and the Impact of Improved Irrigation Management: A Computable General Equilibrium Analysis. *Agric. Econ.* **2011**, *42* (3), 305–323.
4. Pakdel, P. M.; Peighambardoust, S. J. A Review on Acrylic Based Hydrogels and Their Applications in Wastewater Treatment. *J. Environ. Manag.* **2018**, *217*, 123–143.
5. Shah, V.; Daverey, A. Phytoremediation: A Multidisciplinary Approach to Clean up Heavy Metal Contaminated Soil. *Environ. Technol. Innov.* **2020**, *18* (100774), 100774.
6. Ahmed, M. B.; Zhou, J. L.; Ngo, H. H.; Guo, W.; Thomaidis, N. S.; Xu, J. Progress in the Biological and Chemical Treatment Technologies for Emerging Contaminant Removal from Wastewater: A Critical Review. *J. Hazard. Mater.* **2017**, *323*, 274–298.
7. Ranjan, M.; Singh, P. K.; Srivastav, A. L. A Review of Bismuth-Based Sorptive Materials for the Removal of Major Contaminants from Drinking Water. *Environ. Sci. Pollut. Res. Int.* **2020**, *27* (15), 17492–17504.
8. Azizullah, A.; Khattak, M. N. K.; Richter, P.; Häder, D.-P. Water Pollution in Pakistan and Its Impact on Public Health—A Review. *Environ. Int.* **2011**, *37* (2), 479–497.
9. Srivastav, A. L.; Kaur, T.; Rani, L.; Kumar, A. Scientific Research Production of India and China in Environmental Chemistry: A Bibliometric Assessment. *Int. J. Environ. Sci. Technol. (Tehran)* **2019**, *16* (8), 4989–4996.
10. Paul, D. Research on Heavy Metal Pollution of River Ganga: A Review. *Ann. Agrar. Sci.* **2017**, *15* (2), 278–286.
11. Gall, J. E.; Boyd, R. S.; Rajakaruna, N. Transfer of Heavy Metals through Terrestrial Food Webs: A Review. *Environ. Monit. Assess.* **2015**, *187* (4), 201.

12. Tee, P. F.; Abdullah, M. O.; Tan, I. A. W.; Rashid, N. K. A.; Amin, M. A. M.; Nolasco-Hipolito, C.; Bujang, K. Review on Hybrid Energy Systems for Wastewater Treatment and Bio-Energy Production. *Renew. Sustain. Energy Rev.* **2016**, *54*, 235–246.
13. Liu, S.; Yang, B.; Liang, Y.; Xiao, Y.; Fang, J. Prospect of Phytoremediation Combined with Other Approaches for Remediation of Heavy Metal-Polluted Soils. *Environ. Sci. Pollut. Res. Int.* **2020**, *27* (14), 16069–16085.
14. Blaylock, M. *Phytoremediation of Contaminated Soil and Water: Field Demonstration of Phytoremediation of Lead Contaminated Soils*; Lewis Publishers: Boca Raton, FL, USA, 2008.
15. Sarwar, N.; Imran, M.; Shaheen, M. R.; Ishaque, W.; Kamran, M. A.; Matloob, A.; Rehim, A.; Hussain, S. Phytoremediation Strategies for Soils Contaminated with Heavy Metals: Modifications and Future Perspectives. *Chemosphere* **2017**, *171*, 710–721.
16. Kushwaha, A.; Hans, N.; Kumar, S.; Rani, R. A Critical Review on Speciation, Mobilization and Toxicity of Lead in Soil-Microbe-Plant System and Bioremediation Strategies. *Ecotoxicol. Environ. Saf.* **2018**, *147*, 1035–1045.
17. Ali, S.; Abbas, Z.; Rizwan, M.; Zaheer, I.; Yavaş, İ.; Ünay, A.; Abdel-DAIM, M.; Bin-Jumah, M.; Hasanuzzaman, M.; Kalderis, D. Application of Floating Aquatic Plants in Phytoremediation of Heavy Metals Polluted Water: A Review. *Sustainability* **2020**, *12* (5), 1927.
18. Nejad, Z. D.; Jung, M. C.; Kim, K. H. Remediation of Soils Contaminated with Heavy Metals with an Emphasis on Immobilization Technology. *Environ. Geochem. Health* **2018**, *40* (3), 927–953.
19. Tangahu, B. V.; Sheikh Abdullah, S. R.; Basri, H.; Idris, M.; Anuar, N.; Mukhlisin, M. A Review on Heavy Metals (as, Pb, and Hg) Uptake by Plants through Phytoremediation. *Int. J. Chem. Eng.* **2011**, *2011*, 1–31.
20. Kanwar, V. S.; Sharma, A.; Srivastav, A. L.; Rani, L. Phytoremediation of Toxic Metals Present in Soil and Water Environment: A Critical Review. *Environ. Sci. Pollut. Res. Int.* **2020**, *27* (36), 44835–44860.
21. Ansari, A. A.; Naeem, M.; Gill, S. S.; AlZuaibr, F. M. Phytoremediation of Contaminated Waters: An Eco-Friendly Technology Based on Aquatic Macrophytes Application. *Egypt. J. Aquat. Res.* **2020**, *46* (4), 371–376.
22. Pratas, J.; Paulo, C.; Favas, P. J. C.; Venkatachalam, P. Potential of Aquatic Plants for Phytofiltration of Uranium-Contaminated Waters in Laboratory Conditions. *Ecol. Eng.* **2014**, *69*, 170–176.
23. Prasad, M. N. V.; De Oliveira Freitas, H. M. Metal Hyperaccumulation in Plants - Biodiversity Prospecting for Phytoremediation Technology. *Electron. J. Biotechnol.* **2003**, *6* (3). https://doi.org/10.2225/vol6-issue3-fulltext-6.
24. Stoltz, E.; Greger, M. Accumulation Properties of As, Cd, Cu, Pb and Zn by Four Wetland Plant Species Growing on Submerged Mine Tailings. *Environ. Exp. Bot.* **2002**, *47* (3), 271–280.
25. Chorom, M.; Parnian, A.; Jaafarzadeh, N. Nickel Removal by the Aquatic Plant (Ceratophyllum Demersum L.). *Int. J. Environ. Sci. Dev.* **2012**, 372–375.
26. Mishra, S.; Srivastava, S.; Tripathi, R. D.; Trivedi, P. K. Thiol Metabolism and Antioxidant Systems Complement Each Other during Arsenate Detoxification in Ceratophyllum Demersum L. *Aquat. Toxicol.* **2008**, *86* (2), 205–215.
27. Chen, M.; Zhang, L. L.; Li, J.; He, X. J.; Cai, J. C. Bioaccumulation and Tolerance Characteristics of a Submerged Plant (Ceratophyllum Demersum L.) Exposed to Toxic Metal Lead. *Ecotoxicol. Environ. Saf.* **2015**, *122*, 313–321.

28. Mishra, S.; Srivastava, S.; Tripathi, R. D.; Kumar, R.; Seth, C. S.; Gupta, D. K. Lead Detoxification by Coontail (Ceratophyllum Demersum L.) Involves Induction of Phytochelatins and Antioxidant System in Response to Its Accumulation. *Chemosphere* **2006**, *65* (6), 1027–1039.
29. Mishra, S.; Tripathi, R. D.; Srivastava, S.; Dwivedi, S.; Trivedi, P. K.; Dhankher, O. P.; Khare, A. Thiol Metabolism Play Significant Role during Cadmium Detoxification by Ceratophyllum Demersum L. *Bioresour. Technol.* **2009**, *100* (7), 2155–2161.
30. Mechora, S.; Cuderman, P.; Stibilj, V.; Germ, M. Distribution of Se and Its Species in Myriophyllum Spicatum and Ceratophyllum Demersum Growing in Water Containing Se (VI). *Chemosphere* **2011**, *84* (11), 1636–1641.
31. Abdallah, M. A. M. Phytoremediation of Heavy Metals from Aqueous Solutions by Two Aquatic Macrophytes, Ceratophyllum Demersum and Lemna Gibba L. *Environ. Technol.* **2012**, *33* (13–15), 1609–1614.
32. Xue, P.; Yan, C.; Sun, G.; Luo, Z. Arsenic Accumulation and Speciation in the Submerged Macrophyte Ceratophyllum Demersum L. *Environ. Sci. Pollut. Res. Int.* **2012**, *19* (9), 3969–3976.
33. Kamel, A. K. Phytoremediation Potentiality of Aquatic Macrophytes in Heavy Metal Contaminated Water of El-Temsah Lake, Ismailia, Egypt. *Middle–East J. Sci. Res.* **2013**, *14* (12), 1555–1568.
34. Kastratovic, V.; Krivokapic, S.; Bigovic, M.; Djurovic, D.; Blagojevic, N. Bioaccumulation and Translocation of Heavy Metals by Ceratophyllum Demersum from Skadar Lake, Montenegro. *J. Serb. Chem. Soc.* **2014**, *79* (11), 1445–1460.
35. Polechońska, L.; Klink, A.; Dambiec, M.; Rudecki, A. Evaluation of Ceratophyllum Demersum as the Accumulative Bioindicator for Trace Metals. *Ecol. Indic.* **2018**, *93*, 274–281.
36. Aravind, P.; Prasad, M. N. V. Modulation of Cadmium-Induced Oxidative Stress in Ceratophyllum Demersum by Zinc Involves Ascorbate–Glutathione Cycle and Glutathione Metabolism. *Plant Physiol. Biochem.* **2005**, *43* (2), 107–116.
37. Aravind, P.; Prasad, M. N. V.; Malec, P.; Waloszek, A.; Strzałka, K. Zinc Protects Ceratophyllum Demersum L. (Free-Floating Hydrophyte) against Reactive Oxygen Species Induced by Cadmium. *J. Trace Elem. Med. Biol.* **2009**, *23* (1), 50–60.
38. Duman, F.; Koca, F. D. Single and Combined Effects of Exposure Concentration and Duration on Biological Responses of Ceratophyllum Demersum L. Exposed to Cr Species. *Int. J. Phytoremed.* **2014**, *16* (7–12), 1192–1208.
39. Afaj, A. H.; Jassim, A. J.; Noori, M. M.; Schüth, C. Effects of Lead Toxicity on the Total Chlorophyll Content and Growth Changes of the Aquatic Plant Ceratophyllum Demersum L. *Int. J. Environ. Stud.* **2017**, *74* (1), 119–128.
40. Vacula, J.; Komínková, D.; Pecharová, E.; Doksanská, T.; Pechar, L. Uptake of 133Cs and 134Cs by Ceratophyllum Demersum L. under Field and Greenhouse Conditions. *Sci. Total Environ.* **2020**, *720* (137292), 137292.
41. Fawzy, M. A.; Badr, N. E.-S.; El-Khatib, A.; Abo-El-Kassem, A. Heavy Metal Biomonitoring and Phytoremediation Potentialities of Aquatic Macrophytes in River Nile. *Environ. Monit. Assess.* **2012**, *184* (3), 1753–1771.
42. VahdatiRaad, L.; Khara, H. Heavy Metals Phytoremediation by Aquatic Plants (Hyrocotyle Ranocloides, Ceratophyllum Demersum) of Anzali Lagoon. *Int. J. Mar. Sci. Eng.* **2012**, *2* (4), 249–254.

43. Markich, S. J. Water Hardness Reduces the Accumulation and Toxicity of Uranium in a Freshwater Macrophyte (Ceratophyllum Demersum). *Sci. Total Environ.* **2013**, *443*, 582–589.
44. Borisova, G.; Chukina, N.; Maleva, M.; Prasad, M. N. V. Ceratophyllum Demersum L. and Potamogeton Alpinus Balb. from Iset'river, Ural Region, Russia Differ in Adaptive Strategies to Heavy Metals Exposure–A Comparative Study. *Int. J. Phytoremed.* **2014**, *16* (6), 621–633.
45. Jiang, B.; Xing, Y.; Zhang, B.; Cai, R.; Zhang, D.; Sun, G. Effective Phytoremediation of Low-Level Heavy Metals by Native Macrophytes in a Vanadium Mining Area, China. *Environ. Sci. Pollut. Res. Int.* **2018**, *25* (31), 31272–31282.
46. Liu, N.; Wu, Z. Growth and Antioxidant Response in Ceratophyllum Demersum L. under Sodium Dodecyl Sulfate (SDS), Phenol and Joint Stress. *Ecotoxicol. Environ. Saf.* **2018**, *163*, 188–195.
47. Beheary, M.; M. Sheta, B.; Hussein, M.; Nawareg, M.; A. El-Matary, F.; Hyder, A. Environmental Remediation of Tilapia Aquaculture Wastewater Using Ceratophyllum Demersum and Lemna Minor. *Egypt. J. Aquat. Biol. Fish.* **2019**, *23* (2), 379–396.
48. Dogan, M.; Karatas, M.; Aasim, M. Cadmium and Lead Bioaccumulation Potentials of an Aquatic Macrophyte Ceratophyllum Demersum L.: A Laboratory Study. *Ecotoxicol. Environ. Saf.* **2018**, *148*, 431–440.
49. Dogan, M.; Karataş, M.; Aasim, M. Invesitigation of Chromium (III) Accumulation of Ceratophyllum Demersum L. under in Vitro Conditions. *Kahramanmaras Sutcu Imam Univ. J. Natur. Sci.* **2018**, *21*, 277–285.
50. Dogan, M.; Karataş, M.; Aasim, M. Kadmiyum, Krom ve Kurşunun Ceratophyllum Demersum L. Ve Pogostemon Erectus (Dalzeli) Kuntze Üzerine Fitotoksisitesinin Değerlendirilmesi. *Karaelmas Fen ve Mühendislik Dergisi* **2018**, *8*, 543–550.
51. Aasim, M.; Department of Plant Protection, Faculty of Agricultural Sciences and Technologies, Sivas University of Science and Technology, 58140, Sivas, Turkey. Bioaccumulation Potential of In Vitro Regenerated Plants of Ceratophyllum Demersum against Chromium – A Lab Study. *Asian J. Agric. Biol.* **2020**, *8* (3), 233–239.
52. Johnson, U. E.; Adeogun, B. K.; Ugya, A. Y. Efficacy of Aquatic Plants in Industrial Effluent Treatment Using Vertical Subsurface Flow Constructed Wetland: Studies on Ceratophyllum Demersum, Ludwigia Abyssinica and Hydrolea Glabra. *Ann. Fac. Eng. Hunedoara* **2019**, *17* (1), 213–217.
53. Bunluesin, S.; Kruatrachue, M.; Pokethitiyook, P.; Lanza, G. R.; Upatham, E. S.; Soonthornsarathool, V. Plant Screening and Comparison of Ceratophyllum Demersum and Hydrilla Verticillata for Cadmium Accumulation. *Bull. Environ. Contam. Toxicol.* **2004**, *73* (3), 591–598.
54. Patel, D. K.; Kanungo, V. K. Ecological Efficiency of Ceratophyllumdemersuml in Phytoremediation of Nutrients from Domestic Wastewater. *Int. Quart. J. Environ. Sci* **2010**, *4*, 257–262.
55. Suryani, Y.; Cahyanto, T.; Sudjarwo, T.; Panjaitan, D. V.; Paujiah, E.; Jaenudin, M. Chromium Phytoremediation of Tannery Wastewater Using Ceratophyllum Demersum L. (Coontail). *Biosaintifika* **2017**, *9*(2). https://doi.org/10.15294/biosaintifika.v9i2.8716.
56. Kumar, G. P.; Prasad, M. N. V. Cadmium-Inducible Proteins in Ceratophyllum Demersum L. (a Fresh Water Macrophyte): Toxicity Bioassays and Relevance to Cadmium Detoxification. *Bull. Environ. Contam. Toxicol.* **2004**, *73* (1), 174–181.

57. Parneyan, A.; Chorom, M.; Haghighi-Fard, N. J.; Dinarvand, M. Phytoremediation of Nickel from Hydroponic System by Hydrophyte Coontail (Ceratophyllum Demersum L.). *J. Sci. Technol. Greenh. Cult.* **2011**, *2* (6), 75–85.
58. Khang, H. V.; Hatayama, M.; Inoue, C. Arsenic Accumulation by Aquatic Macrophyte Coontail (Ceratophyllum Demersum L.) Exposed to Arsenite, and the Effect of Iron on the Uptake of Arsenite and Arsenate. *Environ. Exp. Bot.* **2012**, *83*, 47–52.
59. Duman, F.; Koca, F. D.; Sahan, S. Antagonist Effects of Sodium Chloride on the Biological Responses of an Aquatic Plant (Ceratophyllum Demersum L.) Exposed to Hexavalent Chromium. *Water Air Soil Pollut.* **2014**, *225* (2), 1–12.
60. Srivastava, S.; Sounderajan, S.; Udas, A.; Suprasanna, P. Effect of Combinations of Aquatic Plants (Hydrilla, Ceratophyllum, Eichhornia, Lemna and Wolffia) on Arsenic Removal in Field Conditions. *Ecol. Eng.* **2014**, *73*, 297–301.
61. Aravind, P.; Prasad, M. N. V. Cadmium-Induced Toxicity Reversal by Zinc in Ceratophyllum Demersum L. (a Free Floating Aquatic Macrophyte) Together with Exogenous Supplements of Amino-and Organic Acids. *Chemosphere* **2005**, *61* (11), 1720–1733.
62. Maleva, M.; Borisova, G.; Chukina, N.; Kumar, A. Urea Increased Nickel and Copper Accumulation in the Leaves of Egeria Densa (Planch.) Casp. and Ceratophyllum Demersum L. during Short-Term Exposure. *Ecotoxicol. Environ. Saf.* **2018**, *148*, 152–159.
63. Karataş, M.; Aasim, M.; Dogan, M. Efficacy of in Vitro Propagated Coontail (Ceratophyllum Demersum L.) on Quality of Different Water Samples. *Fresenius Environ. Bull.* **2016**, *25*, 5113–5119.
64. Saygideğer, S.; Doğan, M. Lead and Cadmium Accumulation and Toxicity in the Presence of EDTA in Lemna Minor L. and Ceratophyllum Demersum L. *Bull. Environ. Contam. Toxicol.* **2004**, *73* (1), 182–189.
65. Poklonov, V. A. Removal of Aluminum by Aquatic Plants Ceratophyllum Demersum and Chara Fragilis from Water of Experimental Ecosystems. *Russ. J. Gen. Chem.* **2016**, *86* (13), 2978–2982.
66. Ewadh, H. M. Removal of Methylene Blue by Coontail (Ceratophyllum Demersum) Using Phytoremediation Concept. *Plant Arch.* **2020**, *20* (1), 2677–2681.
67. Fagbote, O. Evaluation of the Status of Heavy Metal Pollution of Water (Surface and Ground) and Aquatic Macrophyte (Ceratophyllum Demersum) of Agbabu Bitumen Deposit Area, Nigeria. *Br. J. Appl. Sci. Technol.* **2013**, *3* (2), 289–306.
68. Nuwansi, K. K. T. Performance Evaluation and Phytoremediation Efficiency of Selected Aquatic Macrophytes on Aquaculture Effluent. 2018.
69. Ahmad, S. S.; Reshi, Z. A.; Shah, M. A.; Rashid, I.; Ara, R.; Andrabi, S. M. A. Heavy Metal Accumulation in the Leaves of Potamogeton Natans and Ceratophyllum Demersum in a Himalayan RAMSAR Site: Management Implications. *Wetl. Ecol. Manag.* **2016**, *24* (4), 469–475.
70. Ahmed, Z. F.; Ameer, Q. A. A.; Abbas, R. F. Knowing of Accumulation Capacity of [Ceratophyllum Demersum L. and Hydrilla Verticillata Plant] When One Plant Is Used to Remove the Copper Element in a Laboratory-Contaminated Water-Polluting Ecosystem. *Curr. Res. Microbiol. Biotechnol.* **2018**, *6*(1), 1501–1505.
71. Mishra, S.; Wellenreuther, G.; Mattusch, J.; Stärk, H.-J.; Küpper, H. Speciation and Distribution of Arsenic in the Nonhyperaccumulator Macrophyte Ceratophyllum Demersum. *Plant Physiol.* **2013**, *163* (3), 1396–1408.
72. Mishra, S.; Stärk, H.-J.; Küpper, H. A Different Sequence of Events than Previously Reported Leads to Arsenic-Induced Damage in Ceratophyllum Demersum L. *Metallomics* **2014**, *6* (3), 444–454.

73. Zhang, F.-Q.; Wang, Y.-S.; Lou, Z.-P.; Dong, J.-D. Effect of Heavy Metal Stress on Antioxidative Enzymes and Lipid Peroxidation in Leaves and Roots of Two Mangrove Plant Seedlings (Kandelia Candel and Bruguiera Gymnorrhiza). *Chemosphere* **2007**, *67* (1), 44–50.
74. Sharma, P.; Jha, A. B.; Dubey, R. S.; Pessarakli, M. Reactive Oxygen Species, Oxidative Damage, and Antioxidative Defense Mechanism in Plants under Stressful Conditions. *J. Bot.* **2012**, *2012*, 1–26.
75. Mittler, R. Oxidative Stress, Antioxidants and Stress Tolerance. *Trends Plant Sci.* **2002**, *7* (9), 405–410.
76. Grill, E.; Mishra, S.; Srivastava, S.; Tripathi, R. D. Role of Phytochelatins in Phytoremediation of Heavy Metals. In *Environmental Bioremediation Technologies*; Springer Berlin Heidelberg: Berlin, Heidelberg, 2007; pp 101–146.
77. Ohkawa, H.; Ohishi, N.; Yagi, K. Assay for Lipid Peroxides in Animal Tissues by Thiobarbituric Acid Reaction. *Anal. Biochem.* **1979**, *95* (2), 351–358.
78. Devi, S. R.; Prasad, M. N. V. Copper Toxicity in Ceratophyllum Demersum L. *Plant science* **1998**, *138* (2), 157–165.
79. Jambunathan, N. Determination and Detection of Reactive Oxygen Species (ROS), Lipid Peroxidation, and Electrolyte Leakage in Plants. *Methods Mol. Biol.* **2010**, *639*, 292–298.
80. McKersie, B. D.; Leshem, Y. Y. *Stress and Stress Coping in Cultivated Plants*; Kluwer Academic Publishes, 1994.
81. Noctor, G.; Foyer, C. H. Ascorbate and Glutathione: Keeping Active Oxygen under Control. *Annu. Rev. Plant Physiol. Plant Mol. Biol.* **1998**, *49* (1), 249–279.
82. Martin, M. N.; Slovin, J. P. Purified γ-Glutamyl Transpeptidases from Tomato Exhibit High Affinity for Glutathione and GlutathioneS-Conjugates. *Plant Physiol.* **2000**, *122* (4), 1417–1426.
83. Grill, E.; Löffler, S.; Winnacker, E. L.; Zenk, M. H. Phytochelatins, the Heavy-Metal-Binding Peptides of Plants, Are Synthesized from Glutathione by a Specific γ-Glutamylcysteine Dipeptidyl Transpeptidase (Phytochelatin Synthase. *Proc. Natl. Acad. Sci.* **1989**, *86* (18), 6838–6842.
84. Shukla, D.; Kesari, R.; Tiwari, M.; Dwivedi, S.; Tripathi, R. D.; Nath, P.; Trivedi, P. K. Expression of Ceratophyllum Demersum Phytochelatin Synthase, CdPCS1, in Escherichia Coli and Arabidopsis Enhances Heavy Metal (Loid) s Accumulation. *Protoplasma* **2013**, *250* (6), 1263–1272.
85. Shukla, D.; Kesari, R.; Mishra, S.; Dwivedi, S.; Tripathi, R. D.; Nath, P.; Trivedi, P. K. Expression of Phytochelatin Synthase from Aquatic Macrophyte Ceratophyllum Demersum L. Enhances Cadmium and Arsenic Accumulation in Tobacco. *Plant Cell Rep.* **2012**, *31* (9), 1687–1699.

CHAPTER 9

Genetically Modified Plants (GMPs) and Their Potential in Protection, Constraints, Prospects, Challenges, and Opportunities Against Environmental Pollution

WAJIHA ANUM[1], UMAIR RIAZ[2*], LIAQUAT ALI[1], LAL HUSSAIN AKHTER[1], RANA IMTIAZ AHMED[1], ABID ALI[1], and ASAD-UR-RAHMAN[1]

[1]*Regional Agricultural Research Institute, Department of Agriculture, Government of Punjab, Bahawalpur 63100, Punjab, Pakistan*

[2]*Soil and Water Testing Laboratory for Research, Department of Agriculture, Government of Punjab, Bahawalpur 63100, Punjab, Pakistan*

[*]*Corresponding author. E-mail: umairbwp3@gmail.com*

ABSTRACT

Environmental pollution has emerged as a threatening issue in the current era while the increase in population has resulted in the extensive resource-oriented agricultural commodities production creating a disturbance in the biodiversity and health of the ecosystem. But at the same time, the fluctuations in environmental condition have paved ways for downfall of productive soils. The situation is more aggravated by disease, insect pest's incidence as the present crop varieties/plant species loss their resistance with the passage of time. Under such circumstances, the slight modification in the plants genetic set up may result in better confrontation and productivity. In this scenario, the role of genetically modified plants (GMPs) is crucial in controlling the pollution as caused by continuous use of chemicals. These chemicals are often used in the form of fertilizers, insecticides, pesticides,

Environmental Pollution Impact on Plants: Survival Strategies under Challenging Conditions. Tariq Aftab, PhD & Khalid Rehman Hakeem, PhD (Eds.)

plant growth promoting sprays. The GM plants require fewer amounts of agri-inputs as they will be designed stronger and more adaptable to the stressed environmental conditions. Transgenic plants also remediate the contaminated sites by their biological set ups. The environment pollution controlling transgenic plants are inserted with specific genes that have capability of reducing harmful elements, such as heavy metals and inorganic compounds into non/less toxic compounds. In this regard, techniques and novel tools for genetically modifying plants, development processes, gene incorporation process, their suitability and spatial feasibility are considered as crucial things. The aim of this chapter is to know the basics of GM plants and their mechanism in reducing the environmental pollution, how they can prevent and reduce plant protection chemicals and how the contaminated soils can be reclaimed.

9.1 INTRODUCTION

In the present era, the world is liable to many depreciating substances which are generally created or resulted as human actions. With the passage of time, the situation is aggravated and impacts the overall health of the environment consequential in structural alteration of ecosystems. Among the worst results, environmental pollution is causing toxic effects on the well-being of humans and other biota.

Environmental pollution is causing concerns substantial to public interest especially in developing countries. It is also a matter of debate at global level. The extensive urbanization besides industrialization has caused overexploitation of the natural resources which is directly giving upswing to environmental pollution.[3]

In order to mitigate the injurious effects of toxic substances persisting in the environment, in either form, there is a need to identify the possible ways for effective control and management strategies according to the type, sources and origin of the pollutant causing environmental pollution.

Nowadays, the advancement in scientific research leads to identification of ways for making environment healthy. One of these is the use of biotechnology, by implementing the best-suited biotechnology technique; harmful substances can be converted into decomposable forms which further helps in shaping the pollution control methods. Genetically modified organisms (GMOs), both plants and microorganisms play a crucial role in decreasing toxicity of elements, removing industrial wastes and helps to reduce toxic effects of hydrocarbons and petrol leakages.[49]

Predications by World Energy Congress (WEC) about environmental pollution revealed that if fuel reserves are consumed at the present rate, then by 2025, the world can face an irreversible damage.[72]

9.2 ENVIRONMENTAL POLLUTION (TYPES, SOURCES, AND CHARACTERISTICS)

Environmental pollution results from a number of anthropogenic incidents and cause toxicity in soil, water bodies, and atmosphere. The common sources are oil spillage from pipeline sabotage, accidental discharge from oil wells/rigs, use of automobiles, engines, oil bunkering, and raw sewage discharge from petrochemical processing units. They have increased the concentrations of heavy metals and toxic substances especially polycyclic aromatic hydrocarbons (PAHs) in soils. At a broader range, such incidents have altered natural environs and ecosystem.

9.2.1 AGRICULTURE

Agriculture sector of each country has a significant position and people livelihood depends on the produce. However, modern agriculture nowadays is highly dependent on the extensive use of different chemicals for managing the yield decreasing aspects, such as insect pest attacks and nutrient deficiency. But inappropriate usage of such chemicals (insecticide, pesticides, fertilizers) cause environmental pollution because they are washed off and enters into surface and ground water. Furthermore, they are reserved into the soil thus polluting the environment. Fertilizers significantly cause eutrophication (enhanced amounts of N and P) and as a result, marine life is affected. On the other hand, pesticides pose direct impact on human and animal life as their residues (toxic chemicals) persist on the food and enters the human and animal bodies when sprayed food is consumed. Pesticides cause carbon and chlorine and phosphorus pollution. Biomass burning is another pollution source that elevates carbon levels in environment, forested areas are burned for land clearance which is then used for agricultural use and cause particulate pollution.[56]

Macro and micronutrients are added into the agricultural lands in the form of synthetic fertilizers, though they pose threats to environment if not applied in balanced quantities. For instance, nitrogenous fertilizer is converted into nitrous oxide.[54] Being a greenhouse gas, nitrous oxide is causing noteworthy effects in terms of climate change. Similarly, when excess nitrogen and phosphorus fertilizers are added into the water bodies by means of runoff, they give rise to excessive growth of toxic algae. The situation is worsening by combined effects of global warming and fertilizer pollution. Phosphorus runoff results in more bioavailability of glyphosate, and almost each water body nearby extensive farming farms contains huge quantities of glyphosate and its degradation products.[8,13,16,52,60]

9.2.2 DISTILLERY INDUSTRIES

Distillery industries are causing environmental effluence by discharging a dark-colored waste water that is known to contain high amounts of toxic heavy metals, phosphate, phenolics, sulfate, biological oxygen demand (BOD), chemical oxygen demand (COD), and total solids.[14]

Other pollution causing sources are related to municipal and agricultural waste. In such cases, water born bacteria, and arsenic (As) enters the watershed.[72]

9.3 IMPACT ON BIODIVERSITY AND AGRICULTURE

Environmental pollution is altering the balance in ecosystem and with the passage of time, a shift has been observed in invasive and native species. For example, the distillery wastewater when entered into the terrestrial region has resulted in seed germination inhibition, and affects growth and development of native species negatively. It is because, the wastewater reduces soil alkalinity and decrease the Mn availability. On the other hand, when this wastewater enters aquatic bodies, it reduced light penetration thus leading to reduced photosynthetic activities of aquatic plants. It also reduced the amount of dissolved oxygen thus damaging the aquatic bodies.[14]

9.3.1 INSECTICIDES RESIDUES IN SOIL INVERTEBRATES

Orchards and forests are an important component of ecosystem in which many species prevail and a regular web is sustained, however, the problem arises when extensive sprays of insecticides affect the nontargeted organisms. Orchards have almost similar management schedules practiced each year as a consequence of similar plantations for longer time periods. Similarly, insecticide treatments are scheduled same time each year. Furthermore, the soils also receive no or minimal cultivated/tillage. Although, the residue has less persistence if they are not incorporated in soil,[17] however, foliar sprays are popular nowadays, the residues as a result of foliar sprays tend to accumulate and forms a reservoir in soil layers. Deciduous trees also shed their leaves contributing in the reservoir. Previous research has claimed that an amount 167.5 ppm of DDT residues was contained in the leaf sample of treated elm trees.[12] The highest amounts of DDT have been found in orchards soils especially in the top years (Harris et al., 1966; Harris and Sans, 1969).

The insecticide residues are then transferred to the invertebrates that are extant in the soil. The amount of accumulation depends on the substrate on which the insecticide is applied. For example, a study revealed *A. caliginosa* accumulated more Dieldrin as compared with *L. terrestris* in their bodies when both were kept under the same treated soil. The difference was observed because of high ingestion rate observed in *A. caliginosa.*[15]

9.4 WHAT ARE GMOS?

Modern biotechnology has given rise to production of GMOs, it refers to the organisms in which their genetic material is altered by some artificial means under laboratory conditions. This introduction is considered as a breakthrough in agriculture similar to green revolution in the beginning of 70s.[19]

9.4.1 RISKS AND HAZARDS

Genetic modification in organisms has led to a controversy in the world where many are concerned about its effect on the biodiversity, specially, the gene flow into the nontargeted organisms with special emphasis on insect herbivores, natural enemies, and soil biota. Another concern, argued by the ecologist is the development of resistance in the insect pests that can lead to endangering significant natural resources, for example, *Bacillus thuringiensis* that acts like a natural microbial insecticide.[19]

Supporters claim that it is unjust not to use GM crops for reducing the hunger and mal nutrition in developing countries. While the opposers stated that GM crops can be a useful means as a technological advancement for enhancing yield and crop quality. However, they should be introduced after stable and efficient institutions in order to avoid any negative affect on natural system.[67]

9.4.2 *GENETICALLY ENGINEERED MICROORGANISMS (GEMS)*

The true performance of any species is dependent upon the environmental conditions prevailing in the field. Under the adverse environmental conditions, the natural degrader may not survive or work as efficiently as it may under the laboratory conditions.[59]

In this regard an improved degradation of the pollutants persistent in the environment is required. It can be achieved by enhancing the enzymatic activity of the biochemical pathways through incorporation of genetic engineering. For instance, a higher expression of key enzymes will achieve improved and higher degradation rates. Biotechnology can be implemented for developing microorganism for environmental remediation.[64,65] and for specific remediation regarding environmental pollutants. A general description of transgenic plants development is depicted in Figure 9.1.

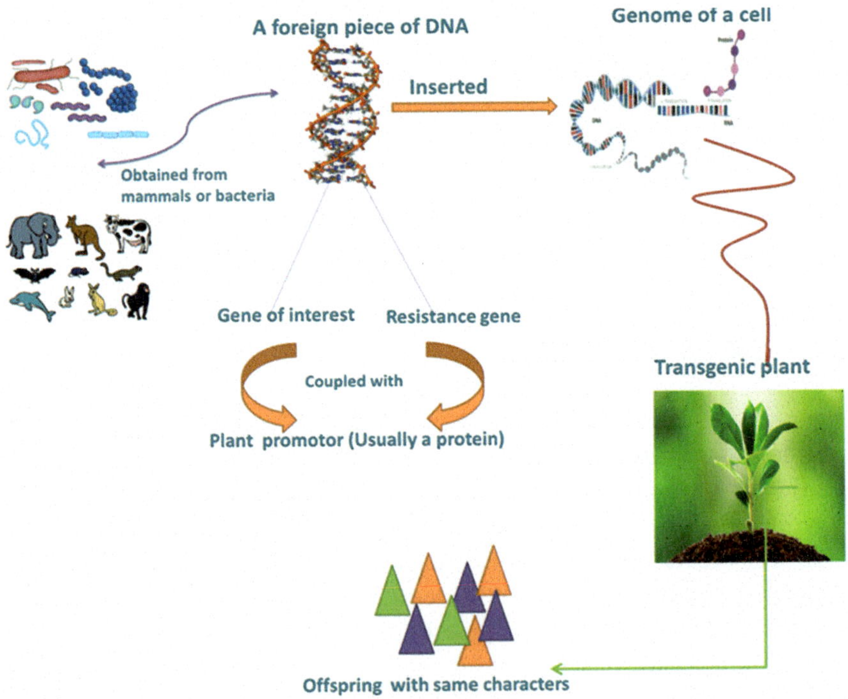

FIGURE 9.1 General procedure for the formation of transgenic plants.

Plants are genetically engineered for purposes, such as enhanced tolerance for biotic and abiotic stresses, efficient nutrient uptake, diminishing the harmful effects of agrochemicals and crop yield increase, tolerance to toxic pollutants, an improved symbiotic relationship with microbes, and their efficiency for phytoremediation processes.[2,18]

Genetic engineering has made it possible for the plants to detoxify the environment by extracting or sequestrating the contaminants.[43]

9.5 GENETIC STRATEGIES FOR MAXIMIZING THE PLANT POTENTIAL AGAINST ENVIRONMENTAL POLLUTION

Agroforestry sector is growing day by day because of increase demand of wood and food products. Moreover, the establishment of windbreaks, biofiltration facilities and feedstock for biofuel production has increased the agroforestry tree plantations.[38,63] In order to fulfil the required demands, high biomass yielding and fast-growing tree varieties, such as eucalyptus, palms, and poplars are preferred for cultivation. However, this leads to high rates of leaf isoprene emissions. Isoprene is a volatile substance that is produced in trees (chloroplast) with high productivity rate during the light-dependent metabolism. The emissions rates are similar to those of methane produced globally[22]). The emitted isoprene is oxidized under a series of photochemically reactions within hours. As a result, the oxidation of isoprene results in pollutants like organic nitrates (e.g., PAN) and tropospheric ozone (O_3).[45]

Genetically modified poplars grown in controlled conditions reveal that stress interactions are complex. Studies have shown that under isoprene emissions, plants are protected from abiotic stress but when CO_2 is increased in the atmosphere, there exists an inversely proportional relationship between the two gases, that is, CO_2 and isoprene cause a reduction in the protection against stresses.[57,71,73] Elevated temperatures (increase in global temperatures) can mitigate the effects due to high CO_2 concentration,[44,50] under this scenario, the protective effect of isoprene is maintained even under increased CO_2 concentrations.[34]

9.6 ROLE OF GMOS IN DECOMPOSITION OF POLYCYCLIC AROMATIC HYDROCARBONS PAHS

PAHs are generally solids compounds with high boiling and melting points, they are less soluble in aqueous and have low vapor pressure.[1,42] PAHs are most commonly produced when there is incomplete combustion of organic compounds like oil, wood, and coal.[1]

PAHs are photolyzed or oxidized with strong oxidizing agents among which the common ones are ozone, nitrogen oxides, and OH radicals,[41] Sometimes heterogeneous oxidation also results when PAHs oxidize with tiny water droplets.[55] Niu et al.[48] reveal that PAHs experience lower photolysis reaction half-lives, when they are present in atmospheric particulate matter as compared with those present in organic solvents or water.[66]

Genetic transformations can increase crop yields upto 16.7% on an average. It means that 800 million more people can be fed if planted wheat, maize, and rice were GM plants considering these three crops present two-thirds of worlds energy intake. GM crops have higher shoot biomass and NUpE responsible for increase in yields. The genetic transformation can increase the yields without additional N inputs, thus decreasing the environmental risks caused by N fertilizer losses and causes increased uptake of N in GM crops, it also causes an increase in the nitrogen use efficiency (NUE) (16.2%) and PFPN (9.47%) in GM crops. The three major crops are applied with 50 Tg N globally and if GM crops are planted, then a minimum of 4.7 Tg N can be saved each year.[33] Moreover, less fossil fuel is consumed if the demand of N is reduced as a result of less demand for transport and production of N fertilizers.[37]

Nitrogen is lost to the environment by means of denitrification, leaching, runoff, volatilization, and microbial consumption.[7] As a result, water eutrophication,[40] soil acidification, air pollution, and environmental degradation occur.[23] Other examples of genetically combined organisms with their specific purposes are presented in Table 9.1.

9.7 GM MAIZE

Among the genetically engineered food crops, maize is at the second position in the world and includes about 24% of total maize production in 2008.[27] The transgenic maize comprises insect resistance and HT, partly as separate and partly also as stacked technologies. For insect resistance, the genes are mostly from the soil bacterium *Bacillus thuringiensis* (Bt), and such genes are responsible for controlling different stem borers specifically corn rootworm and European corn borer (Romeis et al., 2008). GM maize is commonly cultivated in North and South America, South Africa, and Philippines.[51]

Ike et al.[25] carried out an experiment for modulating heavy metal accumulating in leguminous crops with the help of work on root-associated rhizobia. Plant and rhizobia have a symbiotic relationship in which rhizobia forms nitrogen-fixing nodule (contains more than 108 bacterial progenies)

TABLE 9.1 Sources of Transgenic Plants and their Usage.

Source of gene	Transgenic plant	Purpose	Reference
Enterobacter cloacae PB2	*N. tabacum*	Degrade explosives, ability to grow on them	21
Comamonas testosteroni (bph C)	*N. tabacum*	Hydroxyderivatives, increase resistance	20
Human (Metallothionein)	*Nicotiana tabacum*	Increase Cd tolerance	28
Glycine max (Ferritin)	*Oryza sativa*	Increased Fe uptake in seeds	28
E. coli (gshl)	*B. juncea*	High level of Cd in shoot on media (50 μM Cd^{2+})	74
S. cerevisiae (GSH1)	*A. thaliana*	Augmented As and Cd accumulation	23
S. oleracea (OAS-TL)	*N. tabacum*	2.5 times higher biomass on media (300 μM Cd^{2+})	29
		1.3 times higher biomass on media (Ni^{2+})	
		1.5 times higher biomass on media (on medium with 250 μM SeO_4^{2-})	
A. bisulcatus (SMT)	*B. juncea*	No phytotoxicity of 25 μM SeO_3^{2-} on medium (97% growth inhibition with WT). By 40% reduced growth on medium with 25 μM SeO_4^{2-} (60% inhibition with WT)	35
N. tabacum (NtCBP4)	*A. thaliana*	Increased tolerance to Ni^{2+}	10
		20% greater accumulation of Pb in shoots on media (100 μM Pb^{2+})	
		60% reduction in Ni uptake in shoots on media (200 μM Ni^{2+})	
Cynobacteria (glutathione-S –transferase) Bacteria (cyanidase)	Tobacco	Increased cyanide degradation	30
		Strengthened antioxidant system and homeostatic mechanisms	
S. cerevisiae (YCF1)	*A. thaliana*	2.2 times Higher biomass in media (60 μM Cd^{2+})	Song et al., 2003
		1.8 times higher biomass in media (900 μM Pb^{2+})	
Streptococcus thermophilus	*Beta vulgaris L*	More accumulation of Cu, Zn, and Cd ions in shoots	39
		High phytochelatin and GSH levels under different heavy metal stresses	

TABLE 9.1 *(Continued)*

Source of gene	Transgenic plant	Purpose	Reference
Bacteria	*Nicotiana tabacum*	Transgenic accumulated both organic and inorganic Hg at levels surpassing soil concentrations	24
Arabidopsis thaliana	*Brassica juncea*	High Se volatilization (2-- fold) More tolerant to selenite	70
Staphylococcus aureus	*Populus alba* and *P. tremula* var. glandulosa	Transgenic performed better (one of strain produced 4.54.8 times more Hg° than controls when exposed to 50 μM Hg Lowers mercury content in transgenics	26
Escherichia coli (Nitroreductase)	*Arabidopsis thaliana*	Degradation of 2,4,6-trinitrotoluene (TNT)	69
Bacteria (Mercuric reductase (merA) Organomercurial lyase (merB)	*Arabidopsis thaliana*	Detoxification of methylmercury	9
Human (Cytochrome P450)	Poplar (Populus)	Remediation of trichloroethylene (TCE)	68

on root of the plant. When four mammalian MT-coding sequences along with AtPCS1 were genetically fused and were expressed in *Mesorhizobium huakuii* subsp. rengei (strain B_3), it increased the natural capacity of the bacterium to accumulated cadmium ions up to 25-fold from a media containing 30 µM Cd^{2+}. The colonization of leguminous *milkvetch Astragalus sinicum* with the B3 strain in rice paddy soil promoted Cd^{2+} uptake in roots, but not in nodules, by three times. The contribution of these free-living modified rhizobia to the collection of Cd^{2+} in the soil, and the subsequent chemotaxis-mediated transport of accumulated metal for uptake in the legume's roots, seems to be likely. Although the enhanced Cd^{2+} accumulation phenotype of the roots was not accompanied by an increased metal translocation to the shoots, such a strategy would have a beneficial potential for the rhizofiltration or transient phytostabilization of heavy metals in soil.[32]

9.8 HOW IT WORKS—PAHs DEGRADATION

Organic pollutants, such as polychlorinated biphenyls (PCBs), polycyclic aromatic hydrocarbons (PAH) pesticides, 2, 4, 6-trinitrotoluene (TNT), herbicides, and inorganic nutrients are creating havoc for environment safety. The situation has given rise to developed association between plants and microbes which can effectively control/mitigate the polluted sites.[36] For instance, hybrid poplars are used for degrading one of the common pollutant trichloroethylene (TCE).[47] The degradation process involves the production of certain enzymes, such as nitrilase, laccase, dehalogenase, nitroreductase, and laccase which then breakdown the pollutants.[11,46,61] The term "Phytoremediation" covers the processes of contaminants/toxins degradation within the plant, in soil water or surface water through enzymes. It is mostly effective for moderately hydrophobic organic chemicals (octanol–water partition coefficients, log Kow = 0.5 ~ 3.0).[58,61] Such chemicals include short chain aliphatic chemicals, chlorinated solvents, and benzene, toluene, ethyl benzene, xylene (BTEX chemicals). If the log Kow is less than 0.5, the organics cannot get into the plants as they become too hydrophilic and cannot pass through the membranes. While on the other hand, if log Kow is more than 3.0, the hydrophobic chemicals are strongly bound with the plant roots that they cannot get translocated into the plant.[61]

Hydrocarbon contamination in soils is responsible for affecting the natural balance between living organisms and natural environment and affects the natural microflora, microorganisms and bioavailability of resources.[4,62] PAHs pose threats to agriculture and crop production as they can be integrated into the food web, contaminate underground water, fresh water.[5] Moreover, they become part of the plants during the cultivation, hence, threatening food security and pose health hazards as PAHs are known to be neurogenic, carcinogenic, and mutagenic in nature.[6]

9.9 CONCLUSION

Transgenic plants for controlling environmental pollution generally are modified so that there is less need of inputs such as fertilizers and pesticides. This branch of biotechnology has covered a lot of advancement and successfully gene of interests has been incorporated for achieving the desired purpose. Most of the work has been done with microorganism incorporation into plants while a vast range is still unexplored. There is a need to create/modify plants with other plant species and strains of microorganisms which are more stress-resistant and can adapt to a number of environmental conditions. Moreover, a gap prevails and hinders the expansion of GMPs due to stereotypic concepts, which need to be looked after careful considerations of all aspects of society.

KEYWORDS

- **agrochemicals**
- **environmental pollution**
- **genetic engineering**
- **GMO**
- **phytostabilization**
- **toxins removal**

REFERENCES

1. Abdel-Shafy, H. I.; Mansour, M. S. A Review on Polycyclic Aromatic Hydrocarbons: Source, Environmental Impact, Effect on Human Health and Remediation. *Egypt. J. Pet.* **2016,** *25* (1), 107–123.
2. Abiri, R.; Valdiani, A.; Maziah, M.; Shaharuddin, N. A.; Sahebi, M.; Yusof, Z. N. B.; Talei, D. A Critical Review of the Concept of Transgenic Plants: Insights into

Pharmaceutical Biotechnology and Molecular Farming. *Curr. Issues Mol. Biol.* **2015,** *18*, 21–42.

3. Ajibade, F. O.; Adelodun, B.; Lasisi, K. H.; Fadare, O. O.; Ajibade, T. F.; Nwogwu, N. A.; Wang, A. Environmental Pollution and their Socioeconomic Impacts. In *Microbe Mediated Remediation of Environmental Contaminants*; Woodhead Publishing, 2021, pp. 321–354.
4. Amadi, A.; Abbey, S. D.; Nma, A. Chronic Effects of Oil Spill on Soil Properties and Microflora of a Rainforest Ecosystem in Nigeria. *Water Air Soil Poll.* **1996,** *86* (1), 1–11.
5. Asemoloye, M. D.; Jonathan, S. G.; Ahmad, R. Synergistic Plant-Microbes Interactions in the Rhizosphere: A Potential Headway for the Remediation of Hydrocarbon Polluted Soils. *Int. J. Phytoremed.* **2019,** *21* (2), 71–83.
6. Asemoloye, M. D.; Jonathan, S. G.; Jayeola, A. A.; Ahmad, R. Mediational Influence of Spent Mushroom Compost on Phytoremediation of Black-Oil Hydrocarbon Polluted Soil and Response of Megathyrsus Maximus Jacq. *J. Environ. Manag.* **2017,** *200*, 253–262.
7. Basso, B.; Shuai, G.; Zhang, J.; Robertson, G. P. Yield Stability Analysis Reveals Sources of Large-Scale Nitrogen Loss from the US Midwest. *Sci. Rep.* 2019, *9* (1), 1–9.
8. Battaglin, W. A.; Kolpin, D. W.; Scribner, E. A.; Kuivila, K. M.; Sandstrom, M. W. Glyphosate, Other Herbicides, and Transformation Products in Midwestern Streams, 2002. *JAWRA* **2005,** *41* (2), 323–332.
9. Bizily, S. P.; Rugh, C. L.; Meagher, R. B. Phytodetoxification of Hazardous Organomercurials by Genetically Engineered Plants. *Nat. Biotechnol.* **2000,** *18* (2), 213–217.
10. Bizily, S. P.; Rugh, C. L.; Summers, A. O.; Meagher, R. B. Phytoremediation of Methylmercury Pollution: merB Expression in Arabidopsis Thaliana Confers Resistance to Organomercurials. *Proc. Natl. Acad. Sci.* **1999,** *96* (12), 6808–6813.
11. Boyajian, G. E.; Carreira, L. H. Phytoremediation: A Clean Transition from Laboratory to Marketplace? *Nat. Biotechnol.* **1997,** *15* (2), 127–128.
12. Boykins, E. A. The Effects of DDT-Contaminated Earthworms in the Diet of Birds. *BioSci.* **1967,** *17* (1), 37–39.
13. Chang, F. C.; Simcik, M. F.; Capel, P. D. Occurrence and Fate of the Herbicide Glyphosate and its Degradate Aminomethylphosphonic Acid in the Atmosphere. *Environ. Toxicolo. Chem.* **2011,** *30* (3), 548–555.
14. Chowdhary, P.; Yadav, A.; Kaithwas, G.; Bharagava, R. N. Distillery Wastewater: A Major Source of Environmental Pollution and its Biological Treatment for Environmental Safety. In *Green Technologies and Environmental Sustainability*; Springer: Cham, 2017; pp. 409–435.
15. Davis, B. N. K. Laboratory Studies on the Uptake of Dieldrin and DDT by Earthworms. *Soil Biol. Biochem.* **1971,** *3* (3), 221–233.
16. Dubrovsky, N. M.; Hamilton, P. A. *The Quality of our Nation's Water: Nutrients in the Nation's Streams and Groundwater; National Findings and Implications*; US Geological Survey, National Water-Quality Assessment Program, 2010; pp. 2010–3078.
17. Edwards, C. A. Insecticide Residues in Soils. In *Residue Reviews*; Springer: New York, NY, 1966; pp. 83–132.
18. Fernández-Luqueño, F.; López-Valdez, F.; Sarabia-Castillo, C. R.; García-Mayagoitia, S.; Pérez-Ríos, S. R. Bioremediation of Polycyclic Aromatic Hydrocarbons-Polluted Soils at Laboratory and Field Scale: A Review of the Literature on Plants and Microorganisms. In *Enhancing Cleanup of Environmental Pollutants*; Anjum, N., Gill, S., Tuteja, N., Eds.; Springer: Cham, 2017; pp. 43–64.

19. Fontes, E. M.; Pires, C. S.; Sujii, E. R.; Panizzi, A. R. The Environmental Effects of Genetically Modified Crops Resistant to Insects. *Neotrop. Entomol.* **2002,** *31* (4), 497–513.
20. Francova, K.; Sura, M.; Macek, T.; Szekeres, M.; Bancos, S.; Demnerova, K.; Mackova, M. Preparation of Plants Containing Bacterial Enzyme for Degradation of Polychlorinated Biphenyls. *Fresenius Environ. Bull.* **2003,** *12* (3), 309–313.
21. French, C. E.; Rosser, S. J.; Davies, G. J.; Nicklin, S.; Bruce, N. C. Biodegradation of Explosives by Transgenic Plants Expressing Pentaerythritol Tetranitrate Reductase. *Nat. Biotechnol.* **1999,** *17* (5), 491–494.
22. Guenther, A. B.; Jiang, X.; Heald, C. L.; Sakulyanontvittaya, T.; Duhl, T.; Emmons, L. K.; Wang, X. The Model of Emissions of Gases and Aerosols from Nature Version 2.1 (MEGAN2. 1): An Extended and Updated Framework for Modeling Biogenic Emissions. *Geosci. Model Dev.* **2012,** *5* (6), 1471–1492.
23. Guo, J.; Dai, X.; Xu, W.; Ma, M. Overexpressing GSH1 and AsPCS1 Simultaneously Increases the Tolerance and Accumulation of Cadmium and Arsenic in Arabidopsis Thaliana. *Chemosphere* **2008,** *72* (7), 1020–1026.
24. Hussein, H. S.; Ruiz, O. N.; Terry, N.; Daniell, H. Phytoremediation of Mercury and Organomercurials in Chloroplast Transgenic Plants: Enhanced Root Uptake, Translocation to Shoots, and Volatilization. *Environ. Sci. Technol.* **2007,** *41* (24), 8439–8446.
25. Ike, A.; Sriprang, R.; Ono, H.; Murooka, Y.; Yamashita, M. Bioremediation of Cadmium Contaminated Soil Using Symbiosis between Leguminous Plant and Recombinant Rhizobia with the MTL4 and the PCS Genes. *Chemosphere* **2007,** *66* (9), 1670–1676.
26. Im Choi, Y.; Noh, E. W.; Lee, H. S.; Han, M. S.; Lee, J. S.; Choi, K. S. Mercury-Tolerant Transgenic Poplars Expressing Two Bacterial Mercury-Metabolizing Genes. *J. Plant Biol.* **2007,** *50* (6), 658–662.
27. James, C. Global Status of Commercialized Biotech/GM Crops: 2004. *ISAAA Briefs* **2004,** *32*, 1–12.
28. Kärenlampi, S.; Schat, H.; Vangronsveld, J.; Verkleij, J. A. C.; van der Lelie, D.; Mergeay, M.; Tervahauta, A. I. Genetic Engineering in the Improvement of Plants for Phytoremediation of Metal Polluted Soils. *Environ. Poll.* **2000,** *107* (2), 225–231.
29. Kawashima, C. G.; Noji, M.; Nakamura, M.; Ogra, Y.; Suzuki, K. T.; Saito, K. Heavy Metal Tolerance of Transgenic Tobacco Plants Over-Expressing Cysteine Synthase. *Biotechnol. Lett.* **2004,** *26* (2), 153–157.
30. Kebeish, R.; El-Ayouty, Y.; El-Naggar, A.; Saleh, A. M. Synchronous Overexpression of Glutathione-S-Transferase and Cyanidase Maintains the Redox Homeostasis and Improves Cyanide Remediation Capacity in Tobacco. *Environ. Expe. Bot.* **2017,** *137*, 220–227.
31. Kennedy, G. G. Integration of Insect-Resistant Genetically Modified Crops within IPM Programs. In *Integration of Insect-Resistant Genetically Modified Crops within IPM Programs*, 2008; pp. 1–26.
32. Kotrba, P.; Najmanova, J.; Macek, T.; Ruml, T.; Mackova, M. Genetically Modified Plants in Phytoremediation of Heavy Metal and Metalloid Soil and Sediment Pollution. *Biotechnol. Adv.* **2009,** *27* (6), 799–810.
33. Ladha, J. K.; Tirol-Padre, A.; Reddy, C. K.; Cassman, K. G.; Verma, S.; Powlson, D. S.; Pathak, H. Global Nitrogen Budgets in Cereals: A 50-Year Assessment for Maize, Rice and Wheat Production Systems. *Sci. Rep.* **2016,** *6* (1), 1–9.
34. Lantz, A. T.; Solomon, C.; Gog, L.; McClain, A. M.; Weraduwage, S. M.; Cruz, J. A.; Sharkey, T. D. Isoprene Suppression by CO2 is Not due to Triose Phosphate Utilization (TPU) Limitation. *Front. Forests Global Change* **2019,** *2*, 8.

35. LeDuc, D. L.; Tarun, A. S.; Montes-Bayon, M.; Meija, J.; Malit, M. F.; Wu, C. P.; Terry, N. Overexpression of Selenocysteine Methyltransferase in Arabidopsis and Indian Mustard Increases Selenium Tolerance and Accumulation. *Plant Physiol.* **2004,** *135* (1), 377–383.
36. Lee, J. H. An Overview of Phytoremediation as a Potentially Promising Technology for Environmental Pollution Control. *Biotechnol. Bioprocess Eng.* **2013,** *18* (3), 431–439.
37. Li, M.; Xu, J.; Gao, Z.; Tian, H.; Gao, Y.; Kariman, K. Genetically Modified Crops are Superior in their Nitrogen Use Efficiency-A Meta-Analysis of Three Major Cereals. *Sci. Rep.* **2020,** *10* (1), 1–9.
38. Limayem, A.; Ricke, S. C. Lignocellulosic Biomass for Bioethanol Production: Current Perspectives, Potential Issues and Future Prospects. *Prog. Energy Combust. Sci.* **2012,** *38* (4), 449–467.
39. Liu, D.; An, Z.; Mao, Z.; Ma, L.; Lu, Z. Enhanced Heavy Metal Tolerance and Accumulation by Transgenic Sugar Beets Expressing Streptococcus Thermophilus StGCS-GS in the Presence of Cd, Zn and Cu Alone or in Combination. *PLoS One* **2015**, *10* (6), e0128824.
40. Lu, C.; Tian, H. Global Nitrogen and Phosphorus Fertilizer use for Agriculture Production in the Past Half Century: Shifted Hot Spots and Nutrient Imbalance. *Earth Syst. Sci. Data* **2017,** *9* (1), 181–192.
41. Marr, L. C.; Dzepina, K.; Jimenez, J. L.; Reisen, F.; Bethel, H. L.; Arey, J.; Molina, M. J. Sources and Transformations of Particle-Bound Polycyclic Aromatic Hydrocarbons in Mexico City. *Atmospheric Chem. Phys.* **2006,** *6* (6), 1733–1745.
42. Masih, J.; Singhvi, R.; Kumar, K.; Jain, V. K.; Taneja, A. Seasonal Variation and Sources of Polycyclic Aromatic Hydrocarbons (PAHs) in Indoor and Outdoor Air in a Semi Arid Tract of NORTHERN INDIA. *Aerosol Air Qual. Res.* **2012,** *12* (4), 515–525.
43. Meagher, R. B. Phytoremediation of Toxic Elemental and Organic Pollutants. *Curr. Opin. Plant Biol.* **2000,** *3* (2), 153–162.
44. Monson, R. K.; Neice, A. A.; Trahan, N. A.; Shiach, I.; McCorkel, J. T.; Moore, D. J. Interactions Between Temperature and Intercellular CO_2 Concentration in Controlling Leaf Isoprene Emission Rates. *Plant Cell Environ.* **2016,** *39* (11), 2404–2413.
45. Monson, R. K.; Winkler, B.; Rosenstiel, T. N.; Block, K.; Merl-Pham, J.; Strauss, S. H.; Schnitzler, J. P. High Productivity in Hybrid-Poplar Plantations Without Isoprene Emission to the Atmosphere. *Proc. Natl. Acad. Sci.* **2020,** *117* (3), 1596–1605.
46. Morikawa, H.; Erkin, Ö. C. Basic Processes in Phytoremediation and Some Applications to Air Pollution Control. *Chemosphere* **2003,** *52* (9), 1553–1558.
47. Newman, L. A.; Strand, S. E.; Choe, N.; Duffy, J.; Ekuan, G.; Ruszaj, M.; Gordon, M. P. Uptake and Biotransformation of Trichloroethylene by Hybrid Poplars. *Environ. Sci. Technol.* **1997,** *31* (4), 1062–1067.
48. Niu, J.; Sun, P.; Schramm, K. W. Photolysis of Polycyclic Aromatic Hydrocarbons Associated with Fly Ash Particles Under Simulated Sunlight Irradiation. *J. Photochem.. Photobiol. A* **2007,** *186* (1), 93–98.
49. Ozcan, F.; Kahramanogullari, C. T.; Kocak, N.; Yildiz, M.; Haspolat, I.; Tuna, E. Use of Genetically Modified Organisms in the Remediation of Soil and Water Resources. *Fresen. Environ. Bull*, **2012,** *21*, 3443–3447.
50. Potosnak, M. J.; LeStourgeon, L.; Nunez, O. Increasing Leaf Temperature Reduces the Suppression of Isoprene Emission by Elevated CO2 Concentration. *Sci. Total Environ.* **2014,** *481*, 352–359.

51. Qaim, M. The Economics of Genetically Modified Crops. *Annu. Rev. Resour. Econ.* **2009,** *1* (1), 665–694.
52. Quarles, W. Genetic Engineering and Pest Control. *IPM Practitioner* **2016,** *35* (3/4), 1–9.
53. Quarles, W. Glyphosate Toxicity: Smoke or Fire. *IPM Practitioner* **2016,** *35* (5/6), 1–7.
54. Quarles, W. Glyphosate, GMO Soybean Yields and Environmental Pollution. *IPM Practitioner* **2017,** *35* (11/12), 1–9.
55. Raja, S.; Valsaraj, K. T. On the Reactive Uptake of Gaseous PAH Molecules by Micron-Sized Atmospheric Water Droplets. *Atmos. Res.* **2006,** *81* (4), 277–292.
56. Rieuwerts, J. *The Elements of Environmental Pollution*; Routledge, 2017.
57. Rosenstiel, T. N.; Potosnak, M. J.; Griffin, K. L.; Fall, R.; Monson, R. K. Increased CO_2 Uncouples Growth from Isoprene Emission in an Agriforest Ecosystem. *Nature* **2003,** *421* (6920), 256–259.
58. Salt, D. E.; Smith, R. D.; Raskin, I. Phytoremediation. *Annu. Rev. Plant Biol.* **1998,** *49* (1), 643–668.
59. Samanta, S. K.; Singh, O. V.; Jain, R. K. Polycyclic Aromatic Hydrocarbons: Environmental Pollution and Bioremediation. *Trends Biotechnol.* **2002,** *20* (6), 243–248.
60. Sasal, M. C.; Demonte, L.; Cislaghi, A.; Gabioud, E. A.; Oszust, J. D.; Wilson, M. G.; Repetti, M. R. Glyphosate Loss by Runoff and its Relationship with Phosphorus Fertilization. *J. Agric. Food Chem.* **2015,** *63* (18), 4444–4448.
61. Schnoor, J. L.; Light, L. A.; McCutcheon, S. C.; Wolfe, N. L.; Carreia, L. H. Phytoremediation of Organic and Nutrient Contaminants. *Environ. Sci. Technol.* **1995,** *29* (7), 318A–323A.
62. Snowden, R. J.; Ekweozor, I. K. E. The Impact of a Minor Oil Spillage in the Estuarine Niger Delta. *Mar. Pollut. Bull.* **1987,** *18* (11), 595–599.
63. Stolarski, M. J.; Krzyżaniak, M.; Łuczyński, M.; Załuski, D.; Szczukowski, S.; Tworkowski, J.; Gołaszewski, J. Lignocellulosic Biomass from Short Rotation Woody Crops as a Feedstock for Second-Generation Bioethanol Production. *Ind. Crops Prod.* **2015,** *75*, 66–75.
64. Timmis, K. N.; Pieper, D. H. Bacteria Designed for Bioremediation. *Trends Biotechnol.* **1999,** *17* (5), 201–204.
65. Timmis, K. N.; Steffan, R. J.; Unterman, R. Designing Microorganisms for the Treatment of Toxic Wastes. *Annu. Rev. Microbiol.* **1994,** *48* (1), 525–557.
66. Tobiszewski, M.; Namieśnik, J. PAH Diagnostic Ratios for the Identification of Pollution Emission Sources. *Environ. Pollut.* **2012,** *162*, 110–119.
67. Toft, K. H. GMOs and Global Justice: Applying Global Justice Theory to the Case of Genetically Modified Crops and Food. *J. Agric. Environ. Ethics* **2012,** *25* (2), 223–237.
68. Van Aken, B. Transgenic Plants for Phytoremediation: Helping Nature to Clean Up Environmental Pollution. *Trends Biotechnol.* **2008,** *26* (5), 225–227.
69. Van Aken, B. Transgenic Plants for Enhanced Phytoremediation of Toxic Explosives. *Curr. Opin. Biotechnol.* **2009,** *20* (2), 231–236.
70. Van Huysen, T.; Abdel-Ghany, S.; Hale, K.L.; LeDuc, D.; Terry, N.; Pilon-Smits, E. A. Overexpression of Cystathionine-Gamma-Synthase Enhances Selenium Volatilization in Brassica Juncea. *Planta* **2003,** *218* (1), 71–78.
71. Vanzo, E.; Jud, W.; Li, Z.; Albert, A.; Domagalska, M. A.; Ghirardo, A.; Schnitzler, J. P. Facing the Future: Effects of Short-Term Climate Extremes on Isoprene-Emitting and Nonemitting Poplar. *Plant Physiol.* **2015,** *169* (1), 560–575.

72. Vaseashta, A.; Vaclavikova, M.; Vaseashta, S.; Gallios, G.; Roy, P.; Pummakarnchana, O. Nanostructures in Environmental Pollution Detection, Monitoring, and Remediation. *Sci. Technol. Adv. Mater.* **2007,** *8* (1–2), 47.
73. Way, D. A.; Schnitzler, J. P.; Monson, R. K.; Jackson, R. B. Enhanced Isoprene-Related Tolerance of Heat-And Light-Stressed Photosynthesis at Low, but Not High, CO 2 Concentrations. *Oecologia* **2011,** *166* (1), 273–282.
74. Zhu, Y. L.; Pilon-Smits, E. A.; Tarun, A. S.; Weber, S. U.; Jouanin, L.; Terry, N. Cadmium Tolerance and Accumulation in Indian Mustard is Enhanced by Overexpressing γ-Glutamylcysteine Synthetase. *Plant Physiol.* **1999,** *121* (4), 1169–1177.

CHAPTER 10

Detoxification of Sewage Sludge by Natural Attenuation and Application as a Fertilizer

TAQI RAZA[1*], IFTIKHAR ALI[2], NAWAB KHAN[3], NEAL S. EASH[1], MUHAMMAD FARHAN QADIR[4], and HANUMAN SINGH JATAV[5]

[1]*Department of Biosystems Engineering & Soil Science, The University of Tennessee, Tennessee, USA*

[2]*Department of Soil and environmental sciences, The University of Agriculture Peshawar, KPK, Pakistan*

[3]*Climate, Energy and Water Research Institute (CEWRI), National Agricultural Research Centre, Islamabad, Pakistan*

[4]*Institute of Soil and Environmental Sciences, University of Agriculture, Faisalabad, Punjab, Pakistan*

[5]*Sri Karan Narendra Agriculture University Jobner-Jaipur, Rajasthan, India*

[*]*Corresponding author. E-mail: taqiraza85@gmail.com*

ABSTRACT

Sewage sludge is an unpreventable by-product of the wastewater treatment system due to its increasing volume, and the impacts related to its disposal is making it a key issue for many countries. Management and recycling of sewage sludge are the best options, but it requires a high level of characterization because it contains many harmful compounds, which are disturbing the ecosystem, such as heavy metals and organic pollutants. It is considered as an alternative valuable raw material for the production of agricultural fertilizers

Environmental Pollution Impact on Plants: Survival Strategies under Challenging Conditions. Tariq Aftab, PhD & Khalid Rehman Hakeem, PhD (Eds.)

due to the composition of high nutrient and organic matter. The use of sewage sludge as a means of fertilization in the agricultural system seems to be the best possible way for its proper disposal. Due to the blind jump in need of food and feed, a risk exists for shortage supply of fertilizer, and impact on the economy has increased manyfolds. Researches established the methods of utilization of sewage sludge for fertilizer materials manufacturing. The application of sewage sludge in agriculture is attractive from the environment as well as economic points of view. However, the presence of toxic material constrains the use of sewage sludge that may represent a significant hazardous impact on the groundwater, environment, and human health due to the presence of hydrophilic compounds. This chapter will describe the natural method of attenuation for detoxification of sewage slug to reduce its toxicity and use as fertilizer to achieve fertilizer security for potential agriculture production.

10.1 INTRODUCTION

Urban residues are increasing year after year worldwide whose management is necessary to sustain a good environment.[1] During the process of wastewater treatment, a large quantity of sewage sludge is accumulated and it is the most useless component of the total waste.[2] Usually, in sewage sludge, significant quantity of organic matter is present that enhances the fertility of the soil. Hence, sewage sludge application into the agricultural system is an effective alternative for its disposal.[3] Hence the wide variety of sewage sludge contains undesirable inorganic and organic contaminants, in an agricultural fertilizer, the valorization is shown due to the pathogenic microorganisms.[4,5]

Almost all countries are trying to recycle sewage sludge as an alternative source of nutrients for agricultural production. Despite that the negative effects of sewage sludge are more, thus needed to prevent the environment from its harmful effects. Significant part of polycyclic aromatic hydrocarbons, heavy metals, linear alkylbenzene sulphonates, halogenated organic compounds, as well as pathogens are present in sewage sludge. Each country must follow the legislation that describes how to treat and disposal off the sewage sludge in the soil, that is, Regulation 40 CFR Part 503 in USA, Resolution 375 of CONAMA in Brazil. Directive 86/278/EEC in Europe.

Among environmental contaminants, pollutants and endocrine disruptors are causing apprehension health issues due to their high biological activities and their role in disturbing the function of targeted living cells. Endocrine disruptors are of special concern of environmental pollution caused by sewage sludge, as they affect the functioning of the endocrine system.[6]

Dioxins are important class of compounds that are present in sewage sludge that activates to aryl hydrocarbon receptor. Activation of aryl hydrocarbon receptor is the starting part of the metabolic chain that carried the toxic effects of harmful pollutants like co-planar PCBs and benzopyrenes 2,3,7,8-tetrachlorodibenzo-(p)-dioxin.[7] These effects include endocrine disruption, immune dysfunction, reproductive toxicity, cancer in vertebrates, and developmental defects.[8] The interaction of receptors with pollutants can be identified through a number of single-cell-based assays. Yeast-based bioassays are common to be used, in which yeast strains were modified genetically to produce the vertebrate signaling response to either AhR/ER-ligands. These yeast-based assays are the common tools to find ER- and AhR-binding activities of the samples.[9,10] The aquatic life is also under concern due to the effects of sewage sludge that considers the source of soil and contamination, especially leaching and runoff of contaminants, potentially harming the aquatic life.[11]

Therefore, decontamination step is necessary for sustainable sewage sludge management to make better more before disposal into the environment.[12,13] Among various methods suggested for the detoxification of sewage sludge, natural attenuation considered an environment-friendly, cost-effective, and efficient practice. However, its efficiency to remove some highly bioactive contaminants as well as pollutants is yet need to be examined.[14,15] Keeping in view, the potential of sewage sludge carrying toxic substances at different periods of natural attenuation, this study is aimed to comprehensively analyze the effectiveness of different practices and techniques mitigating the toxic effects of sewage sludge and to evaluate the potential of sewage sludge application as a fertilizer in the agricultural system.

10.1.1 ORGANIC WASTE MANAGEMENT AND CIRCULAR ECONOMY

Numerous studies have described the importance and useful benefits derived from treated organic wastes,[16] for example, industrial solid wastes,[17] food wastes[18], and municipal wastes (sewage sludge and solid waste).[19,20]

One of the best ways to manage organic waste including municipal wastes is to reuse it for possible benefits to circulate the economy. One of the alternative uses of sewage sludge is its detoxification and utilization as a fertilizer for agricultural production.[21] Sewage sludge is the result of different types of municipal wastewater treatments and is composed of organic and inorganic contents present in the form of dissolved or suspended particles.[22]

Globally, the use of treated wastewater for agriculture production to replenish the nutrient status of the soil is increasing. In the current era, especially in European countries, the use of sewage sludge as an alternative fertilizer is growing rapidly to replace the cost of chemical fertilizers to attain the economic sustainability. Nowadays, to uplift the economy, countries are generating sewage sludge from wastewater treatment and using it as an alternative in many industries as a raw material. The increasing trend in sewage sludge production is directly linked with the government planning to reuse wastewater, the capacity of a treatment plant, awareness, the introduction of new and advanced treatment techniques, and developing new facilities for reuse.[23] According to a report, in 2020, the sewage sludge production from the treatment of wastewater increased in the Baltic region (14.4%) as compared with 2010. However, the production was found up by 82.3% in Poland, 40% in Belarus, and about 11.1% in Russia. However, the production was found at the same level in Finland, Denmark, Lithuania, Germany, Estonia, and Sweden as in 2010. The reuse of sewage sludge after detoxification of wastewater for the extraction of phosphorus has increased from 25% to 35%.[24] From the last few years, the concept of land application of sewage sludge as a fertilizer increased especially in Europe, and now this trend is also becoming more popular all over the world to cope with the chemical fertilizers to agricultural sustainability as well as powerful economy. Recently, land disposal and storage of sewage sludge is continuously decreasing, and replenishing the land with SS after detoxification is beneficial. The reuse of sewage sludge reduces the pressure on the chemical fertilizer industry but the lack of public awareness and acceptance, its use as an alternative to chemical fertilizer is not common. In the last few decades, researchers giving attention to the recycling and reuse of sewage sludge to increase the environmental and economic sustainability.[25]

10.2 EXPERIMENTAL METHODS AND MATERIALS

10.2.1 CHEMICAL CHARACTERIZATION OF SEWAGE SLUDGE

Land use of sewage sludge as a soil conditioner and fertilizer is a remarkable step unless it is free of toxic substances. Due to the different composition of sewage sludge produced by various methods of treatments, the composition of the sewage sludge should be examined, and on these composition basis, the sewage sludge should be utilized for land use.[26] Characterization in the case of sewage sludge importantly depends on the processes and methods involved in the treatment of wastewater and sewage sludge. In general, sewage sludge consists of organic matter contents, macro, and micronutrients, a number of harmful and toxic contaminants, including pathogens, organic pollutants, and heavy metals.[27] Nutrients and organic matter-rich nature of sewage sludge serves as an alternative soil beneficial conditioner to improve the soil fertility.[28] The chemical composition of sewage sludge is mentioned in Table 10.1, which is reported by the European Directorate and Research & Development Authority, Spain. At the same pace, the chemical composition of sewage sludge in Thialand, Morocco, and India are also reported in Table 10.2. The treatment system not only received domestic sewage sludge but also all the other sludges, including industrial, municipal, and runoff water. Therefore, sewage sludge is composed of various types of harmful materials (salts, detergents, hormone disrupter, agrochemicals residue, and pathogens) in addition to organic matter and nutrients. According to a survey report conducted by the US-EPA (United State Environmental Protection Agency), sewage sludge additionally consisted of volatile compounds, benzene, organic chemicals, furans and dioxins, sorbed compounds, hydrophobic chemical, and bioaccumulative toxic elements. According to a report published by US-EPA in 2003, sewage sludge contained almost 803 types of different chemicals, and most of them were found to be toxic for human health, and ecological risks are associated with it.

TABLE 10.1 Chemical Characterization of Soil Fertilized with Sewage Sludge as a Fertilizer and Legal limitation Established by European Directorate and Research & Development Authority Spain on Land Use of Sewage Sludge.

Properties	Unit	Range
pH	–	7.47
Organic matter	%	20.5
Nitrogen	%	3.5
Phosphorus	%	1.8
Potassium	%	0.4
Calcium	%	2.72
Magnesium	%	0.83
Sodium	%	0.16
Iron	%	1.8
Zinc	ppm	1248.6
Copper	ppm	143
Manganese	ppm	6.2
Cadmium	ppm	0.71
Chromium	ppm	38.34
Lead	ppm	83.9

European Directorate 86/278 and Spain R.D. 1310/1990.

TABLE 10.2 Chemical Composition Ranges of Sewage Sludge Samples Collected from Spain, Thailand, Morocco and India respectively.[31–34]

Properties	Unit	Range
pH	–	6.8–8.6
Organic matter	%	19.7–43.3
Total nitrogen	%	2.4–3.44
Total phosphorus	%	1–03–1.4
Exchangeable Calcium	%	0.8332
Exchangeable Potassium	%	0.0870
Zinc	ppm	444–1901
Manganese	ppm	400–2621
Copper	ppm	173–702
Cadmium	ppm	1–1.23

In addition, lots of new chemicals are introduced into commerce, so there is a need for a recent survey to assess their risk concentration and characterization. These include pharmaceuticals, chlorinated and without chlorinated paraffin, odorants, surfactant, flame retardants, polycyclic and nitro musk, and all the other chemicals that are used for treating the wastewater sludge. These chemicals should be separately characterized for future assessment of risk.

Chemical characterization of water taken from eight different locations of Indiana state, USA showed that the composition of sewage sludge over 2 years contained almost 1–4% inorganic carbon and 50% organic matter. It was also found that the organic N and inorganic P formed the major content of total N and P, respectively in sewage sludge. It was also found that alkaline earth metals (Mg and Ca), organic P and N, and inorganic C and N were present constantly during the sampling period. Similarly, it was found that the inorganic N and organic K and P were gently varied throughout the period. However, significant variations were noted in heavy metals (Pb, Ni, Zn, Cu, and Cd) during this period.[29] Analysis of sewage sludge from Calcutta, India as a fertilizer assessment showed that its pH varied from 7 to 7.5, and it contained a higher concentration of salt in winter season and cation exchange capacity (CEC) was higher in monsoon season. The sludge was found as a rich source of available nitrogen and organic carbon.[30]

10.2.2 THE PROCESS INVOLVED IN DETOXIFICATION OF SEWAGE SLUDGE BY NATURAL ATTENUATION

Natural attenuation is an eco-friendly process that involves a transformation of harmful contaminants into a less harmful product by a natural process, in other words detoxification. Detoxification of sewage sludge (SS) sample can be carried out through anaerobic conditions, and dewatering of SS is done by centrifugation of the sample. SS samples should be taken from different depths of the sludge pile and composite sludge samples were prepared and then packed in a plastic bag having 0.5 mm holes in diameter with 1 cm spacing between holes for perforation. A hole is dug in a contamination-free environment by following the protocol given by the Mazzeo group of scientists and a sewage sludge sample packed bag should be buried in it.[35] The buried samples should be kept in such an environment for a long time for maximum detoxification by natural attenuation. Previous studies showed that the minimum period for natural attenuation is 2 months and the maximum period is 12 months. During this period, alteration will occur in the samples which reduced the toxicity of the SS samples. Maximum temperature ensures the supreme detoxification of the sewage sludge sample. The various analytical techniques can be used to determine the toxicity concentration of the final detoxified samples, but the liquid and gas chromatography techniques are the two most important techniques which were considered best to determine the pollutant and toxicity capacity of the sewage sludge sample. The maximum concentration of m- and p-cresol and minimum concentration of furans and dioxins of sewage sludge samples should be recorded periodically.[35] The schematic diagram of natural attenuation process is presented in Figure 10.1 which is convenient to understand. After the natural attenuation period, the sewage sludge sample should be freeze-dried and sieved through an 80 mm sieve. Now, SS samples should be stored in dark with minimum temperature for the next process as organic amendments and mineral fertilizer source. Furthermore, it can be evaluated for a hydrosoluble substance-using protocol (NBR10.006 ABNT, 2004). Similarly, the microorganism diversity can be carried out through dilution-plate technique containing blood agar and amended with MacConkey agar, Sabouraud agar, and SS agar. colony-forming units (CFU) will be carried out for microorganism quantification.

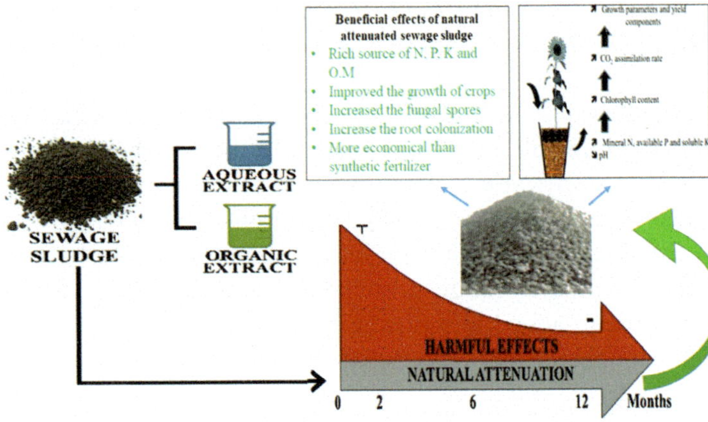

FIGURE 10.1 Schematic diagram of process involved in natural attenuation of sewage sludge.

10.3 RESULTS AND DISCUSSION

10.3.1 LAND APPLICATION OF SS-BASED FERTILIZER

Significant sewage sludge production demands some specific additional steps for its safe disposal and utilization to reduce the associated dangerous effects of heavy metals and other micropollutants on human health and its surrounding environment. Keeping in view the nutrients value, sewage sludge can be used to replenish the soil nutrients status, but on the parallel way to nutritive importance, a non-monitored concentration of heavy metals and other toxic compounds are also reported.[36] Thus, the best application of detoxified sewage sludge as organic fertilizers for agricultural production not only reduces the environmental contamination but also lowers the demand for chemical fertilizer application. As fertilizers are recovered from biowaste as well as from their by-products, they have enormous environmental and economic benefits, but in the end, they have some limitations with direct application like the presence of toxic compounds.

Application of properly treated sewage sludge on agricultural land is the only best and alternative option to manage the sewage sludge. Recycling options for circular economy and EU regulations boost the application of sewage sludge for agriculture production, because it is considered the low-cost alternative to restore the nutrient and organic matter of the land. Many scientists reported that the detoxification of sewage sludge by natural attenuation is the most suitable method to produce environmentally friendly fertilizer from wastewater. It was also reported that the natural attenuated detoxified fertilizer contains more nutrients and organic matter contents as compared with fertilizer produced by other methods, and also it significantly improves the fertility level of the soil. Land application of detoxified sewage sludge as fertilizer for crops and vegetable production is a cost-effective alternative to enhance soil productivity. Application of detoxified sewage sludge collected from small and large municipal station as a fertilizer to low-quality soil improved the performance of plants (Willow) and soil characterization.[37]

In Europe (EU), especially the Mediterranean area of EU, due to intensive cultivation and high temperature, the organic matter content of soil is reducing consistently.[38] Every year about 40% of the detoxified sewage sludge is added to soils which improves the soil's organic matter content. All the developed countries are focusing on sewage sludge as an amendment to increase the organic matter contents in their soil, and the application percentage varies among countries. Almost half or more than half of all the produced sewage sludge treated in some European countries, such as UK, Belgium, Ireland, Denmark, Spain, and France is used as an amendment or as fertilizer in agriculture. Similarly, less than 5% of the total sludge produced is used as organic fertilizer in agricultural soil in Netherland, Greece, Slovenia, Romania, and Slovakia.[39]

The nutrient and organic matter status of the sludge vary greatly depending upon the source, composition, and other environmental factors. The application of SS-based fertilizer is considered more important concerning phosphorus recycling sources which are limited in most of the soil.[40] As the global population is increasing steadily, this leads to an increase in food demand, putting pressure on soil and reducing the nutrients from it. Therefore, it is necessary to find an alternative resource to fulfill the nutrients requirement of soil for maximum crop production, while phosphorus reserves are expected to be diminishing in the upcoming 50–100 years. According to a study, sewage sludge can cover almost 20–21% of the phosphorus demand. Land application of SS-based fertilizer enhanced the activities of microorganisms and increased the enzyme production. It was recorded that catalase, protease, and phosphatase activities are increased by the application of SS-fertilizer, but no changes were recorded in the activity of dehydrogenase..

A 3-year-research study on the microbiological activities of soil by the application of SS-fertilizer showed that microbial activities increased and mostly varied from soil to soil type and the highest activities of enzymes were recorded at the application of 800 m^3/ha except for dehydrogenasic activity.[41] It was also recorded that the soil physical properties are also influenced by the application of SS fertilizer, but it is also reported that in many cases, changing trend varies among the soils. For example, the field capacity of Neurosis soil increased 41% but decreased about 9% in most Strassfeld soils.

In several cases, changes in the physical properties resulted from the sewage sludge application have been recorded. It can be concluded that soil type is an important factor in the application of SS fertilizer.[42]

10.3.2 SOIL PROPERTIES AFFECTED BY THE APPLICATION OF SEWAGE SLUDGE

Land application of sewage sludge becoming more popular because it plays an important role in the soil to enhance soil productivity due to the presence of valuable recycling components, such as organic matter, nitrogen, phosphorus, and many other micronutrients.[43] Soil application of sewage sludge may also reduce the need for inorganic fertilizers for agricultural production.[29] Sludge is the organic fertilizer that increased the fertility of the soil for a long time[44] and changes in biophysicochemical properties of soil are affected gradually which are mentioned in Table 10.3. Otherwise, when untreated sludge bears high concentrations of metals and other toxic constituents, it deteriorates the properties of soil.[45]

10.3.2.1 EFFECT OF SS-FERTILIZER ON SOIL PHYSICAL PROPERTIES

Soil physical properties are considered most important for crop production and their role in soil sustainability cannot be negotiated. All the soil properties depend on each other in some way. The most important physical properties of soil include soil structure, pore size, soil aggregation, stability, soil compaction, density, hydrolytic conductivity, nutrient, and water-holding capacity. The water movement, exchange of oxygen, and process of nutrients absorption, root formation are highly influenced by the physical properties of soil. Sewage sludge used as organic fertilizer or amendment should be considered as the best strategy to reuse the wastewater. The use of sewage sludge as a soil amendment is considered a sustainable and alternate strategy to improve the physical properties. Most importantly, it improves the soil organic contents that improve the soil properties especially porosity, bulk density, and water-holding capacity. A study of sewage sludge at rate of 0.5% on soil water-holding, aggregate stability, and hydraulic conductivity showed that digested sludge as well as raw sludge enhanced the total soil water-holding capacity with the highest increase in the raw sludge amendment.[45] In sludge, the higher organic matter may increase the aggregate stability and may decrease the bulk density. Many other structural properties of soil are directly linked with the bulk density of soil and can be used as an index to measure the changes in soil structure. The improvements in physical properties of soil increased the water-retention capacity by promoting the greatest water retention in sludge-amended soils.[46] Another study showed that the application of sewage sludge improves the physical properties in calcareous and sandy soils. They found that the bulk density decreased and organic matter, water saturation percentages are increased by the application SS organic fertilizer.[47] Sludge application improves the physical properties of soil through enhancing water retention, aeration, aggregation, permeability, infiltration, and decreasing the crusting on the soil surface. Johnson reported that no significant differential effect of sewage sludge and farm yard manure (FYM) on soil physical properties and organic matter.[48] The application of digested sludge to soil enhances the water saturation and holding capacity more as compared with FYM.[49] It is reported that application of 50,000–60,000 kg sewage sludge increase 1.6%, 0.33–15 bars, 12–18.5% organic matter, soil tension, and field capacity, respectively.[45,50,51] Continued application of SS for 9 years to soil increases about 35% available water capacity and 16–33% soil aggregation stability.[45,51] The reported evidence showed that the positive effect of sewage sludge application is likewise the conventional organic fertilizers, but more than inorganic fertilizers in improving soil physical properties and crop yield.[52] Detoxification of sewage sludge adds up in the beneficial effect on soil nutrients status and crop production. Thus, wise management of sewage sludge is a resource, not a burden.

10.3.2.2 EFFECT OF SS-FERTILIZER ON SOIL CHEMICAL PROPERTIES

Soil chemical properties are also an important indicators of soil health because these properties are directly interconnected with soil nutrients availability and plant growth. Soil chemical properties are directly linked with the capacity of soil to release or store the nutrients and availability to plant from the soil solution. Importantly, soil structure is significantly affected by the chemical properties of soil. Therefore, it is necessary to understand the importance of the chemistry of soil in relation to soil amendment.[53] Most important chemical characteristics of soil include pH, Eh, EC, CEC, total carbon and nitrogen, heavy metal concentration, and macro and micronutrients. Nutrient availability varied remarkably according to soil chemical properties.

Sewage sludge amendment to soil significantly changes the chemical properties of soils.[47] It was also suggested that chemical properties especially soluble cations and anions, EC, available micronutrients, phosphorus, and heavy metals were increased along with HCO-3 and pH decreases with an increase in the rate of sewage sludge amendment. It has also been reported that in soils, the pH is increased with the application of municipal sewage sludge.[54] Sludge decomposition in soil produces acid and calcium carbonate resulted in change in soil pH.[29] Sewage sludge having an abundance of trace metal plays a crucial role in the consideration of soil pH.

The bioavailability of the sludge-borne metals is influenced by the microorganism and the chemical properties of soil, such as redox potential (Eh), pH, sesquioxide content, and organic matter,[55] as well as application rate of sludge are also affected.[56] A difference in pH and CEC was also recorded between surface and subsurface soil amendment with sewage sludge. The surface soil had lower pH (7.3) and higher CEC (18.4–22.8 cmolc kg^{-1}), meanwhile, the subsurface had higher pH (7.5) and low CEC (15.1–19.1 cmolc kg^{-1}).

Similarly, subsurface soil contains a lower amount of organic carbon (1.16%) than the surface (1.31%) on sewage amendment. Available phosphorus and nitrogen were at moderate to high levels.[70] It was reported that the pH of soil is also affected when nickel, copper, and zinc become available in soil either from sludge or soil bonded due to the release of supernatant liquid high metal concentration as pH decreased below the threshold value, which is 4.5 for Cu, 6.3 for Ni, and 5.8 for Zn-loaded sludge.[57,58] In another study, it has been also recorded that the CEC of the soil significantly increased on the application of sewage sludge having high salt contents. An increase in heavy metals, N-NH4+ content, and total soluble salts was observed on application to soil.[59–61]

A study was conducted by Hernández et al.[62] to evaluate the effect of sewage sludge on the availability of soil macronutrients (nitrogen, phosphorus, and potassium) as well as heavy metals (Zn, Cu, Fe, Mn, Cd, Ni, Cr, and Pb). Total nitrogen and extractable potassium were monitored and nitrogen content was high in the sludge-amended soil, whereas the extractable potassium remained unchanged. The extractability of Zn, Cu, Fe, Mn, and Pb increased due to the application of sludge as compared with the control. Relatively cation exchange capacity was reported high due to the application of high rates of sewage sludge, which helps to hold the nutrients due to additional cation binding sites within the rooting zone.[63] Thus, the use of sewage sludge as a fertilizer is a good source to increase the nutrient availability of soil and may improve crop production.

10.3.2.3 EFFECT OF SS-FERTILIZER ON SOIL BIOLOGICAL ACTIVITIES

Soil biodiversity is a very important component of soil and cannot be ignored when directed with soil properties, soil and plant functions, and agricultural crop production. A number of biological activities are responsible for various types of transformations in soil. All the reactions that involve the movement and recycling of nutrients are notably influenced by the biological activities of soil. Thus, it is very important to estimate the biological activities of soil microbiomes in terms of system functions that are directly related to environmental sustainability. Sewage sludge for agricultural utilization is gaining importance due to the high contents of plant nutrients and organic matter, because it can improve the soil biological activities that result in a surge of soil productivity. The microbial diversity of sewage sludge is highly affected by the composition of sewage sludge especially by the concentration of heavy metals contaminations and the source of sewage sludge. The impact of heavy metals on microbial activities of soil is getting emphasis and has been reported many times in the literature.[64] The enzymatic activities in soil are indirectly influenced by the heavy metals concentration in sewage sludge.[65] It is also reported that the treated sewage sludge contained a low concentration of heavy metals that increases microbial activities, soil organic matter, enzymatic activities, and soil carbon biomass, but these activities have been decreased when high-level heavy metal-containing sewage sludge was applied to soil.[66] Meanwhile, it is concluded that the microbial diversity is highly affected by the high concentrated sewage sludge. Application of a recommended dose of sewage sludge increases the microbial activities by tying up the heavy metals in soil pores and making them unavailable at low concentrations.[67] As enzymatic activities play a major role to improve the soil fertility, metals present in the sewage sludge hinder the enzymatic activities.[65] The effect of sewage sludge may be used as an indicator of soil pollution for biological activity.[67] The amendment of sewage sludge increases the microbial activity, the soil enzyme activities as well as soil respiration.[68]

Incorporation of sewage sludge into soil stimulates enzymatic activities, especially lowers the dehydrogenase activity. The reason behind the stimulation of enzymic activities is the presence of intra- and extracellular enzymes, thus it may increase the microbial activities of soil. Author also reported that phosphatase alkaline, arylsulfatase, and urease activities were higher in sewage sludge-amended soil as compared with unamended soil.[69] Organic carbon was also found 2.5 times more and microbial biomass monitored was high as compared with unamended soil.[70] Microorganism also plays a vital role in the decomposition of sewage sludge in the soil, and it is reported that mineralization rate of organic N and C increases which stimulates the activities of bacterial population and protease activity the first 3 days of application and afterward declined quickly.[71]

TABLE 10.3 The Amendments of Soil (Physical, Chemical and Biological) Properties Through the Sewage Sludge.

Properties	Effect	References
Physical		
Soil aggregate stability	Increase	46
Water-holding capacity	Increase	72
Bulk density	Decrease	46
Erosion	Decrease	46
Porosity	Increase	72
Chemical		
Humus content	Increase	27
pH	Decrease	45, 73
Toxic elements	Increase	27
Electrical conductance	Increase	74
Soil organic carbon	Increase	75
Cation exchange capacity	Increase	72
N and P	Increase	43
Biological		
Pathogenic organisms	Increase	27
Aerobic bacteria	Increase	72
Yeast population	Increase	27

10.3.2.4 IMPACT OF SS FERTILIZER ON AMOUNT OF N, P, AND K

Sewage sludge-based fertilizer application for crop production either directly or blended with other nutrients material like compost is a well-heated discussion for fertilizer requirements due to its rich sources of organic matter and essential nutrients especially nitrogen (N) and phosphorus (P).[37,76] Sometimes, it also contains other nutrients, such as Mg, Ca, and K, but the concentration of K is low in it.[70] When it is applied to soil, it increases the organic matter in soil which indirectly improves the soil physicochemical and microbial properties.[77–79] In long term, it improves the soil nutrients status and fertility level.[80] A higher quantity of organic matter and increased level of nutrients in soil improve the soil's organic carbon, microbial, and enzyme activities.[70] There are many factors that influence the bioavailability of nutrients from SS fertilizer to soil including the biological and chemical nature of the waste, composition, and processes of detoxification of sewage sludge.[81–83] It is very necessary to consider the SS fertilizer to soil ratio before application to soil.[70] The nutrient contents varied and depend upon the treatment process and availability of water.[78] Inorganic phosphorus and organic nitrogen are the major part of total phosphorus and total nitrogen in SS fertilizer.[70] It is reported that the percentage of three main nutrients in sewage sludge after treatment of 1500 wastewater plant is potassium, 0.27%; total phosphorus, 2.68%; and total nitrogen, 9.76% of total solids.[84]

The availability of N and P from sewage sludge occurs through the mineralization process of the organic matter, and this process of mineralization depends upon the crop cultivated previously and soil type which varied from climate to climate. Mineralization increases the availability of nitrogen in the soil and decreases the requirement of fertilizers for crop production. In the first year of application, it makes the nutrients available up to 50%. The availability of phosphorus from SS varied from 40 to 80% and is considered as the renewable reservoir of P contents.[79] The application of SS fertilizer also influences the availability of nutrients in plant as well as in the soil. It is reported in the literature that the application of sewage sludge increases the number of nutrients (N, P, and K) in different parts of soybean, alfalfa, and faba bean.[85] Another study concluded that the application of sewage sludge as fertilizer or organic amendment increases the concentration of macro (N, P, K, Mg and Ca) and micronutrients (Mn, Fe, Zn, and Cu) in cotton, corn, barley, and sugarbeet. Thus, the nutritional composition of sewage sludge can be considered a renewable resource to overcome the nutrients losses from the soil and maintain the health of the soil for maximum crop productivity.

10.3.2.5 EFFECT OF SS FERTILIZER ON GROWTH, YIELD, AND ACCUMULATION OF HEAVY METAL IN PLANTS

The quality criteria of sewage sludge for agriculture use are indicated by the concentration of heavy metals, a number of pathogens, bacterial colonies, and toxic elements. The regulatory authority of the US also recommended that the permissible rate of sewage sludge is 2–3 years for agriculture purposes and reclamation of nonagricultural land while landfilling is forbidden due to leaching reasons and related environmental issues.[86] Large cities are continuously producing sludge and its incineration is a costly and environmentally controversial way for its management. However, the agricultural use or reclamation purpose of sewage sludge is considered the best alternative strategy for disposed of, and recycling method of nutrients and organic matter. On average, sludges from municipal wastes contain nitrogen 2.6%, phosphorus 1.83% as a dry matter.[87] Assuming the forecasted sludge production amount of P would be sufficient to replace mineral P fertilizers at the area of 618 thousand hectares of arable land (6.2% of total arable land) since the average application rate of P in Poland is 20.9 kg ha^{-1}. Several controversies are always there on the wider use of sewage sludge either as a soil reclamation source or as a fertilizer in agriculture. These are due to inadequate information regarding environmental hazards or benefits of the use of sewage sludge. Such lack of information arises regarding soil functions, soil biodiversity, nutrient cycling, and a cycle of organic matter.

On a sustainability basis, the management of sewage sludge into a valuable alternative resource is important instead of considering it as a burden. Agriculture application drags our immediate attention to the usage of SS-based fertilizer for crop production all over the world. In the European Union (36–37%), China (44%), US (60–65%), and in Egypt (85%) of the total produced sewage sludge has been used for agriculture production.[88–90] Agricultural use of SS has a couple of positive effects on the soil as it improves the soil health and enhances the soil productivity through increasing N, P, and many other micronutrients. Sometimes, it also improves the physiochemical and biological activities of soil that ultimately improve the crop production.[70,90]

It was estimated that almost 13 million tons of sewage sludge were produced in E27 countries by the end of the 2020 year.[33] It is very challenging to manage this huge sewage sludge due to its pollutant nature including toxic compounds and heavy metals.[79,91] It is also a rich source of pathogens and microorganisms that cause many environmental issues.[92] Thus, the land use of sewage sludge after natural attenuation for agriculture production would be a good alternative use of this waste material. All the developed countries are promoting the land application of detoxified sewage sludge as a fertilizer for managing and recycling the resources efficiently. Recycling of wastes is the efficient and eco-friendly strategy for smooth growth of country economy circle. Similarly, developing countries should also invest in sewage sludge production to reduce waste accumulation, and it should be used in agriculture as it decreases the cost and dependence on natural resources which are near to end. Hence, ultimately, it will stimulate the economic growth. Keeping this point in view, sewage sludge can be detoxified through different methods and used as alternative sources of mineral nutrients and organic matter.

The plant cycle from germination to maturity is significantly influenced by the application of SS-based fertilization. A study of sunflower production under the fertilization of SS-based fertilizer at different levels showed that germination and seedlings are significantly affected by its application as compared with control. Plant height, root and aerial part, dry biomass, growth, and yield attributes were significantly increased as compared with control (unamended) treatment. It was also recorded that SS increases transpiration rate, stomatal conductance, and net assimilation rate. Similarly, grain yield was recorded 7–8 times more than unamended.[90]

10.3.2.6 USE OF SS FERTILIZER IN ENHANCING CROP PRODUCTION

A profound amount of nitrogen, phosphorus, and organic matter content makes sewage sludge a potential fertilizer that can be applied to agricultural soils.[94] It has been reported that 1 ton of dried sludge contains organic matter, nitrogen, phosphorus, and a variety of soluble salts up to 200, 6, 8, and 10 kg respectively.[95] As sewage sludge is a potential source of organic matter and other plant essential nutrients, sewage sludge has been reported as a vital source of fertilization.

In literature, there are numerous examples of the application of SS fertilizers to agronomic crops, forest trees, horticulture crops, and plants grown on reclaimed lands.[96,97] Furthermore, the plant cycle from germination to maturity significantly affected by the application of SS-based fertilization. A positive response has been shown by papaya plants in terms of increase in their biomass with the application of sewage sludge. These authors also found an agronomic efficiency index of 198% for the application of 60 t ha^{-1} of sewage sludge compared with chemical fertilization.[98] The residual effect of sewage sludge applied at the rate of 30 t ha^{-1} enhanced the maize yield by 77% from the first cycle which was 22% more efficient than the application of mineral fertilizers, which points out the importance of using sewage sludge as a residual source of nutrients.[99] As it has been evident from many studies that SS fertilizers are rich sources of plant nutrients and enhance plant growth and yield, not only the direct application of SS fertilizers are effective, but the residual effect of SS fertilizers also had a profound effect in enhancing plant growth and yield.[99,100] Yield enhancement of wheat crop with the application of sewage sludge fertilizers was recorded over the year. Similar results were reported by Jamil et al.[101] Further examples of increased crop production with the direct application of SS fertilizer as well as its residual effect are depicted in Table 10.4.

TABLE 10.4 Examples of Crop Production Enhanced with SS Fertilizer.

Crop name	Scientific name	Yield status	Reference
Wheat	*Triticum aestivum*	Increased	100, 101
Corn/ Maize	*Zea mays*	Increased	41, 43, 102, 103
Barley	*Hordeum vulgare*	Increased	41
Faba bean	*Vicia faba*	Increased	103
Sunflower	*Helianthus*	Increased	104

10.3.3 ENVIRONMENTAL RISK OF NON-DETOXIFIED SEWAGE SLUDGE APPLICATION

Application of sewage sludge as a fertilizer or soil amendment is a common practice all over the world because it contains high concentrations of macro and micronutrients as well as other organic compounds that improve plant growth significantly. However, continuous use of sewage sludge for agricultural purposes may cause different types of contaminations, such as toxic compounds (organic and inorganic), heavy metals accumulation, and harmful pathogen accumulation.[105-107] Continuous accumulation of heavy metals in soil may cause a harmful and dangerous effect on the human food system through the consumption of crops and vegetables. The concentrations of metals in the SS depend on various factors, such as (1) sewage treatment processes, (2) sewage origin, and (3) sludge treatment processes.[56] The low amount of heavy metals and their restricted mobility with soil depth suggest the possible use of this waste material for agricultural purposes without imposing a toxic effect on plants.

There are a wide number of other dangerous substances part of sewage sludge, such as polycyclic hydrocarbons, aromatic hydrocarbons, biphenyls, ethers, detergents, hormones, steroids, and pharmaceuticals compounds found in it.[70,108,109] These toxic compounds cause the adsorption of heavy metals in soil. The adsorption process of heavy metals in soil is influenced by the organic matter, pH, CEC, and mobility of metals within sewage sludge.[70,110] It is also reported that heavy metal contents will be lower depending upon the rate of application of sewage sludge into the soil.

In European Union countries, the application of sewage sludge for agriculture purpose is under the control of Directorate directive number 86/278/EEC which provides the legal limitation for different heavy metals present in the sewage sludge used for agriculture. But, many countries are still not considering the environmental prospect of its application and deteriorating the soil health.

The number of organic contaminants is increasing with time, there is a need for proper planning and advancement in the sector of sewage sludge management. In the future, it needs to initiate monitoring programs and technological changes in sewage sludge management to control the concentration of organic contaminant within the permissible level[111] To reduce the adverse effect and to improve the sewage sludge management considering the nutrients as well as organic matter recycling to the soil, sewage sludge can be used as a component in chemical fertilizers production, which can help to meet the fertilizer requirement and may reduce the contaminants. Fertilizers that are derived from biological waste are usually cost-effective and eco-friendly, but they have no free access to European Union's market. This is the problem associated with various legislation in the members' state.[112] The European Commission proposed a new regulation to expedite the utilization of organic-based fertilizers. The commission also proposed to allow the organomineral fertilizers to be competitive with mineral fertilizers through the harmonization of legislation. Besides, the commission updates the requirements of the fertilizer, simplifies the mechanisms and procedures for bringing fertilizers to the market.

10.4 CONCLUSIONS

After the brief and comprehensive review of sewage sludge as organic fertilizer through the natural attenuation process, now it is concluded that municipal solid waste as well as sewage sludge affected the biological and physiochemical properties of the soil. The positive effect on the biological properties of soil could be attributed to the organic amendments that stimulate the microbial population or it may be attributed to the addition of microbial cells to the soil along with organic amendments, which can mitigate the adverse effect of salinity and heavy metal contaminated sites. Numerous studies have revealed that the application of sewage sludge to the agricultural system enhances nutrients available to plants, as sewage sludge is a rich source of soil macronutrients, such as nitrogen, phosphorus, and potassium as well as some micronutrients. It was also demonstrated by many studies that the application of sewage sludge to croplands enhanced the crop yield and growth attributes. Proper management of hazardous effects of SS and their application into the agricultural system will the fire-up the economy of a country by the win–win situation as it will lessen the hazardous effects of SS and should be used as valuable fertilizer, therefore, there is a need across the board of legislation to legalize and remove the obstacles in the way of using sewage sludge fertilizers into the agricultural system, and to introduce them as a commercial fertilizer into markets worldwide after complete treatments.

KEYWORDS

- **sewage sludge**
- **organic fertilizers**
- **soil properties**
- **environmental risks**
- **crop production**

REFERENCES

1. Zotos, G.; Karagiannidis, A.; Zampetoglou, S.; Malamakis, A.; Antonopoulos, I. S.; Kontogianni, S.; Tchobanoglous, G. Developing a Holistic Strategy for Integrated Waste Management Within Municipal Planning: Challenges, Policies, Solutions and Perspectives for Hellenic Municipalities in the Zero-Waste, Low-Cost Direction. *Waste Manag.* **2009,** *29*, 1686–1692.
2. Pathak, A.; Dastidar, M. G.; Sreekrishnan, T. R. Bioleaching of Heavy Metals from Sewage Sludge: A Review. *J. Env. Manage.* **2009,** *90*, 2343–2353.
3. Zuloaga, O.; Prieto, A.; Ahmed, K.; Sarkar, S. K.; Bhattacharya, A.; Chatterjee, M.; Bhattacharya, B. D.; Satpathy, K. K.; Distribution of Polycyclic Aromatic Hydrocarbons in Recent Sediments of Sundarban Mangrove Wetland of India and Bangladesh: A Comparative Approach. *Envir. Earth Sci.* **2013,** *68*, 355–367.
4. Harrison, E. Z.; Oakes, S. R.; Hysell, M.; Hay, A. Organic Chemicals in Sewage Sludges. *Sci. Total Envir.* **2006,** *367*, 481–497.
5. Gibson, R. S. *Principles of Nutritional Assessment.* Oxford UniversityPress: USA; 2005.
6. Giudice, B. D.; Young, T. M. Mobilization of Endocrine-Disrupting Chemicals and Estrogenic Activity in Simulated Rainfall Runoff from Land-Applied Biosolids. *Envir. Toxic. Chem.* **2011,** *30*, 2220–2228.
7. Engwall, M.; Hjelm, K. Uptake of Dioxin-Like Compounds from Sewage Sludge into Various Plant Species—Assessment of Levels Using a Sensitive Bioassay. *Chemosphere* **2000,** *40*, 1189–1195.
8. Nebert, D. W.; Puga, A.; Vasiliou, V. Role of the Ah Receptor and the Dioxin-Inducible [Ah] Gene Battery in Toxicity, Cancer, and Signal Transduction. *Ann. N. Y. Acad. Sci.* **1993,** *685*, 624–640.

9. Noguerol, T. N.; Boronat, S.; Casado, M., Raldúa, D.; Barceló, D.; Piña, B. Evaluating the Interactions of Vertebrate Receptors with Persistent Pollutants and Antifouling Pesticides Using Recombinant Yeast Assays. *Anal. Bioanal. Chem.* **2006,** *385*, 1012–1019.
10. Noguerol, T. N.; Boronat, S.; Jarque, S.; Barceló, D.; Piña, B. Detection of Hormone Receptor Ligands in Yeast by Fluorogenic Methods. *Talanta* **2006,** *69*, 351–358.
11. Eriksen, S.; Aldunce, P.; Bahinipati, C. S.; Martins, R. D. A.; Molefe, J. I.; Nhemachena, C.; O'brien, K.; Olorunfemi, F.; Park, J.; Sygna, L.; Ulsrud, K. When Not Every Response to Climate Change is a Good One: Identifying Principles for Sustainable Adaptation. *Clim. Dev.* **2011,** *3*, 7–20.
12. Domene, X.; Alcañiz, J. M.; Andrés, P. Comparison of Solid-Phase and Eluate Assays to Gauge the Ecotoxicological Risk of Organic Wastes on Soil Organisms. *Environ. Pollut.* **2008,** *151* (3), 549–558.
13. Roig, N.; Sierra, J.; Martí, E.; Nadal, M.; Schuhmacher, M.; Domingo, J. L. Long-term Amendment of Spanish Soils with Sewage Sludge: Effects on Soil Functioning. *Agri. Ecosys. Environ.* **2012,** *158*, 41–48.
14. Bhupathiraju, V. K.; Krauter, P.; Holman, H. Y.; Conrad, M. E.; Daley, P. F.; Templeton, A. S.; Hunt, J. R.; Hernandez, M.; Alvarez-Cohen, L. Assessment of In-Situ Bioremediation at a Refinery Waste-Contaminated Site and an Aviation Gasoline Contaminated Site. *Biodegradation* **2002,** *13*, 79–90.
15. Robinson, B.; Green, S.; Mills, T.; Clothier, B.; van der Velde, M.; Laplane, R.; Fung, L.; Deurer, M.; Hurst, S.; Thayalakumaran, T.; van den Dijssel, C. Phytoremediation: Using Plants as Biopumps to Improve Degraded Environments. *Soil Res.* **2003,** *41*, 599–611.
16. Martínez-García, C.; Eliche-Quesada, D.; Pérez-Villarejo, L.; Iglesias-Godino, F. J.; Corpas-Iglesias, F. A. Sludge Valorization from Wastewater Treatment Plant to its Application on the Ceramic Industry. *J. Environ. Manag.* **2012,** *95*, 343–348.
17. Vieira, C. M. F.; Andrade, P. M.; Maciel, G. S.; Vernilli F. Jr.; Monteiro, S. N. Incorporation of Fine Steel Sludge Waste into Red Ceramic. *Mater. Sci. Eng. A* **2006,** *427*, 142–147.
18. Lin, C. S. K.; Pfaltzgraff, L. A.; Herrero-Davila, L.; Mubofu, E. B.; Abderrahim, S.; Clark, J. H.; Koutinas, A. A.; Kopsahelis, N.; Stamatelatou, K.; Dickson, F.; Luque, R.; Food Waste as a Valuable Resource for the Production of Chemicals, Materials and Fuels. Current Situation And Global Perspective. *Energy Environ. Sci.* **2013,** *6*, 426–464.
19. Menikpura, S. N. M.; Sang-Arun, J.; Bengtsson, M. Integrated Solid Waste Management: An Approach for Enhancing Climate Co-benefits Through Resource Recovery. *J. Clean. Prod.* **2013,** *58*, 34–42.
20. Cusidó, J. A.; Cremades, L. V. Environmental Effects of Using Clay Bricks Produced with Sewage Sludge: Leachability and Toxicity Studies. *Waste Manag.* **2012,** *32*, 1202–1208.
21. Eliche-Quesada, D.; Martínez-García, C.; Martínez-Cartas, M. L.; Cotes-Palomino, M. T.; Pérez-Villarejo, L.; Cruz-Pérez, N.; Corpas-Iglesias, F. A. The Use of Different Forms of Waste in the Manufacture of Ceramic Bricks. *Appl. Clay Sci.* **2011,** *52*, 270–276.
22. Przewrocki, P.; Kulczycka, J.; Wzorek, Z.; Kowalski, Z.; Gorazda, K.; Jodko, M. Risk Analysis of Sewage Sludge-Poland and EU Comparative Approach. *Pol. J. Environ. Stud.* **2004,** *13*, 237–244.
23. Lone, A. H.; Raverkar, K.; Pareek, N. In-Vitro Effects of Herbicides on Soil Microbial Communities. *Bioscan* **2014,** *9*, 11–16.
24. Werle, S.; Wilk, R. K. A Review of Methods for the Thermal Utilization of Sewage Sludge: The Polish perspective. *Renew. Energy* **2010,** *35*, 1914–1919.

25. Buratti, C.; Barbanera, M.; Testarmata, F.; Fantozzi, F.; Life Cycle Assessment of Organic Waste Management Strategies: An Italian Case Study. *J. Clean. Prod.* **2015,** *89*, 125–136.
26. Zitomer, D. H.; Speece, R. E. Sequential Environments for Enhanced Biotransformation of Aqueous Contaminants. *Environ. Sci. Techn.* **1993,** *27*, 226–244.
27. Kulling, D.; Stadelmann, F.; Herter, U. *Sewage Sludge–Fertilizer or Waste.* In UKWIR Conf. Brussels, 2001, pp. 9–11.
28. Logan, T. J.; Harrison, B. J. Physical Characteristics of Alkaline Stabilized Sewage Sludge (N-Viro Soil) and their Effects on Soil Physical Properties. *J. Environ. Qual.* **1995,** *24*, 153–164.
29. Sommers, L. E.; Nelson, D. W.; Yost, K. J. Variable Nature of Chemical Composition of Sewage Sludges. *J. Environ. Qual.* **1976,** *5*, 303–306.
30. Maiti, P. S.; Sah, K. D.; Gupta, S. K.; Banerjee, S. K. Evaluation of Sewage Sludge as a Source of Irrigation and Manure. *Eval. Sewage Sludge Source Irrigation Manure* **1992,** *40*, 168–172.
31. Parkpain, P.; Sirisukhodom, S.; Carbonell-Barrachina, A. A. Heavy Metals and Nutrients Chemistry in Sewage Sludge Amended Thai Soils. *J. Environ. S. Health A* **1998,** *33* (4), 573–597.
32. Lugo, M. D. L.; Martínez, G. A.; Rivera, L. E.; González-Vélez, A. Weed Incidence after Application of Sewage Sludge Compost. *J. Agri. Uni. Puerto Rico.* **2002,** *86*, 67–68.
33. Alis, H.; Bozkurt, M. A.; Oner, O. Z.; Dolay, K.; Turhan, A. N.; Uçar, A.; Inci, E. Aygun, E. Case Report: Acute Pancreatitis Caused by Postcholecystectomic Hemobilia. *BMC Gastroenterol.* **2010,** *10*, 1–4.
34. Nandakumar, K.; Ramamurthy, S.; Rajarajan, A.; Savarirnuthu, E. Suitability of Dindigul Town's Sewage Sludge for Field Application: Nutritional Perspective. *Pollut. Res.* **1998,** *17*, 61–63.
35. Mazzeo, D. E. C.; Fernandes, T. C. C.; Levy, C. E.; Fontanetti, C. S.; Marin-Morales, M. A.; Monitoring the Natural Attenuation of a Sewage Sludge Toxicity Using the Allium Cepa Test. *Ecol. Indic.* **2015,** *56*, 60–69.
36. Oleszczuk, P.; Malara, A.; Jośko, I.; Lesiuk, A. The Phytotoxicity Changes of Sewage Sludge-Amended Soils. *Water Air Soil Pollut* **2012,** *223*, 4937–4948.
37. Urbaniak, M.; Wyrwicka, A.; Tołoczko, W.; Serwecińska, L.; Zieliński, M. The Effect of Sewage Sludge Application on Soil Properties and Willow (Salix sp.) Cultivation. *Sci. Total Environ.* **2017,** *586*: 66–75.
38. Martínez-Sánchez, A.; Allende, A.; Luna, C.; Selma, M. V.; Tudela, J. A.; Gil, M. I. In *Impact of Climate Change and Global Trends on the Microbial Quality of Leafy Greens*, Intl. Conference Quality Manag. Fresh Cut Produce: Convenience Food Tasteful Life 2011, 1209, 51–56.
39. Kominko, H.; Gorazda, K.; Wzorek, Z.; Wojtas, K. Sustainable Management of Sewage Sludge for the Production of Organo-Mineral Fertilizers. *Waste Biomass Valori* **2018,** *9*, 1817–1826.
40. Mambelli, S.; Bird, J. A.; Gleixner, G.; Dawson, T. E.; Torn, M. S. Relative Contribution of Foliar and Fine Root Pine Litter to the Molecular Composition of Soil Organic Matter after In Situ Degradation. *Organic Geochem.* **2011,** *42* (9), 1099–1108.
41. Suess A. Use of Sewage Sludge as a Fertilizer for Increasing Soil Fertility and Crop Production. 1997.

42. Beck, A. J.; Johnson, D. L.; Jones, K. C. The Form and Bioavailability of Non-Ionic Organic Chemicals in Sewage Sludge-Amended Agricultural Soils. *Sci. Total Environ.* **1996,** *185*, 125–149.
43. Vaca, R.; Lugo, J.; Martinez, R.; Esteller, M. V.; Zavaleta, H. Effects of Sewage Sludge and Sewage Sludge Compost Amendment on Soil Properties and Zea Mays L. Plants (heavy metals, Quality and Productivity). *Intl. J. Environ. Pollut.* **2011,** *27*, 304–311.
44. Archie, S. G.; Smith, M. I. L. T. O. N. Survival and Growth of Plantations in Sewage Sludge Treated Soil and Older Forest Growth Study. *Municipal Sludge Applic. Pacific North-West Forest Lands.* **1981,** 105–113.
45. Epstein, E.; Taylor, J.; Chaney, R. *Effects of Sewage Sludge and Sludge Compost Applied to Soil on Some Soil Physical and Chemical Properties*; Wiley Online Library, 1976.
46. Ojeda, G.; Alcañiz, J. M.; Ortiz, O. Runoff and Losses by Erosion in Soils Amended with Sewage Sludge. *Land Degrad. Develop.* **2003,** *14*, 563–573.
47. Hussein, A. Impact of Sewage Sludge as Organic Manure on Some Soil Properties, Growth, Yield and Nutrient Contents of Cucumber Crop. *J. Appl. Sci.* **2009,** *9*, 1401–1411.
48. Johnston, A. Research Seminar on the Manurial Value of Sewage Sludge. *Directorate Water Engin. Res. Develop. Divis. Tech. Note* **1976,** *1977*, 3–11.
49. Bunting, A. H. Experiments on Organic Manures. *J. Agric. Sci.* **1963,** *60*, 121–140.
50. Gupta, S. C.; Dowdy, R.; Larson, W. Hydraulic and Thermal Properties of a Sandy Soil as Influenced by Incorporation of Sewage Sludge. *Soil. Sci. Soc. Am. J.* **1977,** *41*, 601–605.
51. Salter, P.; Williams, J.; Harrison, D.; Effects of bulky organic manures on the available water capacity of a fine sandy loam. *Exper. Hort.* **1965,** *13*, 69-75.
52. Catroux, G.; L'hermite, P.; Suess, E. *The Influence of Sewage Sludge Application on Physical and Biological Properties of Soils*, 2012.
53. Sparks, D. L.; Page, A.; Helmke, P.; Loeppert, R.H.; *Methods of Soil Analysis, Part 3: Chemical Methods*; Vol 14; John Wiley Sons; 2020.
54. Tsadilas, C. D.; Matsi, T.; Barbayiannis, N.; Dimoyiannis, D.; Influence of Sewage Sludge Application on Soil Properties and on the Distribution and Availability of Heavy Metal Fractions. *Commun. Soil Sci. Plant Anal.* **1995,** *26* (15–16), 2603–2619.
55. Kalra, S.; Kelkar, D.; Galwankar, S. C.; Papadimos, T. J.; Stawicki, S. P.; Arquilla, B. The Emergence of e bola as a Global Health Security Threat: from 'Lessons Learned' to Coordinated Multilateral Containment Efforts. *J. Glob. Infec. Dis. 6* (4), **2014,** 164–167.
56. Hue, N. V.; Ranjith, S. A. Sewage Sludges in Hawaii: Chemical Composition and Reactions with Soils and Plants. *Water Air Soil Pollut.* **1994,** *72*, 265–283.
57. Korboulewsky, N.; Dupouyet, S.; Bonin, G. Environmental Risks of Applying Sewage Sludge Compost to Vineyards: Carbon, Heavy Metals, Nitrogen, and Phosphorus Accumulation. *J. Environ. Quality* **2002,** *31*, 1522–1527.
58. Adams, T. M.; Sanders, J. R. The Effects of pH on the Release to Solution of Zinc, Copper and Nickel from Metal-Loaded Sewage Sludges. *Environ. Pollut. Ser. B Chem. Phy.* **1984,** *8*, 85–99.
59. Abdel-Nasser, G.; Harhash, M.; Effect of Organic Manures in Combination with Elemental Sulphur on Soil Physical and Chemical Characteristics, Yield, Fruit Quality, Leaf Water Contents and Nutritional Status of Flame Seedless Grapevine I. Soil Physical and Chemical Characteristics. *J. Agric. Sci. Mansoura Univ.* **2000,** *25*, 3541–3558.

60. Antolín, M. C.; Pascual, I.; García, C.; Polo, A.; Sánchez-Díaz, M. Growth, Yield and Solute Content of Barley in Soils Treated with Sewage Sludge Under Semiarid Mediterranean Conditions. *Field Crops Res.* **2005,** *94*, 224–237.
61. Mendoza, J.; Garrido, T.; Castillo, G.; San-Martin, N. Metal Availability and Uptake by Sorghum Plants Grown in Soils Amended with Sludge from Different Treatments. *Chemosphere* **2006,** *65*, 2304–2312.
62. Hernández, T.; Moreno, J. I.; Costa, F. Influence of Sewage Sludge Application on Crop Yields and Heavy Metal Availability. *Soil Sci. Plant Nutr.* **1991,** *37* (2), 201–210.
63. Soon, Y. K.; Solubility and Sorption of Cadmium in Soils Amended with Sewage Sludge. *J. Soil Sci.* **1981,** *32*, 85–95.
64. Obbard, J. Ecotoxicological Assessment of Heavy Metals in Sewage Sludge Amended Soils. *Appl. Geochem.* **2001,** *16*, 1405–1411.
65. Kandeler, E.; Tscherko, D.; Bruce, K.; Stemmer, M.; Hobbs, P. J.; Bardgett, R. D. Structure and Function of the Soil Microbial Community in Microhabitats of a Heavy Metal Polluted Soil. *Bio. Fert. Soils* **2000,** *32*, 390–400.
66. Fliessbach, A.; Martens, R.; Reber, H. Soil Microbial Biomass and Microbial Activity in Soils Treated with Heavy Metal Contaminated Sewage Sludge. *Soil Bio. Biochem.* **1994,** *26*, 1201–1205.
67. Siebielec, G.; Siebielec, S.; Lipski, D. Long-Term Impact of Sewage Sludge, Digestate and Mineral Fertilizers on Plant Yield and Soil Biological Activity. *J. Clean. Prod.* **2018,** *187*, 372–379.
68. Banerjee, M.; Burton, D.; Depoe, S.; Impact of Sewage Sludge Application on Soil Biological Characteristics. *Agric. Ecosys. Environ.* **1997,** *66*, 241–249.
69. Liang, Y.; Si, J.; Nikolic, M.; Peng, Y.; Chen, W.; Jiang, Y. Organic Manure Stimulates Biological Activity and Barley Growth in Soil Subject to Secondary Salinization. *Soil Biol. Biochem.* **2005,** *37*, 1185–1195.
70. Singh, R.; Agrawal, M. Potential Benefits and Risks of Land Application of Sewage Sludge. *Waste Manag* **2008,** *28*, 347–358.
71. Hattori, H. Microbial Activities in Soil Amended with Sewage Sludges. *Soil Sci. Plant Nutri.* **1988,** *34*, 221–232.
72. Sree-Ramulu, U. S. *Reuse of Municipal Sewage and Sludge in Agriculture*. Sci. Publish: India, 2001.
73. Nielson, R. R. *Sampling and Analysis Plan for the Former Atomic Energy Commission bus Lot Property (No. PNNL--11923)*. Pacific Northwest National Lab, 1998.
74. Walter, I.; Martínez, F.; Alonso, L.; Cuevas, J. D. G. G. Extractable Soil Heavy Metals Following the Cessation of Biosolids Application to Agricultural Soil. *Environ. Pollut.* **2002,** *117*, 315–321.
75. Kladivko, E. J.; Nelson, D. W. Changes in Soil Properties from Application of Anaerobic Sludge. *J. Water Pollut. Cont. Federat.* **1979,** *51* (2), 325–332.
76. Fytili, D.; Zabaniotou, A. Utilization of Sewage Sludge in EU Application of Old and New Methods—A Review. *Renew. Sustain. Energy Rev.* **2008,** *12*, 116–140.
77. Shuman, L. M. Effect of Organic Waste Amendments on Cadmium and Lead in Soil Fractions of Two Soils. *Commun. Soil Sci. Plant Anal.* **1998,** *29*, 2939–2952.
78. Moss, L. H.; Epstein, E.; Logan, T. Comparing the Characteristics, Risks and Benefits of Soil Amendments and Fertilizers Used in Agriculture. *Proceed. Water Environ. Fed.* **2002,** *3*, 602–623.

79. Andreoli, C. V.; Von-Sperling, M.; Fernandes, F.; Ronteltap, M. *Sludge Treatment and Disposal*. IWA Publishing; 2007.
80. Draeger, K. J.; Pundsack, J. W.; Jorgenson, M. D.; Mulloy, P. J. Watershed Effects of Biosolids Land Application: Literature Review. 1999.
81. Bertran, E.; Sort, X.; Soliva, M.; Trillas, I. Composting Winery Waste: Sludges and Grape Stalks. *Biores. Technol*. **2004,** *95*, 203–208.
82. Mantovi, P.; Baldoni, G.; Toderi, G.; Reuse of Liquid, Dewatered, and Composted Sewage Sludge on Agricultural Land: Effects of Long-Term Application on Soil and Crop. *Water Res.* **2005,** *39*, 289–296.
83. Warman, P. R.; Termeer, W. C. Evaluation of Sewage Sludge, Septic Waste and Sludge Compost Applications to Corn and Forage: Yields and N, P and K Content of Crops and Soils. *Biores. Technol.* **2005,** *96*, 955–961.
84. Uggetti, E.; Ferrer, I.; Llorens, E.; García, J. Sludge Treatment Wetlands: A Review on the State of the Art. *Biores. Technol.* **2010,** *101* (9), 2905–2912.
85. Elsokkary, I. H. Biogeochemical Cycles of Specific Pollutants: Mercury. 1989.
86. Kijo-Kleczkowska, A.; Otwinowski, H.; Środa, K.; Properties and Production of Sewage Sludge in Poland with Reference to the Methods of Neutralizing. *Archives Waste Manag. Environ. Protect.* **2012,** 14.
87. Siebielec, G.; Stuczynski, T. Trace Metal Aspects of Biosolids Use. *Proceed. Protect. Restor. Environ.* **2008,** 7.
88. Ghazy, M.; Dockhorn, T.; Dichtl, N. Sewage Sludge Management in Egypt: Current Status and Perspectives Towards a Sustainable Agricultural Use. *World Acad. Sci. Eng. Technol.* **2009,** *57*, 492–500.
89. Spinosa, L.; Ayol, A.; Baudez, J. C.; Canziani, R.; Jenicek, P.; Leonard, A.; Rulkens, W.; Xu, G.; Van Dijk, L. Sustainable and Innovative Solutions for Sewage Sludge Management. *Water* **2011,** *3*, 702–717.
90. Mohamed, B.; Mounia, K.; Aziz, A.; Ahmed, H.; Rachid, B.; Lotfi, A.; Sewage Sludge Used as Organic Manure in Moroccan Sunflower Culture: Effects on Certain Soil Properties, Growth and Yield Components. *Sci. Total Environ.* **2018,** *627*, 681–688.
91. Mtshali, J. S.; Tiruneh, A. T.; Fadiran, A. O. Characterization of Sewage Sludge Generated from Wastewater Treatment Plants in Swaziland in Relation to Agricultural Uses. *Resour. Environ.* **2014,** *4*, 190–199.
92. Latare, A. M.; Kumar, O.; Singh, S. K.; Gupta, A.; Direct and Residual Effect of Sewage Sludge on Yield, Heavy Metals Content and Soil Fertility Under Rice–Wheat System. *Ecologic. Engin.* **2014,** *69*, 17–24.
93. Kotowska, U.; Żalikowski, M.; Isidorov, V. A. HS-SPME/GC–MS Analysis of Volatile and Semi-Volatile Organic Compounds Emitted from Municipal Sewage Sludge. *Environ. Monitor. Assess.* **2012,** *184*, 2893–2907.
94. Metcalf, E. Wastewater Engineering Treatment, Disposal and Reuse. 1991.
95. Lixandru G. *The Use of Sewage Sludge as a Fertilizer in Agriculture, Pedogenic Factors and Processes in the Temperate Zone*; Al, I., Ed.; Cuza University, 2005.
96. Larney, F. J.; Angers, D.A. The Role of Organic Amendments in Soil Reclamation: A Review. *Can. J. Soil Sci.* **2012,** *92*, 19–38.
97. Gubišová, M.; Horník, M.; Hrcˇková, K.; Gubiš, J.; Jakubcová, A.; Hudcovicová, M.; Ondreicˇková, K. Sewage Sludge as a Soil Amendment for Growing Biomass Plant Arundo donax L. *Agronmy* **2020,** *10*, 678.

98. Nascimento, A. L.; Sampaio, R. A.; Brandão Júnior, D. D. S.; Junio, G. R. Z.; Fernandes, L. A. Productivity of Castor Bean in Response to Fertilization with Sewage Sludge. *Revista Caatinga* **2011,** *24,* 145–151.
99. Lemainski, J.; Silva, J. E. D. Use of Biosolids for Corn (Zea mays, L.) Production in the Federal District. *Braz. J. Soil Sci.* **2006,** *30* (4), 741–750.
100. Cocarta, D. M.; Subtirelu, V. R.; Badea, A. Effect of Sewage Sludge Application on Wheat Crop Productivity and Heavy Metal Accumulation in Soil and Wheat Grain. *Environ. Engin. Manag. J.* **2017,** *16* (5):1093–1100.
101. Jamil, M.; Qacim, M.; Umar, M. Utilization of Sewage Sludge as Organic Fertilizer in Sustainable Agriculture. *J. Appl. Sci.* **2006,** *6,* 531–535.
102. Arroyo, M. D. M. D.; Cots, M. Á. P.; HORNEDO, R. M. D. I.; Rodríguez, E. M. B.; Beringola, L. B.; Sánchez, J. V. M. Sewage Sludge Compost Fertilizer Effect on Maize Yield and Soil Heavy Metal Concentration. *Intl. J. Environ. Pollut.* **2002,** *18,* 147–150.
103. Elsalam, H. E. A.; El-Sharnouby, M. E.; Mohamed, A. E.; Raafat, B. M.; El-Gamal, E. H.; Effect of Sewage Sludge Compost Usage on Corn and Faba Bean Growth, Carbon and Nitrogen Forms in Plants and Soil. *Agronomy* **2021,** *11,* 628.
104. Albuquerque, H. C. D.; Junio, G. R. Z.; Sampaio, R. A.; Fernandes, L. A.; Prates, F.; Amaral Sobrinho, N.; Residual Effect of Sewage Sludge Fertilization on Sunflower Yield and Nutrition. *Braz. J. Agric. Environ. Engin.* **2015,** *19,* 1005–1011.
105. Langenkamp, H.; Part, P.; Erhardt, W.; Pruess, A. Organic Contaminants in Sewage Sludge for Agricultural Use. *Insti. Environ. Sustain. Soil Waste Unit* **2001,** 1–73.
106. Ottaviani, M.; Santarsiero, A.; De Fulvio, S. Heavy Metals in Sewage Sludge Utilized in Agriculture. *Ann. Ist. Sanità* **1989,** *25* (3), 525–530.
107. Dudley, T. C. Sludge Digestion and Gas Utilization in the Metropolitan Public Health Division of the Thames Water Authority. 1980.
108. Smith, S. R. A Critical Review of the Bioavailability and Impacts of Heavy Metals in Municipal Solid Waste Composts Compared to Sewage Sludge. *Environ. Intl.* **2009,** *35,* 142–156.
109. Fijalkowski, K.; Rorat, A.; Grobelak, A.; Kacprzak, M. J. The Presence of Contaminations in Sewage Sludge—The Current Situation. *J. Environ. Manag.* **2017,** *203,* 1126–1136.
110. Silveira, M. L. A.; Alleoni, L. R. F.; Guilherme, L. R. G. Sewage Sludge and Heavy Metals in Soils. *Agric. Sci.* **2003,** *60* (4), 793–806.
111. Boix, C.; Ibáñez, M.; Fabregat-Safont, D.; Morales, E.; Pastor, L.; Sancho, J. V.; Sánchez-Ramírez, J. E.; Hernández, F.; Behaviour of Emerging Contaminants in Sewage Sludge after Anaerobic Digestion. *Chemosphere* **2016,** *163,* 296–304.
112. Buckwell, A.; Nadeu, E.; A Review of the Issues, Opportunities, and Actions. *Nutri. Rec. Reuse Euro. Agric.* **2016.**

Index

R

S

U

W